Lecture Notes in Management and Industrial Engineering

Series editor

Adolfo López-Paredes, Valladolid, Spain

This bookseries provides a means for the dissemination of current theoretical and applied research in the areas of Industrial Engineering & Engineering Management. The latest methodological and computational advances that both researchers and practitioners can widely apply to solve new and classical problems in industries and organizations constitute a growing source of publications written for and by our readership.

The aim of this bookseries is to facilitate the dissemination of current research in the following topics:

- Strategy and Enterpreneurship
- Operations Research, Modelling and Simulation
- Logistics, Production and Information Systems
- Quality Management
- Product Management
- Sustainability and Ecoefficiency
- Industrial Marketing and Consumer Behavior
- Knowledge and Project Management
- Risk Management
- Service Systems
- Healthcare Management
- Human Factors and Ergonomics
- Emergencies and Disaster Management
- Education

More information about this series at http://www.springer.com/series/11786

Pablo Cortés · Elvira Maeso-González
Alejandro Escudero-Santana
Editors

Enhancing Synergies in a Collaborative Environment

 Springer

Editors
Pablo Cortés
Escuela Técnica Superior de Ingeniería
Universidad de Sevilla
Seville
Spain

Alejandro Escudero-Santana
Escuela Técnica Superior de Ingeniería
Universidad de Sevilla
Seville
Spain

Elvira Maeso-González
Escuela Técnica Superior de Ingeniería
 Industrial
Universidad de Málaga
Málaga
Spain

ISSN 2198-0772　　　　　　　ISSN 2198-0780　(electronic)
Lecture Notes in Management and Industrial Engineering
ISBN 978-3-319-14077-3　　　ISBN 978-3-319-14078-0　(eBook)
DOI 10.1007/978-3-319-14078-0

Library of Congress Control Number: 2014958014

Springer Cham Heidelberg New York Dordrecht London
© Springer International Publishing Switzerland 2015
This work is subject to copyright. All rights are reserved by the Publisher, whether the whole or part of the material is concerned, specifically the rights of translation, reprinting, reuse of illustrations, recitation, broadcasting, reproduction on microfilms or in any other physical way, and transmission or information storage and retrieval, electronic adaptation, computer software, or by similar or dissimilar methodology now known or hereafter developed.
The use of general descriptive names, registered names, trademarks, service marks, etc. in this publication does not imply, even in the absence of a specific statement, that such names are exempt from the relevant protective laws and regulations and therefore free for general use.
The publisher, the authors and the editors are safe to assume that the advice and information in this book are believed to be true and accurate at the date of publication. Neither the publisher nor the authors or the editors give a warranty, express or implied, with respect to the material contained herein or for any errors or omissions that may have been made.

Printed on acid-free paper

Springer International Publishing AG Switzerland is part of Springer Science+Business Media (www.springer.com)

Preface

The Joint CIO-ICIEOM-IIIE 2014 event, comprising the "8th International Conference on Industrial Engineering and Industrial Management," the "XX International Conference on Industrial Engineering and Operations Management," and the "International IIE Conference 2014," was a joint conference resulting from an institutional agreement between Asociación para el Desarrollo de la Ingeniería de Organización (ADINGOR), Associaçâo Brasileira de Engenharia de Produçâo (ABEPRO), and Institute of Industrial Engineers (IIE).

Association for the Development of Industrial Engineering (ADINGOR) is the leading Spanish association dedicated to the development and dissemination of knowledge related to Engineering and Business Organizations (management sciences, social and economic issues, quantitative methods, statistical, and computer applications) in order to improve management processes. ADINGOR brings together professionals of Management and R&D, from organizations, companies, and universities geared to these objectives. The ADINGOR website is www.adingor.es.

Brazilian Association of Industrial Engineering (ABEPRO) is an institution representing teachers, students, and professionals of Production Engineering. The association, founded more than 25 years ago, has the following functions: to clarify the role of the Production Engineer in the society and in its market, to be a link to the government institutes related to organizing, evaluating (MEC e INEP), and promoting courses, as well as to private organizations as CREA, Confea, SBPC, Abenge, and other nongovernmental organizations, which deal with research, teaching, and different extensions of engineering. The ABEPRO website is www.abepro.org.br/.

IIE is the world's largest professional society dedicated solely to the support of the industrial engineering profession and individuals involved with improving quality and productivity. Founded in 1948, IIE is an international nonprofit association that provides leadership for the application, education, training, research, and development of industrial engineering. IEs figure out a better way to do things and work in a wide array of professional areas, including management, manufacturing, logistics, health systems, retail, service, and ergonomics. They influence policy and implementation issues regarding topics such as sustainability, innovation, and Six Sigma. And like the profession, IEs are rooted in the sciences of engineering, the analysis of systems, and the management of people. The IIE website is www.iienet2.org/.

These three institutions have joined their efforts toward bringing together an ongoing basis for discussion, analysis, and scientific collaboration. In fact, Industrial Engineering is an expertise of the Engineering that appears strongly linked to the concepts of Industrial Management, Operations Management, Management Science, Operations Research, Systems Engineering and Engineering Management, among others. However, the Industrial Engineering concept in the twenty-first century is overpassing these concepts and is extending to a systemic vision of all the fields and departments of the productive and service systems in an integrated way.

Industrial Engineering thus appears as a toolbox of key-enabling technologies providing the most efficient management and optimization of complex processes or systems. It relates to the capabilities of designing, developing, and implementing integrated solutions for complex systems associated to the economic, human, knowledge, information, equipment, energy, and materials subsystems.

The Conference motto was "enhancing synergies in a collaborative environment". The Conference Topics received much attention from the scientific and professional communities. They covered: operations research, modeling and simulation, computer and information systems, operations research, scheduling and sequencing, logistics, production and information systems, supply chain and logistics, transportation, lean management, production planning and control, production system design, reliability and maintenance, quality management, sustainability and eco-efficiency, marketing and consumer behavior, business administration and strategic management, economic and financial management, technological and organizational innovation, strategy and entrepreneurship, economics engineering, enterprise engineering, global operations and cultural factors, operations strategy and performance, management social responsibility, environment, and sustainability.

The Conference received 294 papers, and finally only 47 were accepted for publication in this volume. It means a 17 % acceptance ratio showing the high level of scientific exigency in the conference.

We would like to express our gratitude to the authors, coming from Spain, Brazil, United States, Iran, Colombia, Portugal, Turkey, Germany, Italy, Thailand, Great Britain, Saudi Arabia, Slovenia, or South Corea among others, making this Conference a truly international one.

Our special thanks to the Governments of Málaga and Andalucía, to the Universities of Málaga and Sevilla, and to the Sponsors: EMT, ACOTRAL, Andalucía Dry Port, Jorge Ordoñez Winery, Industrial Engineering and Polytechnic Schools and Andalucía Tech, for their support.

We also acknowledge the invited speakers and reviewers. And we also recognize the great effort of the Steering, Scientific and Organizing Committee, particularly to Milton Vieira Junior (President of ABEPRO), Bopaya Bidanda (IIE Senior VP International) and Luis Onieva (President of ADINGOR).

September 2014

Pablo Cortés
Elvira Maeso-González
Alejandro Escudero-Santana

Contents

Part I OR, Modelling and Simulation

Mixed-Model Sequencing Problem Improving Labour Conditions 3
Joaquín Bautista, Rocío Alfaro-Pozo, Cristina Batalla-García
and Sara María Llovera-Laborda

**A Heuristic Approach for Integrated Storage and Shelf-Space
Allocation** 11
Nazanin Esmaili, Bryan A. Norman and Jayant Rajgopal

**Evaluation of Pull Control Strategies and Production Authorisation
Card Policies Recovery Period in a Multi-product System** 19
Chukwunonyelum Emmanuel Onyeocha, Joseph Khoury
and John Geraghty

**Smart Decision in Industrial Site Selection: What's New
in the Case of a Steel Mill in Brazil?** 29
Alexandre de Oliveira Gomes and Carlos Alberto Nunes Cosenza

**Dynamic Multi-objective Maximal Covering Location Problem
with Gradual Coverage** 39
Mohammad Forghani Youshanlo and Rashed Sahraeian

Part II Logistics, Production and Information Systems

**Drivers and Stages in "Packaging Logistics": An Analysis
in the Food Sector** 51
Jesús García-Arca, José Carlos Prado-Prado
and A. Trinidad González-Portela Garrido

A Tabu Search Approach for Production and Sustainable Routing Planning Decisions for Inbound Logistics in an Automotive Supply Chain 61
David Peidro, Manuel Díaz-Madroñero, Josefa Mula and Abraham Navalón

Clothing Industry: Main Challenges in the Supply Chain Management of Value Brand Retailers 69
Sandra Martínez, Ander Errasti, Martin Rudberg and Miguel Mediavilla

Structural Equation Modeling for Analyzing the Barriers and Benefits of E-procurement 77
Peral Toktaş-Palut, Ecem Baylav, Seyhan Teoman and Mustafa Altunbey

Ergonomic Risk Minimisation in Assembly Line Balancing 85
Joaquín Bautista, Cristina Batalla-García and Rocío Alfaro-Pozo

Reverse Logistics Barriers: An Analysis Using Interpretive Structural Modeling 95
Marina Bouzon, Kannan Govindan and Carlos Manuel Taboada Rodriguez

Mixed Trips in the School Bus Routing Problem with Public Subsidy 105
Pablo Aparicio Ruiz, Jesús Muñuzuri Sanz and José Guadix Martín

Analysis of the Criteria Used by Organizations in Supplier Selection 115
Joan Ignasi Moliné, Anna Maria Coves Moreno and Anna Rubio

Event Monitoring Software Application for Production Planning Systems 123
Andrés Boza, Beatriz Cortes, Maria del Mar Eva Alemany and Eduardo Vicens

Optimization of Public Transport Services in Small and Medium Size Towns. A Case Study on Spain 131
Elvira Maeso-González and Juan Carlos Carrasco-Giménez

Methodology to Manage Make-to-Order and Make-to-Stock Decisions in an Electronic Component Plant 139
Julien Maheut, David Rey and José P. Garcia-Sabater

Effects of the Implementation of Antequera Dry Port in Export and Import Flows 147
Guadalupe González-Sánchez, Mª. Isabel Olmo-Sánchez and Elvira Maeso-González

An Iterative Stochastic Approach Estimating the Completion Times of Automated Material Handling Jobs 155
Gunwoo Cho and Jaewoo Chung

Part III Strategy and Entrepreneurship

Transportation Infrastructure and Economic Growth Spillovers 165
Herick Fernando Moralles and Daisy Aparecida do Nascimento Rebelatto

Influence of the Environmental Factors in the Creation, Development and Consolidation of University Spin-offs in the Basque Country 173
E. Zarrabeitia Bilbao, P. Ruiz de Arbulo López and P. Díaz de Basurto Uraga

Conceptual Model for Associated Costs of the Internationalisation of Operations 181
Ángeles Armengol, Josefa Mula, Manuel Díaz-Madroñero and Joel Pelkonen

Relations Between Costs and Characteristics of a Process: A Simulation Study 189
J. Fortuny-Santos, P. Ruiz de Arbulo-López and M. Zarraga-Rodríguez

Critical Success Factors on Implementation of Customer Experience Management (CEM) Through Extended Marketing Mix 197
A. Arineli and H. Quintella

Financing Urban Growth in Aging Societies: Modelling the Equity Release Schemes in the Welfare Mix for Older Persons ... 205
David Bogataj, Diego Ros-McDonnell and Marija Bogataj

Sustainable Urban Growth in Ageing Regions: Delivering a Value to the Community 215
David Bogataj, Deigo Ros McDonnell, Alenka Temeljotov Salaj and Marija Bogataj

The Role of the Entrepreneur in the New Technology-Based
Firm (NTBF)... 225
Juan Antonio Torrecilla García, Agnieszka Grazyna Skotnicka
and Elvira Maeso-González

Part IV Quality and Product Management

Increasing Production and Minimizing Costs During Machining
by Control of Tool's Wears and or Damages 235
Nivaldo Lemos Coppini and Ivair Alves dos Santos

Impact of 5S on Productivity, Quality, Organizational Climate
and IS at Tecniaguas S.A.S 247
Paloma Martínez Sánchez, Natalia Rincón Ballesteros
and Diana Fuentes Olaya

Structuring a Portfolio for Selecting and Prioritizing Textile
Products.. 257
Maurício Johnny Loos and Paulo Augusto Cauchick-Miguel

Methodology for Evaluating the Requirements of Customers
in a Metalworking Company Automotive ISO/TS 16949 Certified 267
Gil Eduardo Guimarães, Roger Oliveira, Luiz Carlos Duarte,
Milton Vieria and Daniel Galeazzi

Improving Resilience and Performance of Organizations
Using Environment, Health and Safety Management
Systems. An Empirical Study in a Multinational Company 275
Barbara Seixas de Siqueira, Mercedes Grijalvo
and Gustavo Morales-Alonso

Part V Knowledge and Project Management

Effectiveness of Construction Safety Programme Elements.......... 285
Antonio López-Arquillos, Juan Carlos Rubio-Romero,
Jesús Carrillo-Castrillo and Manuel Suarez-Cebador

Operational Issues for the Hybrid Wind-Diesel Systems:
Lessons Learnt from the San Cristobal Wind Project 291
Yu Hu, Mercedes Grijalvo Martín, María Jesús Sánchez
and Pablo Solana

**Points of Convergence Between Knowledge Management
and Agile Methodologies in Information Technology (IT)** 299
Eliane Borges Vaz and Maria do Carmo Duarte Freitas

Recent Advances in Patent Analysis Network 307
Javier Gavilanes-Trapote, Rosa Río-Belver, Ernesto Cilleruelo
and Jaso Larruscain

**Information Enclosing Knowledge Networks: A Study
of Social Relations**... 315
L. Sáiz-Bárcena, J.I. Díez-Pérez, M.A. Manzanedo del Campo
and R. Del Olmo Martínez

Part VI Service Systems

**Cost System Under Uncertainty: A Case Study in the Imaging
Area of a Hospital**... 325
Victor Jiménez, Carla Duarte and Paulo Afonso

Innovation in Consulting Firms: An Area to Explore 335
Isaac Lemus-Aguilar and Antonio Hidalgo

**Innovation Capability and the Feeling of Being
an Innovative Organization** 343
Marta Zárraga-Rodríguez and M. Jesús Álvarez

**Minimizing Carbon-Footprint of Municipal Waste Separate
Collection Systems**.. 351
Giancarlo Caponio, Giuseppe D'Alessandro, Salvatore Digiesi,
Giorgio Mossa, Giovanni Mummolo and Rossella Verriello

A MILP Model for the Strategic Capacity Planning in Consultancy... 361
Carme Martínez Costa, Manuel Mateo Doll and Amaia Lusa García

**Forecasting the Big Services Era: Novel Approach Combining
Statistical Methods, Expertise and Technology Roadmapping** 371
Iñaki Bildosola, Rosa Rio-Bélver and Ernesto Cilleruelo

Part I
OR, Modelling and Simulation

Mixed-Model Sequencing Problem Improving Labour Conditions

Joaquín Bautista, Rocío Alfaro-Pozo, Cristina Batalla-García and Sara María Llovera-Laborda

Abstract It is presented an extension of the mixed-model sequencing problem that considers some working conditions agreed between companies and trade unions. In particular, it is formulated a mathematical model with saturation limits which an operator can have throughout his workday and with the possibility of increasing the work pace of the operators at certain times of the workday. In this way, it is possible to improve labour conditions and line productivity simultaneously. In fact, the proposed model is evaluated by means of a computational experience that allows to observe that an increment of 3.3 % on the work pace factor of processors reduces the work overload by 62.6 % while the saturation conditions imposed by collective agreements are satisfied.

Keywords Mixed-model assembly line · Work overload · Saturation · Work pace

---

This work is supported by the Spanish Ministerio de Economía y Competitividad under Project DPI2010-16759 (PROTHIUS-III) including EDRF fundings.

J. Bautista (✉) · R. Alfaro-Pozo · C. Batalla-García · S.M. Llovera-Laborda
Research Group OPE-PROTHIUS. Dpto. de Organización de Empresas, Universitat Politècnica de Catalunya, Avda. Diagonal, 647, 7th Floor, 08028 Barcelona, Spain
e-mail: joaquin.bautista@upc.edu

R. Alfaro-Pozo
e-mail: rocio.alfaro@upc.edu

C. Batalla-García
e-mail: cristina.batalla@upc.edu

S.M. Llovera-Laborda
e-mail: sara.maria.llovera@upc.edu

© Springer International Publishing Switzerland 2015
P. Cortés et al. (eds.), *Enhancing Synergies in a Collaborative Environment*,
Lecture Notes in Management and Industrial Engineering,
DOI 10.1007/978-3-319-14078-0_1

1 Introduction

Currently, many productive systems have mixed-model assembly lines. These lines are capable of processing a set of product types (I) that although similar, they differ in the use of resources and consumption of components at the set (K) of workstations of the line. These differences make necessary sequencing problems. A clear example of such problems is the Mixed-Model Sequencing Problem with Work overload Minimisation (*MMSP-W*) (Boysen et al. 2009). This problem focuses on obtaining product sequences that minimise either the work overload, or uncompleted work, or maximise the line productivity.

However, the selection of a sequence is not the only decision that affects directly the productivity of companies in production systems with ideologies Just In Time (*JIT*) and Douky Seisan (DS^1). In fact, large companies, in particular automotive companies, negotiate a set of labour conditions with trade unions. Some of these conditions are set once the operation processing times of the different products at workstations ($p_{i,k}, \forall i \in I, \forall k \in K$) have been established through the Methods Time Measurement (*MTM*) system considering a standard[2] activity (*MTM_100* in centesimal scale); among them, there are: (1) the effective length of the working day; (2) the maximum and average saturation to which a worker may be subject throughout his effective workday; and (3) the normal level of activity or work pace considered by the company and with which the processing times of operations are set from the standard processing times pre-established.

Thus, taking into account the aim of the *MMSP-W* (Bautista and Cano 2011), in this work it is presented a variant of the problem that allows to assess, using a case study linked to the engine plant from Nissan in Barcelona, the influence of the agreed conditions, between company and trade unions, on the total work overload of the line. In particular, the extension incorporates the limitation of the maximum and average saturation of the processors and the variability of the processing times of operations according to the work pace of the processors.

2 Labour Conditions

Commonly companies and trade unions from the European industry set the work schedule and effective length of a work shift by means of collective agreements.

The work schedule, with one-year horizon, is segmented into numbered weeks (52 or 53) and it serves to set the days of work of the year for the collective. In this environment, it is usual to have an annual work schedule of 225 industrial days (Llovera et al. 2014). Moreover, these work days may have as many as 3 shifts of

---

[1] Synchronised production ideology developed by Nissan.

[2] Work pace that can be kept throughout the workday. It is neither too fast nor too slow. It is the pace of an average skilled worker.

8 h, although the number of effective hours of a work shift is between 6 h 15′ and 7 h, because time is left to operators to meet their physiological needs.

On the other hand, European factories set limits regarding the relationship between effective work time and total available time assigned to the processors to develop their work. That is, they establish limits on the maximum and average saturation of the processors of the workstations.

The maximum saturation limit corresponds to the highest allowed level of saturation of a processor at a given moment. In fact, if this limit is not satisfied, it will be necessary to go to the department of methods and times in order to improve the process and reduce times. Usually its value is $\eta_{max}^{\infty} = 1.2$.

The average saturation is the occupation of a processor along its workday. Normally, the admissible value is $\eta_{med}^{\infty} = 0.95$ and it should be respected by all the demand plans.

Regarding the normal work pace, normally, the automotive companies set their processing times ($p_{i,k} : i \in I, k \in K$) considering as normal activity the resultant of the direct application of the standard times (*MTM*_100) to activity 110 (*MTM*_100) (Llovera et al. 2014). To do this it is necessary to change the time scale from *MTM*_100 to *MTM*_110 as follows:

$$p_{i,k} = p_{i,k}(MTM\_110) = p_{i,k}(MTM\_100) \cdot \frac{100}{110} \quad (i = 1, \ldots, |I|\,;\ k = 1, \ldots, |K|) \quad (1)$$

In such conditions, given a set of processing times of tasks at workstations ($\hat{p}_{i,k} : i \in I, k \in K$) obtained by observation and timekeeping, it is easy to determine the work pace with which a processor carries out its work. This work pace (activity) is represented through the activity factor of a product and a workstation ($\alpha_{i,k} : i \in I, k \in K$). That is: $\alpha_{i,k} \cdot \hat{p}_{i,k} = \alpha^N \cdot p_{i,k} : i \in I, k \in K$. In short, if $\alpha^N = 1.0$: $\alpha_{i,k} = p_{i,k}/\hat{p}_{i,k} : i \in I, k \in K$.

Thus, there are three possible work pace factors: the standard, $\alpha_{i,k}^\circ = 0.90$, normal, $\alpha_{i,k}^N = 1.0$, and optimal[3], $\alpha_{i,k}^* = 1.2$, whom correspond to the scale times *MTM*_100, *MTM*_110 and *MTM*_132, respectively.

3 Incorporating Saturation and Work Pace Conditions into the *MMSP-W*

Robert Yerkes and John Dodson showed that efficiency increases when stress increases in 1908. However, this relationship only holds true up to a certain point of stress level, after which the performance drastically decreases. This relationship,

---

[3] Maximum work pace factor that an operator can bear when he works 8 h a day; it is 20 % higher than the normal work pace.

named the Yerkes-Dodson law, emphasizes that the operator is more productive on the border between the minimum and the maximum stress and it is represented by an inverted U-shaped curve where the optimum stress lies between the excess and deficiency of activation. This idea together with the consideration that workers vary their work pace throughout the workday allows us to establish a relationship between the performance of an operator and his level of "activation" which is reflected in his level of stress and which follows a concave function (Muse et al. 2003).

In this way, it is possible to establish a stepped function for the work pace factor along the workday whereas the adaptation period of the start of the working day, the period of fatigue corresponding to the end of the day, and that the work pace factor depends on the moment of the workday ($\dot{\alpha}_{k,t}$) and not the type of product. Thus, all processors of workstations will process the first ($1 \leq t \leq t_0$) and last ($t_\infty + 1 \leq t \leq T + |K| - 1$) periods with a normal work pace factor ($\alpha^N = 1.0$), and the intermediate periods ($t_0 + 1 \leq t \leq t_\infty$) with a greater work pace factor ($\alpha^{max} = 1.1$) (Bautista et al. 2013). That is, the processing times of intermediate product units will be reduced and, therefore, the processors will complete more quantity of required work.

On the other hand, the incorporation of the saturation limits into the *MMSP-W* has two possible considerations: (1) the static saturation which is due to the processing times of the operations ($p_{i,k}(\forall i, \forall k)$), with a specific work pace factor ($\alpha_{i,k}$), and the production mix ($d_i, (\forall i)$); and (2) the dynamic saturation which is due to the sequence ($\pi(T) = \{\pi_1, \pi_2, \ldots, \pi_T\}$) of T products and the completed work by the processors ($v_{k,t}$, with a dynamic work pace factor $\dot{\alpha}_{k,t}$). That is:

$$\eta_{med}^\circ(k) = \frac{1}{c \cdot T} \cdot \sum_{i=1}^{|I|} \frac{p_{i,k}}{\alpha_{i,k}} \cdot d_i \quad (k = 1, \ldots, |K|) \tag{2}$$

$$\eta_{max}^\circ(k) = \frac{1}{c} \cdot \max_{i \in I} \left\{ \frac{p_{i,k}}{\alpha_{i,k}} \right\} \quad (k = 1, \ldots, |K|) \tag{3}$$

$$\eta_{med}(k) = \frac{1}{c \cdot T} \cdot \sum_{t=1}^{T} \frac{v_{k,t}}{\dot{\alpha}_{k,t}} \quad (k = 1, \ldots, |K|) \tag{4}$$

$$\eta_{max}(k) = \frac{1}{c} \cdot \max_{1 \leq t \leq T} \left\{ \frac{v_{k,t}}{\dot{\alpha}_{k,t}} \right\} \quad (k = 1, \ldots, |K|) \tag{5}$$

It should be noted that the maximum saturation cannot be exceeded, because its violation requires going to the department of methods and times to look for alternatives in the assembly of the products. However, the fact to overtake the average saturation may be compensated with some measures, such as: (1) process improvement; (2) increase in line capacity; and (3) processor rotations between workstations. This fact causes that given a demand plan d_i ($\forall i$) and the processing times $p_{i,k}$, with a specific activity factor $\alpha_{i,k}$, the average saturation, both static and dynamic, can exceed the η_{med}^∞ value. If this occurs, the processors will not complete all the required work and, therefore, the work overload of the line increases. In fact, if we consider the average static saturation, we can determine the inevitable work overload by processor and line, as follows:

$$\omega_0(k) = c \cdot T \cdot \max\{0, \eta_{med}^{\circ}(k) - \eta_{med}^{\infty}\} \quad (k = 1, \ldots, |K|) \tag{6}$$

$$W_0 = \sum_{k=1}^{|K|} b_k \cdot \omega_0(k) \tag{7}$$

Obviously, the inevitable overload calculated through static saturation, will always be less or equal than the work overload due to dynamic saturation, because the effect of the variation of the processing times and the sequence must be added.

Considering the work pace factor and the maximum allowed saturation we formulate the $M4 \cup 3\_\dot{\alpha}I\_\eta$ model from the reference $M4 \cup 3$ model (Bautista et al. 2012). The parameters and variables are:

Parameters	
K	Set of workstations $(k = 1, \ldots, |K|)$
b_k	Number of homogeneous processors at workstation k $(k = 1, \ldots, |K|)$
I	Set of product types $(i = 1, \ldots, |I|)$
d_i	Programmed demand of product type i $(i = 1, \ldots, |I|)$
$p_{i,k}$	Processing time required by a unit of type i at workstation k for each homogeneous processor (at normal pace or activity)
T	Total demand. Obviously, $\sum_{i=1}^{|T|} d_i = T$
t	Position index in the sequence $(t = 1, \ldots, T)$
c	Cycle time. Standard time assigned to workstations to process any product unit
l_k	Time window. Maximum time that each processor at workstation k is allowed to work on any product unit, where $l_k - c > 0$ is the maximum time that the work in progress is held at workstation k $(k = 1, \ldots, |K|)$
$\dot{\alpha}_{k,t}$	Dynamic work pace factor associated with the tth operation of the product sequence $(t = 1, \ldots, T)$ at workstation k $(k = 1, \ldots, |K|)$. Note that if we associate the same dynamic factor to each moment of workday in all workstations, we have: $\dot{\alpha}_{k,t} = \dot{\alpha}_{t+k-1}$ $(k = 1, \ldots, |K|; t = 1, \ldots, T)$
η_{max}^{∞}	Maximum allowable saturation for each processor at workstations $(k = 1, \ldots, |K|)$
η_{med}^{∞}	Allowable average saturation for each processor at workstations $(k = 1, \ldots, |K|)$, during all the workday $(t = 1, \ldots, T)$
Variables	
$x_{i,t}$	Binary variable equal to 1 if product type i $(i = 1, \ldots, |I|)$ is assigned to the position t $(t = 1, \ldots, T)$ of the sequence, and 0 otherwise
$\hat{s}_{k,t}$	Positive difference between the start instant and the minimum start instant of the tth operation at workstation k
$\hat{v}_{k,t}$	Processing time applied and reduced by a work pace factor $\dot{\alpha}_{k,t}$. We impose: $v_{k,t} = \dot{\alpha}_{k,t} \cdot \hat{v}_{k,t}$ $(k = 1, \ldots, |K|; t = 1, \ldots, T + |K| - 1)$, considering $v_{k,t}$ the time applied to the tth unit of the product sequence at station k for each homogeneous processor at normal work pace or activity
$w_{k,t}$	Work overload generated for the unit of the product sequence at station k for each homogeneous processor (at normal activity); measured in time

$M4 \cup 3\_\dot{\alpha}I\_\eta$

$$\min W = \sum_{k=1}^{|K|} \left(b_k \sum_{t=1}^{T} w_{k,t} \right) \Leftrightarrow \max V = \sum_{k=1}^{|K|} \left(b_k \sum_{t=1}^{T} v_{k,t} \right) \tag{8}$$

Subject to:

$$\sum_{t=1}^{T} x_{i,t} = d_i \quad (i = 1, \ldots, |I|) \tag{9}$$

$$\sum_{i=1}^{|I|} x_{i,t} = 1 \quad (t = 1, \ldots, T) \tag{10}$$

$$\dot{\alpha}_{t+k-1} \cdot \hat{v}_{k,t} + w_{k,t} = \sum_{i=1}^{|I|} p_{i,k} \cdot x_{i,t} \quad (k = 1, \ldots, |K|); \ (t = 1, \ldots, T) \tag{11}$$

$$\hat{v}_{k,t} \leq \eta_{\max}^{\infty} \cdot c \quad (k = 1, \ldots, |K|); \ (t = 1, \ldots, T) \tag{12}$$

$$\sum_{t=1}^{T} \hat{v}_{k,t} \leq \eta_{med}^{\infty} \cdot c \cdot T \quad (k = 1, \ldots, |K|) \tag{13}$$

$$\hat{s}_{k,t} \geq \hat{s}_{k,t-1} + \hat{v}_{k,t-1} - c \quad (k = 1, \ldots, |K|); \ (t = 2, \ldots, T) \tag{14}$$

$$\hat{s}_{k,t} \geq \hat{s}_{k-1,t} + \hat{v}_{k-1,t} - c \quad (k = 2, \ldots, |K|); \ (t = 1, \ldots, T) \tag{15}$$

$$\hat{s}_{k,t} + \hat{v}_{k,t} \leq l_k \quad (k = 1, \ldots, |K|); \ (t = 1, \ldots, T) \tag{16}$$

$$\hat{s}_{k,t}, v_{k,t}, \hat{v}_{k,t}, w_{k,t} \geq 0 \quad (k = 1, \ldots, |K|); \ (t = 1, \ldots, T) \tag{17}$$

$$x_{i,t} \in \{0, 1\} \quad (i = 1, \ldots, |I|); \ (t = 1, \ldots, T) \tag{18}$$

$$\hat{s}_{1,1} = 0 \tag{19}$$

In the model, the objective function (8) expresses the equivalence between the minimisation of the work overload, W, and the maximisation of the completed work, V. The constraints (9) and (10) represent the demand plan satisfaction, $d_i(\forall i)$, and the unique assignment of each product unit to a position of the sequence, respectively. The set (11) establishes the relationship between the required processing time, $p_{i,k}$, the applied processing time reduced by the work pace factor, $\hat{v}_{k,t}$, and the work overload generated by the units at the workstations, $w_{k,t}$, considering the same work pace factor for all workstations. The constraints (12) and (13) limit the maximum and average saturation of the processors. The sets (14)–(16)

determine the possible start instants of the operations. The constraints (17) establish the non-negativity of the variables and the set (18) imposes the binary condition to the assignment variables. Finally, the equality (19) sets the start instant of the operations.

4 Computational Experience

To evaluate the impact of the saturation and the work pace factor in the *MMSP-W*, we perform a computing experience based on a demand plan linked to the plant of engines of Nissan from Barcelona. Specifically, the production plan corresponds to a total demand of $T = 200$, divided equally in 9 engine types and belonging to three families (4 × 4 vehicles, vans and trucks). The line, where the engines are manufactured, consists of $|K| = 21$ workstations with one homogeneous processor each one. These processors consist of a team of two equivalent operators and they have a cycle time of $c = 175$ s and a time window of $l_k = 195$ s. On the other hand, based on data from Nissan, the activity factor function used is (see Fig. 1).

After running the model $M4 \cup 3\_\dot{\alpha}I\_\eta$ through Gurobi v4.6.1 solver, on a Apple Macintosh iMac computer with an Intel Core i72.93 GHz processor and 8 GB of RAM using MAC OS × 10.6.7, with a limit of CPU time of 2 h and considering a normal work pace factor, ($\dot{\alpha}^L : \dot{\alpha}_t = 1, \forall t$), and a stepped work pace factor, ($\dot{\alpha}^S$), we have obtained the results showed in Table 1.

In particular, the new model is more competitive than its reference model. In fact, the $M4 \cup 3\_\dot{\alpha}I\_\eta$ reaches the optimal solution, in both cases ($\dot{\alpha}^L$ and $\dot{\alpha}^S$), without using up the limit of CPU time (2.90 and 1.78 s, respectively). However, the improvement of working conditions increases the work overload considerably.

On the other hand, if we only consider the saturation limits (linear function for the work pace factor, $\dot{\alpha}^L$) the inevitable work overload is equal than the actual work

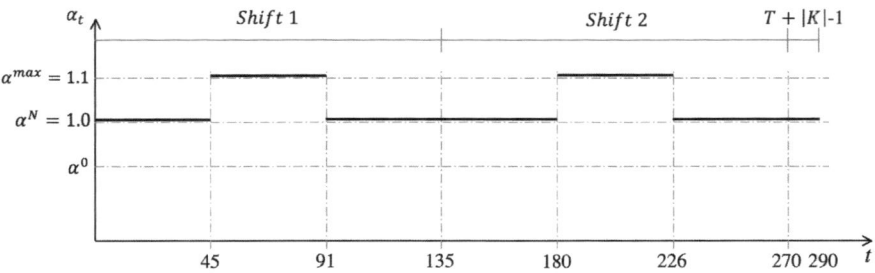

Fig. 1 Stepped function for the work pace factor considering the case study from Nissan

Table 1 Work overload values obtained with $M4 \cup 3$ and $M4 \cup 3\_\dot{\alpha}I\_\eta$, inevitable work overload and workstations with processors over-saturated considering a linear (normal work pace throughout the workday) and stepped function for the work pace factor

ε	$W_{4\cup 3}$	W_0		$W_{4\cup 3\_\dot{\alpha}I\_\eta}$		$k \in K : \eta_{med}(k) > \eta_{med}^{\infty}$	
		$\overline{\alpha^L}$	$\overline{\alpha^S}$	$\dot{\alpha}^L$	$\dot{\alpha}^S$	$\dot{\alpha}^L$	$\dot{\alpha}^S$
#*1*	187	12,315	4,220.6	12,315*	4,601.8*	4, 9, 10, 16, 17, 18	9, 10, 16, 17, 18

Optimal solutions with*

overload, being the total uncompleted work 1,2315 s. However, with the increase of the activity factor of processors by $\widehat{3.3}$ %, the inevitable workload is reduced by 65.7 % (which corresponds to the maximum reduction that could be obtained) and the actual work overload by 62.6 %.

5 Conclusions

A new mathematical model for the *MMSP-W* has been formulated. This model, in addition to minimise the work overload, considers the labour conditions agreed between companies and trade unions. In fact, thanks to a case study from Nissan, we have seen that the saturation limits get worse line productivity. Therefore, in order to improve productivity and labour conditions is recommended increasing the activity factor throughout the workday because in this way the saturation conditions are respected and the actual work overload is reduced by 62.6 %.

References

Bautista J, Cano A (2011) Solving mixed model sequencing problem in assembly lines with serial workstations with work overload minimisation and interruption rules. Eur J Oper Res 210 (3):495–513

Bautista J, Cano A, Alfaro R (2012) Models for MMSP-W considering work station dependencies: A case study of Nissan's Barcelona Plant. Eur J Oper Res 223(3):669–679

Bautista J, Alfaro R, Batalla C, et al (2014) Incorporating the work pace concept into the MMSP-W. In: Hernández, C. et al. (Eds.) Managing complexity- challenges for industrial engineering and operations management. Series: Lecture Notes in Management and Industrial Engineering. Springer International Publishing, ISBN: 978-3-319-04704-1, pp 261–268

Boysen N, Fliedner M, Scholl A (2009) Sequencing mixed-model assembly lines: survey, classification and model critique. Eur J Oper Res 192(2):349–373

Llovera S, Bautista J, Llovera J, Alfaro R (2014) Tiempo efectivo de trabajo: Un análisis normativo de la Jornada Laboral en el Sector de Automoción. doi:10.13140/2.1.1946.3680, Report number: OPE-WP.2014/06 (http://hdl.handle.net/2117/24508), Universitat Politècnica de Catalunya

Muse L, Harris S, Field H (2003) Has the inverted-u theory of stress and job performance had a fair test? Hum Perform 16(4):349–364

A Heuristic Approach for Integrated Storage and Shelf-Space Allocation

Nazanin Esmaili, Bryan A. Norman and Jayant Rajgopal

Abstract We address the joint allocation of storage and shelf-space, using an application motivated by the management of inventory items at Outpatient Clinics (OCs). OCs are limited health care facilities that provide patients with convenient outpatient care within their own community, as opposed to having them visit a major hospital. Currently, patients who are prescribed a prosthetics device during their visit to an OC must often wait for it to be delivered to their homes from a central storage facility. An alternative is the use of integrated storage cabinets at the OCs to store commonly prescribed inventory items that could be given to a patient immediately after a clinic visit. We present, and illustrate with an actual example, a heuristic algorithm for selecting the items to be stocked, along with their shelf space allocations. The objective is to maximize total value based on the desirability of stocking the item for immediate dispensing. The heuristic model considers cabinet characteristics, item size and quantity, and minimum and maximum inventory requirements in order to arrive at the best mix of items and their configuration within the cabinet.

Keywords Shelf space allocation · Heuristics · Two-dimensional packing · Healthcare applications

1 Introduction

This paper addresses a combined packing and shelf-space allocation problem. The problem of packing a set of squares or rectangles into a larger square or rectangle has been widely studied. It has been shown that these problems are strongly

---

N. Esmaili · B.A. Norman · J. Rajgopal (✉)
Department of Industrial Engineering, University of Pittsburgh, Pittsburgh, PA, USA
e-mail: rajgopal@pitt.edu

N. Esmaili
e-mail: nae22@pitt.edu

B.A. Norman
e-mail: banorman@pitt.edu

© Springer International Publishing Switzerland 2015
P. Cortés et al. (eds.), *Enhancing Synergies in a Collaborative Environment*,
Lecture Notes in Management and Industrial Engineering,
DOI 10.1007/978-3-319-14078-0_2

NP-complete (Li and Cheng 1990; Leung et al. 1990), and therefore numerous heuristics have been proposed. General surveys on heuristic method for packing problem can be found in Hopper and Turton (2001), and Ntene and Vuuren (2009). For general surveys on packing problems the reader is referred to Lodi et al. (2002). Similarly, there are many research articles on the shelf space allocation problem to deal with how to optimally allocate shelf space among multiple items so as to maximize profit, minimize inventory costs, or minimize wasted space. For a comprehensive study of shelf space allocation the reader is referred to Yang and Cheng (1999). As noted by Yang (2001) well-managed shelf space can improve the return on inventory investment, raise consumer satisfaction by reducing out-of-stock occurrences, and increase sales and profit margins.

We consider an integrated storage cabinet in the management of prosthetics inventory items at Outpatient Clinics (OCs). OCs are health care facilities that provide patients with a limited suite of services, but the convenience of outpatient care within their own community, obviating the need to visit a relatively distant hospital. Currently, patients who are prescribed a prosthetics device during an OC visit must often wait for it to be delivered to their homes from a central storage facility. An alternative is the use of integrated storage cabinets at the OCs to store commonly prescribed inventory items that could be given to a patient immediately after a clinic visit. The objective is to maximize the total value of items stored in the cabinet, based on the desirability of stocking an item for immediate dispensing and the savings in shipping costs to the patient from the central storage facility.

To our knowledge the specific problem studied has not been considered in the literature and our model and method are new. We consider a restricted version of the combined shelf space allocation and packing problem, where the objective is to pack one or more units of a number of rectangular items into a rectangular cabinet so that the total value of items packed in the cabinet is maximized. We also consider different configurations for each item and propose novel methods to pick the best one for each item. Here, an item's value is defined as the benefit from stocking it in the cabinet as opposed to having it shipped from a central facility. Given that the packing problem by itself is NP-hard in a strong sense, our problem is hard to solve optimally. We therefore develop a new heuristic algorithm to solve it. The remainder of the paper is organized as follows. Section 2 describes the model. Section 3 first introduces three different methods to remove dominated configurations and then presents a heuristic algorithm to solve the problem being modeled. Section 4 illustrates the heuristic with an actual example and examines its performance. Finally, Sect. 5 provides some concluding remarks.

2 Model Description

Integrated storage cabinets come in a wide range of configuration options including open or closed shelving, secure cabinets with several drawers, and specialty storage options. We consider the most general configuration, which is a rectangular form with dimensions H, W and D for the height, width, and depth respectively. The cabinet has a series of shelves arranged vertically, each of which could be of a different height from the set $(h, 2h, 3h, \ldots)$. Thus the maximum number of possible shelves within an integrated storage cabinet is given by $J = \lfloor \frac{H}{h} \rfloor$. Along each shelf are a series of lanes of equal height but possibly varying width that run the depth of the cabinet. Consequently, the maximum number of lanes possible along one shelf is given by $I = \lfloor \frac{W}{w} \rfloor$, where w is the minimum width of each possible lane. The horizontal coordinate of the bottom left corner of a possible lane is indexed by $i = 0, \ldots, I-1$, while its vertical coordinate is indexed by $j = 0, \ldots, J-1$. Note that the each unit on the vertical coordinate is equal to one "shelf height" unit h, and each unit on the horizontal coordinate is equal to one "shelf width" unit w; all of the subsequent discussion will use these units for the height and width of items. Only one item type can be stored in a lane, and units of an item cannot be stacked on top of each other.

Given a set of K items, each of which is a rectangular solid of varying dimensions, the packing and space allocation problem is to decide on the number of each item to pack into a cabinet, along with its configuration in the lane(s) it is assigned, so that the total value across all items in the cabinet is maximized. Each item k can be placed on a shelf in any of up to six configurations (two for each pair of facial dimensions). For each $k = 1, \ldots, K$, we are given its minimum (L_k) and maximum (U_k) number required, along with its storage value V_k. Corresponding to configuration c, item k is characterized by height H_k^c, width W_k^c, and depth D_k^c, $k = 1, \ldots, K$ and $c = 1, \ldots, 6$. The minimum required height of a lane for item k is given by $h_k^c = \lceil \frac{H_k^c}{h} \rceil$, $h_k^c \leq J$, the minimum required width of a lane for item k is given by $w_k^c = \lceil \frac{W_k^c}{w} \rceil$, $w_k^c \leq I$, and the maximum number of item k in one lane is given by $n_k^c = \lfloor \frac{D}{D_k^c} \rfloor$. Then the maximum number of lanes possible for item k is given by $u_k^c = \lceil \frac{U_k}{n_k^c} \rceil$, the minimum number of lanes required for item k is given by $l_k^c = \lceil \frac{L_k}{n_k^c} \rceil$, and the value of having a full lane of item k in the cabinet using configuration c is given by $v_k^c = V_k \times n_k^c$. We assume that a lane dedicated to item k will be filled with n_k^{c*} units, where $c^* \in (1, \ldots, 6)$ is the final configuration chosen for item k; thus $v_k^{c*} = V_k \times n_k^{c*}$. Let us define an integer decision variable x_k as the number of lanes in the cabinet that are to be dedicated to item k. Then the objective of the model is to maximize $\sum_{k=1}^{K} v_k^{c*} x_k$.

3 Description of Heuristic

We first describe three different methods that reduce the number of configurations the algorithm considers, by eliminating similar or dominated configurations for each item. Then, we propose a heuristic algorithm to solve the problem.

3.1 Methods to Remove Dominated Configurations

Once an item's configuration has been selected, any remaining shelf space above and behind that item is unusable by other items. Consequently, one might as well fill up any extra space in the lane with as much as possible of the same item. Note that alternative configurations of the same item may differ in the amount of space wasted. Therefore, we first evaluate the local space efficiency of a configuration using the following methods to remove inefficient or dominated configurations.

The first method removes similar configurations for each item. The maximum number of possible configurations of an item is 6. But if two dimensions of an item have the same value then the number of possible configurations decreases to 3. If all three dimensions are identical, then there is only one unique configuration.

The second method separately removes dominated configurations for each dimension. We divide the three actual dimensions of an item by the minimum possible shelf height, h and then round these up to determine the minimum required height of a lane for that item with the different configurations. If these three values are all different then we don't have any dominated configurations for the height, but if we have a tie, then the one with highest original dimension dominates others as it causes less wasted space. Thus, any dominated dimension need not be considered as a height in any configuration for that item. We use the same procedure to determine dominated configurations for width with the only difference that instead of h we divide all dimensions by w. We use similar logic to find dominated configurations for depth by dividing the depth of the cabinet by the dimensions of the item, but now rounding down instead of up.

The third method eliminates those configurations that are multiples of some other configuration. Suppose that for some item k, (1) the minimum required height of a lane for configurations c and c' are identical (i.e., $h_k^c = h_k^{c'}$), and (2) the minimum required width of a lane, the maximum number in one lane, and the value of a lane of item k for configuration c are all an integer multiple (say m) of configuration c' (i.e., $\frac{w_k^c}{w_k^{c'}} = \frac{v_k^c}{v_k^{c'}} = m$). Then configuration c is equivalent to m parallel lanes of configuration c'. We may therefore eliminate configuration c.

3.2 Algorithm

The algorithm consists of five steps. First, a preprocessing step checks for feasibility. Second, the configuration selection step chooses the best configuration for each item. Third, the priority step builds two different sets of priority indices. Fourth, the allocation step allocates available space to items one by one according to their priority indices. This step is divided into two sub-steps which respectively ensure that the lower and upper bounds for the amount of item k are not violated. Finally, in the fifth step the objective value of the final solution is calculated. Let

s denote the most recently created new shelf, R_s the index set of items on shelf s, and. Start with $s = 1$.

1. *Preprocessing step*: This step checks for problem feasibility by ensuring that the total available space in a cabinet is not less than minimum amount of space required. In particular, $\sum_{k=1}^{K} \min_c \{l_k^c (w_k^c h_k^c)\} > IJ$, then the problem is definitely infeasible, so stop. Otherwise, there may still be a feasible solution so go to the configuration selection step.
2. *Configuration selection step*: In this step, we select the best configuration (c^*) to use for each item in the rest of the algorithm. To do this we execute the next three steps until all items are configured.

    2.1. Consider the ratio $\dfrac{(H_k^c W_k^c D_k^c)\left\lfloor \frac{D}{D_k^c} \right\rfloor}{(hh_k^c)(ww_k^c)D}$ for each configuration of each item. This ratio gives us the fraction of the available space actually being used in one lane of item k with configuration c. If there is a unique maximum, we denote the corresponding c by c^*.

    2.2. Use Sect. 3.1 to remove any dominated configurations from the ones remaining for each item. If there is only one remaining configuration for the item denote it by c^*, and go on to the priority step; else, go to step 2.3.

    2.3. We still have more than one possible configuration, so assign c^* to the one with minimum $w_k^c \times h_k^c$. If there is a tie assign c^* to the one with minimum $w_k^c + h_k^c$, and if there is still a tie assign c^* to the configuration with the lower index value for c; then move on to the priority step.

3. *Priority step*: In this step we build two different priority indices.

    3.1. Sort the elements of $\{1, \ldots, K\}$ in descending order of the item heights, break any ties based on widths and then on item index k. Define this new sorted set as S_1. Then go to step 3.2.

    3.2. Sort the elements of $\{1, \ldots, K\}$ in descending order of the quantity $\dfrac{v_k^{c^*}}{w_k^{c^*} h_k^{c^*}}$, and define this new sorted set as S_2, then go to the allocation step.

4. *Allocation step*:

    4.1. For successive $k \in S_1$, allocate the available space in the cabinet from lower left of the bottom shelf to meet the minimum requirement ($l_k^{c^*}$) of item k using a first-fit decreasing height heuristic (FFDH):

    4.1.1. Is S_1 empty? If yes go to step 4.2; else select the next $k \in S_1$ and go to step 4.1.2.

    4.1.2. Is $x_k^{c^*}$ less than its lower bound $l_k^{c^*}$? If yes go to step 4.1.3; else remove item k from set S_1 and go to step 4.1.1.

    4.1.3. Create a lane for item k (left justified) on the first partially filled shelf where it fits. If no shelf can accommodate item k then go to step 4.1.4; else, increment $x_k^{c^*}$ by 1 and go to the step 4.1.2.

4.1.4. Is there any space left to create new shelf for a lane of k? If yes, create new shelf ($s := s + 1$) and create a lane for k (left justified) on it, then increment $x_k^{c^*}$ by 1 and go to step 4.1.2; else stop, the problem is not solvable with this algorithm.

4.2. Create additional lanes (up to $u_k^{c^*}$ of them) for each k if possible: proceeding from the bottom shelf up:

4.2.1. Identify the first available location (left justified) for a new lane. If this corresponds to an existing (partially filled) shelf (s') go to step 4.2.3; otherwise go to step 4.2.2.

4.2.2. This step creates a new shelf if possible:

4.2.2.1. Is S_2 empty? If yes go to step 5; else select the next $k \in S_2$ and check if $x_k^{c^*}$ is less than its upper bound $u_k^{c^*}$. If yes go to step 4.2.2.2; else remove k from set S_2 and repeat this step

4.2.2.2. Is there any space left to create a new shelf for k? If yes, create a new shelf ($s := s + 1$) and assign the first lane on the shelf to item k, increment $x_k^{c^*}$ by 1 and go to step 4.2.1; else remove k from S_2 and go to step 4.2.2.1.

4.2.3. This step creates a new lane on existing shelf s' if possible:

4.2.3.1. Sort the current elements of S_2 in descending order of the quantity $\frac{v_k^{c^*}}{\sum_{k' \in R_{s'}} (h_{s'} - h_{k'}^{c^*}) w_{k'}^{c^*} + h_{s'} w_k^{c^*}}$, where $h_{s'} = \max_{k' \in R_{s'}} h_{k'}^{c^*}$ is the height of shelf s'. Define this new ordered set as S_3 and go to step 4.2.3.2.

4.2.3.2. Is S_3 empty? If yes go to step 4.2.1; else select the next $k \in S_3$ and check if $x_k^{c^*}$ is less than its upper bound $u_k^{c^*}$? If yes go to step 4.2.3.3; else remove item k from set S_2, S_3 and repeat this step

4.2.3.3. Try to create a lane for item k (left justified) in the available space on shelf s'. If it fits then increment $x_k^{c^*}$ by 1 and go to the step 4.2.1; else, remove k from S_3 and go to step 4.2.3.2

5. *Termination step*: Calculate the objective function $\sum_{k=1}^{K} v_k^{c^*} x_k^{c^*}$

4 Numerical Example

We now solve an actual example from a clinical setting with the proposed heuristic algorithm. We are given a cabinet with height 74 in., width 22 in. and depth 22 in. The height of each possible shelf and the width for each possible lane are 4.66 and 1.00 in. respectively. Therefore the cabinet can have up to 15 shelves, and up to 22 lanes per shelf (each with depth 22 in.). Table 1 provides the item characteristics for the items that we want to allocate in this cabinet.

Table 1 Items parameters

Item name	No.	H_k^{c*}	W_k^{c*}	D_k^{c*}	V_k	L_k^{c*}	U_k^{c*}	h_k^{c*}	w_k^{c*}	v_k^{c*}	l_k^{c*}	u_k^{c*}
Accucheck	1	3.75	6.50	5.50	11	10	30	1	7	44	3	8
BP cuff	2	8.25	4.00	5.50	8	10	20	2	4	32	3	5
Mask	3	7.00	6.50	2.00	2	10	20	2	7	22	1	2
Nebulizer	4	9.25	8.00	8.00	20	5	10	2	8	40	3	5
Thermopress	5	10.00	6.00	2.00	15	7	20	3	6	165	1	2
Heating pad	6	12.00	8.00	3.00	10	7	20	3	8	70	1	3
Humidifier	7	13.00	11.50	8.00	25	1	2	3	12	50	1	1

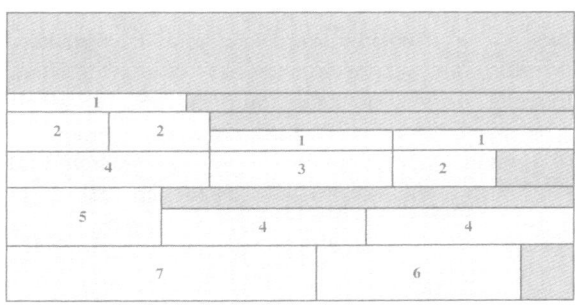

Fig. 1 Example configuration after step 4.1

Solving this example with the proposed algorithm yields an objective value of 1180. Figure 1 shows the algorithm at the end of step 4.1 when the minimum allocations have been completed and Fig. 2 shows the final allocation.

Since we do not know the optimal solution we propose an upper bound for empirically evaluating the heuristic solution. This bound, z_{up} is obtained by solving the following knapsack problem considering the overall volume of the cabinet $\max\{\sum_{k=1}^{K} v_k^{c*} x_k^{c*} : \sum_{k=1}^{K} h_k^{c*} w_k^{c*} x_k^{c*} \leq IJ, l_k^{c*} \leq x_k^{c*} \leq u_k^{c*} \ \forall k, x_k^{c*} \in \mathbb{Z}_+\}$. For our problem, $z_{up} = 1292$; therefore, it follows that the optimum value lies in the interval $[1180, 1292]$. The ratio $1292/1180 = 1.09$ indicates that the best we could possibly do is about 9 % better than our heuristic solution. In fact, it is likely that the

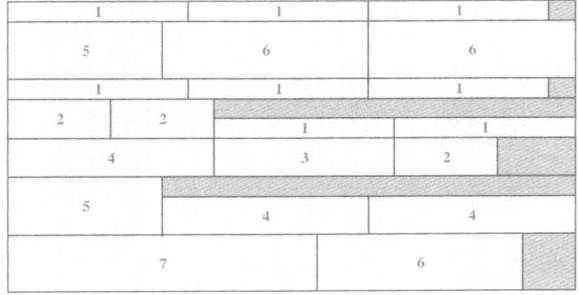

Fig. 2 Example final configuration

improvement that we could get with the optimal solution is actually much smaller since the upper bound is approximate and not necessarily a very tight one.

5 Concluding Remarks

This paper proposes a restricted version of the combined shelf space allocation and packing problem for integrated storage cabinets. Using an FFDH approach, consideration of different configurations and a new prioritization scheme, a heuristic algorithm is proposed to solve this problem. We also introduce different methods to choose the best configuration for each item. We empirically evaluate the heuristic solution by proposing an upper bound. This bound is obtained by considering the overall volume of the cabinet and solving a knapsack problem. The algorithm is illustrated with an actual example from a clinical setting and the solution is shown to be a very good one. We are currently in the process of evaluating the algorithm on a wider range of problems and on developing an integer programming formulation to find the optimal solution for this class of problems.

References

Hopper E, Turton B (2001) An empirical investigation of meta-heuristic and heuristic algorithms for a 2D packing problem. Eur J Oper Res 128(1):34–57
Leung JY, Tam TW, Wong C, Young GH, Chin FY (1990) Packing squares into a square. J Parallel Distrib Comput 10(3):271–275
Li K, Cheng K (1990) On three-dimensional packing. SIAM J Comput 19(5):847–867
Lodi A, Martello S, Vigo D (2002) Recent advances on two-dimensional bin packing problems. Discrete Appl Math 123(1):379–396
Ntene N, Van Vuuren JH (2009) A survey and comparison of guillotine heuristics for the 2D oriented offline strip packing problem. Discrete Optim 6(2):174–188
Yang M (2001) An efficient algorithm to allocate shelf space. Eur J Oper Res 131(1):107–118
Yang M, Chen W (1999) A study on shelf space allocation and management. Int J Prod Econ 60:309–317

Evaluation of Pull Control Strategies and Production Authorisation Card Policies Recovery Period in a Multi-product System

Chukwunonyelum Emmanuel Onyeocha, Joseph Khoury and John Geraghty

Abstract Pull production control strategies have demonstrated poor responses to large variations in demand volume. A recently developed pull production control strategy referred to as Base Stock-Kanban-CONWIP control strategy is capable of minimising WIP while maintaining high volume flexibility. This paper investigates the recovery period of pull production control strategies and production authorisation card policies after lumpy demands in a multi-product manufacturing system. Discrete event simulation and multi-objective optimisation approach were adopted to develop sets of non-dominated optimal solutions for configuring suitable decision set to the pull control strategies. It was shown that the shared Kanban allocation policy outperforms the dedicated Kanban allocation policy. Additionally the Basestock-Kanban CONWIP control strategy outperforms the alternatives.

Keywords Pull control strategies · Production authorisation card policies · Multi-product system · Lumpy demand

1 Introduction

Pull production control strategy has shown to be successful particularly in non-repetitive and single-product manufacturing environments resulting to the recognition of pull as a superior strategy over the push strategy (Womack et al. 1990;

C.E. Onyeocha (✉) · J. Geraghty
Enterprise Process Research Centre, School of Mechanical
and Manufacturing Engineering, Dublin City University, Dublin 9, Ireland
e-mail: chuks247@gmail.com

J. Geraghty
e-mail: John.geraghty@dcu.ie

J. Khoury
Methode Electronics Malta Limited, Industrial Estate Mriehel, Qormi BRK 3000, Malta
e-mail: Joseph.khoury@methodegermany.com

© Springer International Publishing Switzerland 2015
P. Cortés et al. (eds.), *Enhancing Synergies in a Collaborative Environment*,
Lecture Notes in Management and Industrial Engineering,
DOI 10.1007/978-3-319-14078-0_3

Marek et al. 2001; Krishnamurthy et al. 2004). Krishnamurthy et al. (2004) reported that the effective implementation of pull production control strategies in certain original equipment manufacturers led to the proposal that pull production control strategy is valuable in all manufacturing environments. Their findings showed that pull control strategies perform poor in multi-product systems with high varying demands and highly engineered products in small batches. According to Marek et al. (2001), large variations in demand volume and product mix adversely affects pull production control strategies.

Demand variability in multi-product manufacturing systems poses a difficult challenge and requires implementation of appropriate production control strategy to manage such variations. Lumpy and intermittent product volume or mix in a system undermines the principles of pull and results in line congestion, low throughput rate and long lead times in manufacturing systems (Bamber and Dale 2000). Various studies have evaluated the performance of pull production strategies in differs manufacturing conditions however, not much interest has been shown on the time of recovery of pull production control strategy after a lumpy demand. This paper examines the time of recovery of GKCS, EKCS, HK-CONWIP and Basestock Kanban CONWIP (BK-CONWIP) control strategies in both D-KAP and S-KAP modes in a four product five stage assembly line with sequence-dependent changeover times, and finite buffer. Simulation and multi-objective optimisation techniques were used to generate Pareto-frontier for selection of the control parameters for the experiments. The rest of this paper is structured as follows: Sect. 2 provides a brief review of the relevant studies on multi-product pull production control strategies, while a description of the manufacturing system and the experimental set-up is presented in Sect. 3. The simulation results and analysis is provided in Sect. 4, while the research conclusion is presented in Sect. 5.

2 Background

Pull control strategy starts production of parts in a system based on actual demand and it attempts to control the Work-In-Process (WIP) inventory in a system, while observing the throughput. Some pull strategies uses signal cards known as production authorisation cards to release parts in a system, for instance Kanban control strategy (Sugimori et al. 1977), while others do not, such as Basestock control strategy (Kimball 1988). Several variations of pull strategies exist based on the need to improve the performance of pull in certain manufacturing environments (Onyeocha and Geraghty 2012). Some of these variations integrated push and pull control strategies for better performance to certain manufacturing conditions (Spearman et al. 1990; Onyeocha and Geraghty 2012). The ability of pull strategies to control WIP in a system gave pull added advantage over push strategies, as manufacturers prefer to limit the amount of materials released into a system while allowing demands to be on paper than physical materials in a system (Marek et al. 2001). However in multi-product pull controlled systems, large volume of production authorisation cards are planned to respond to multi-products and demand variations, causing proliferation of WIP. Large WIP in a system results to line congestion, long lead time and low throughput.

Proliferation of WIP in multi-product systems undermines the principle of pull control strategies and has not been adequately resolved as a majority of studies on pull control strategies concentrated on single product manufacturing environments (Deleersnyder et al. 1992; Spearman et al. 1990; Onyeocha and Geraghty 2012). On the other hand, the widely studied areas in multi-product manufacturing environments are the planning and scheduling and optimisation (see, Onyeocha and Geraghty 2012). In those studies, the production authorisation cards are dedicated to a product type. The need to reduce the WIP level in multi-product pull control systems led to the findings of Baynat et al. (2002), showing that the control mechanism of production authorisation cards in pull strategies can be dedicated or shared. S-KAP was reported in the literature to have the capability of achieving lower WIP at any given delivery performance with fewer production authorisation cards under steady or low environmental and system variability (Olaitan and Geraghty 2013; Onyeocha et al. 2013). In addition, some of the pull control strategies that are successful in single product manufacturing environments for instance KCS, CONWIP, BSCS and HK-CONWIP are incapable to operate in S-KAP mode (Baynat et al. 2002; Onyeocha and Geraghty 2012). Onyeocha and Geraghty (2012) proposed a modification approach to improve the control mechanisms of pull control strategies to facilitate their operations in S-KAP mode. Furthermore they developed a new pull control strategy (BK-CONWIP) with capability of quick responds manufacturing and high flexibility.

The innovation of developing production control strategies with quick responds manufacturing capability is valuable, however, it is necessary to understand the behaviour of these control strategies and the time of their recovery when lumpy demand occurs in a system.

3 Experimental Conditions

A brief description of the system modelled and the experimental set-up is provided in this section. The system is an automotive electronics components production plant, which produces four product-types. These four products are classified into two product families (A and B) and are assembled in a five stage line (see Fig. 1). Numerous complex operations are performed simultaneously in the system, such as minimisation of the number of set-ups, minimisation of the run/batch quantity and dynamic priority control. The processing times, machine unreliability, machine capacity, and buffer sizes are modelled based on the data from the company (see Table 1).

The demands are made once a week and the due date for an order is one production week. There are two demand datasets (A and B), with each of the four products having demand profile for a six week period as shown in Table 2.

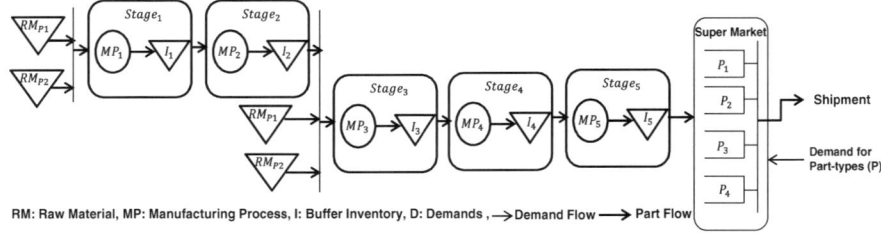

Fig. 1 Five-stage multi-product manufacturing system modelled (*Source* Authors' production)

The demand dataset B has a lumpy demand at its first week demand and backlogs occurs in the system when demands are not met within a production week period. The backlogs are first served in the next production week before commencement of the week's production. The ability to fully clear the backlogs and the demands of the current production week is considered as the recovery period of a pull strategy.

The simulation experiments in this work were conducted using ExtendSim simulation software application and were carried out over thirty replications, four weeks warm-up period and ten weeks run length for each of the models. The control parameters of the models were optimised using the demand dataset A via a multi-objective optimisation application block developed by Kernan and Geraghty (2004). The Pareto optimiser configurations were set as follows: the mutation rate set to 10 %, the number of generations set to 200 and 30 replications per iteration. Table 3 provides a description of the search space of the pull control strategies and production authorisation card policies at 100 % Service Level (SL). It was observed that S-KAP maintained lower production authorisation cards than D-KAP.

4 Simulation Results and Analysis

The optimal settings of demand dataset A that achieved 100 % service level was used in the configuration of the control parameters of a specified Kanban Allocation Policy and specified Production Control Strategy combination (KAP+PCS) and was kept unchanged while simulating the demand dataset B to examine the rate of backlog and time of recovery of each KAP+PCS. The average Total Work-In-Process (TWIP) inventory, average Total Service Level (TSL) and average Total Backlog (TBL) for each of KAP+PCS were plotted in Fig. 2, while Table 4 provides a description of screening and selection of a superior strategy based on the least WIP level achieved by KAP+PCS using the Nelson's ranking and selection procedure (Nelson et al. 2001). In addition, the weekly cumulative backlogs obtained from demand dataset B with lumpy demand in week 1was plotted in Fig. 3 to describe the position of backlogs in the system. Due to space limitations only the first and fifth plots are presented in Fig. 3.

Table 1 System configuration

Stage	Product 1 (Boxes)	Product 2 (Boxes)	Product 3 (Boxes)	Product 4 (Boxes)	Maintenance: exponential distribution mean				Set-up normally distributed
					Product 1 and product 2		Product 3 and product 4		
	Process time (h)	Process time (h)	Process time (h)	Process time (h)	MTBF[c] (h)	MTTR[d] (h)	MTBF[c] (h)	MTTR[d] (h)	Set-up times (h)
1	0.13	0.13	N/A[a]	N/A[a]	29.90	0.75	N/A[a]	N/A[a]	N/A[a]
2	0.11	0.11	N/A[a]	N/A[a]	3.50	0.23	N/A[a]	N/A[a]	N/A[a]
3	0.10	0.10	0.19	0.18	6.10	0.23	6.10	0.23	N^b(0.33, 0.11)
4	0.10	0.10	0.13	0.13	6.10	0.23	6.10	0.23	N^b(0.33, 0.11)
5	0.08	0.08	0.11	0.11	6.10	0.23	6.10	0.23	N^b(0.33, 0.11)

Source Authors' production
[a] Not applicable
[b] Normally distributed
[c] Mean time before failure
[d] Mean time to repair

Table 2 Demand profile

Demand dataset	Product	Production week (Box quantity)					
		Week 1	Week 2	Week 3	Week 4	Week 5	Week 6
A	Product 1	542	452	404	503	247	483
	Product 2	130	224	142	118	129	114
	Product 3	130	184	131	159	125	147
	Product 4	110	138	147	71	61	39
Total demand volume	Box quantity	912	998	824	851	562	783
B	Product 1	1,133	542	452	404	503	247
	Product 2	220	130	224	142	118	129
	Product 3	211	130	184	131	159	125
	Product 4	154	110	138	147	71	61
Total demand volume	Box quantity	1,718	912	998	824	851	562

Source Authors' production

Table 3 Search space and optimal values of KAP+PCS at 100 % service level

KAP+PCS	Search range	CONWIP	Kanban	Basestock
D-KAP HK-CONWIP	4–350	285	44	205
S-KAP HK-CONWIP	4–350	262	26	180
D-KAP GKCS	4–350	N/A[a]	302	254
S-KAP GKCS	4–350	N/A[a]	285	248
D-KAP EKCS	4–350	N/A[a]	290	245
S-KAP EKCS	4–350	N/A[a]	279	245
D-KAP BK-CONWIP	4–350	303	43	0
S-KAP BK-CONWIP	4–350	225	25	0

Source Authors' production
[a] Not applicable

Figure 2 shows that in demand dataset A, all KAP+PCS achieved 100 % average total service level and zero average total backlogs at different work-in-process inventory levels. S-KAP outperformed D-KAP in terms of WIP levels, while BK-CONWIP outperformed HK-CONWIP, GKCS and EKCS. S-KAP BK-CONWIP is superior to all the alternatives as it maintained the least WIP level. The performance in demand dataset B is similar, such that the shared policy outperformed the dedicated policy. BK-CONWIP outperformed HK-CONWIP, EKCS and GKCS, while S-KAP BK-CONWIP outperformed the alternatives. However, no KAP+PCS achieved 100 % average total service level and zero average total backlogs.

The average total work-in-process inventory was screened using Nelson's combined procedure to determine the best KAP+PCS. The control parameters of

Evaluation of Pull Control Strategies and Production ...

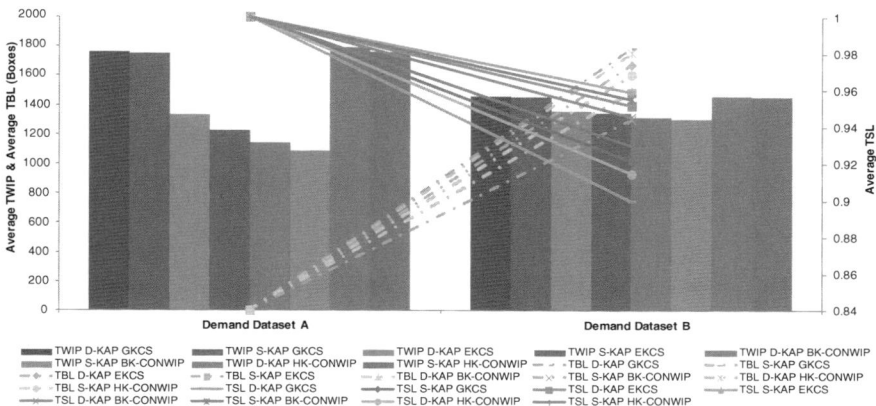

Fig. 2 Results of WIP levels, service levels and backlogs of KAP+PCS (*Source* Authors' production)

Table 4 Application of Nelson's combined procedure for best KAP+PCS

KAP+PCS	i	n_o	$\bar{Y}_i$	S_i^2	Keep?	N_i
D-KAP HK-CONWIP	1	30	1,791.50	18,893.80	Eliminate	330
S-KAP HK-CONWIP	2	30	1,760.80	18,021.01	Eliminate	329
D-KAP GKCS	3	30	1,761.32	18,987.38	Eliminate	417
S-KAP GKCS	4	30	1,750.17	18,859.01	Eliminate	474
D-KAP EKCS	5	30	1,337.20	20,985.78	Eliminate	521
S-KAP EKCS	6	30	1,225.59	20,182.04	Eliminate	496
D-KAP BK-CONWIP	7	30	1,141.32	17,824.63	Eliminate	284
S-KAP BK-CONWIP	8	30	1,085.44	17,830.12	Keep	172

Source Authors' production

the Nelson's combined procedure used are as follows: k = 8, n_o = 30, the overall confidence level (α) is 90 % for the combined procedure, that is α = 0.1, also confidence level of 95 % for each of the two stage sampling procedures is given as $\alpha_0 = \alpha_1 = 0.5$. A practical significant difference of 30 boxes is reasonable, provided by the company's personnel. Rinott's integral h = 4.253, where n_o is the initial number of replications for each strategy. N_i is the number of additional simulation needed for further screening. $S^2 i$ is the variance of the sample data and $\bar{Y}_i$ is the average of the sampled data. The results of the application of Nelson' combined procedure for demand dataset A shows that S-KAP BK-CONWIP is the best KAP +PCS.

Figure 3 shows that BK-CONWIP recovered better than the alternatives. BK-CONWIP in both D-KAP and S-KAP modes recovered on the week 5 (600 h) period. While HK-CONWIP, EKCS and GKCS recovered on week 6 (720 h) period. It was shown that S-KAP BK-CONWIP has the least total backlog in week 1, while D-KAP

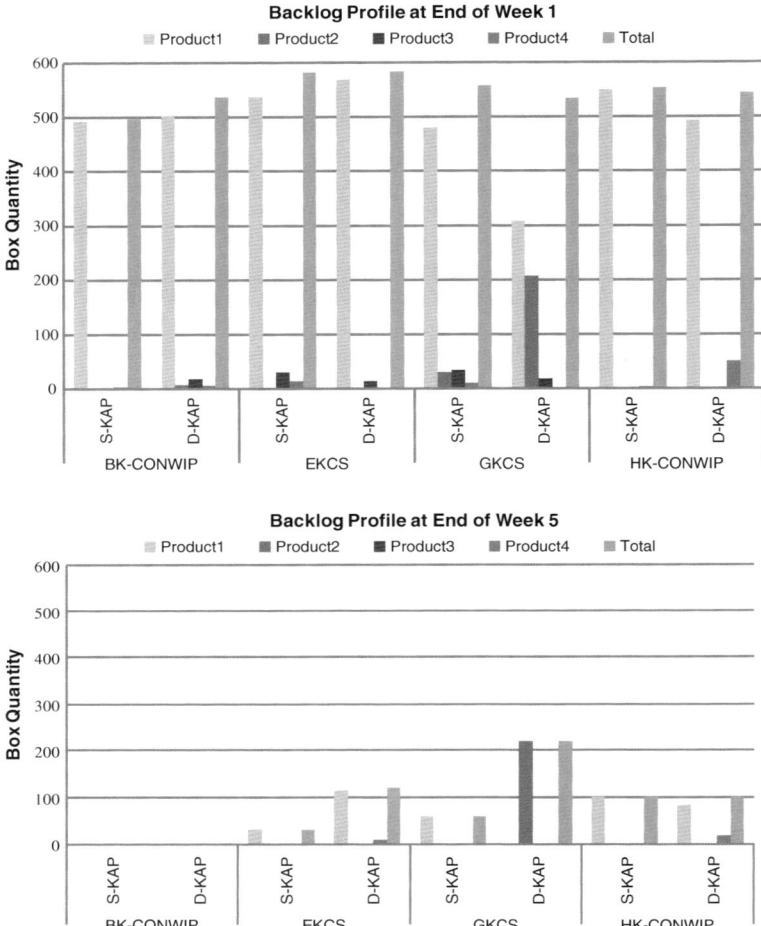

Fig. 3 End of the week 1 backlog positions (*Source* Authors' production)

EKCS has the most backlogs in the system. S-KAP BK-CONWIP maintained the least backlogs in all week. In week 5, BK-CONWIP in both S-KAP and D-KAP modes achieved total recovery and had zero backlogs, followed by S-KAP EKCS with 30 boxes (backlogs), then S-KAP GKCS with 59 boxes (backlogs), next is D-KAP HK-CONWIP with 100 boxes (backlogs), followed by S-KAP HK-CONWIP with 101 boxes, and D-KAP EKCS with 121 boxes, while D-KAP GKCS has the most backlogs (220 boxes) in the system. In week 6, the remaining KAP+PCS recovered and achieved zero backlogs in the system. S-KAP was observed to outperform D-KAP in quick recovery, while BK-CONWIP outperformed the alternatives as it recovered earlier than the rest.

5 Conclusion

We have demonstrated that S-KAP BK-CONWIP provides the best flexibility in terms of minimising WIP in the presence of unsteady demands. While the performances of all systems degraded significantly when a surge in demand volume occurred in the week 1 of demand dataset B. The shared policy of BK-CONWIP and EKCS maintained good service levels, above 95 %. When lumpy demand occurred, all KAP+PCS accumulated significant proportion of the end of week backlog (see Fig. 3). However, S-KAP BK-CONWIP maintained the least quantity of the total backlogs. Similarly, we showed that BK-CONWIP provides quick response with regards to backlog recovery period. Shared Kanban allocation policy recovered quicker than the dedicated policy, while the best KAP+PCS is the S-KAP BK-CONWIP by maintaining the least total end of week backlog. Further research on the robustness of these KAP+PCS systems with more complexity in terms of number of products and number of stages would provide clearer understanding of appropriateness of these strategies for multi-product environments with lumpy demand.

References

Bamber L, Dale BG (2000) Lean production: a study of application in a traditional manufacturing environment. Prod Plan Control 11(3):291–298

Baynat B, Buzacott JA, Dallery Y (2002) Multi-product kanban-like control systems. Int J Prod Res 40:4225–4255

Deleersnyder JL, Hodgson TJ, King RE, O'Grady PJ, Savva A (1992) Integrating kanban type pull systems and MRP type push systems: insights from a Markovian model. IIE Trans 24:43–56

Kernan B, Geraghty J (2004) A multi-objective genetic algorithm for extend. In: Proceedings of the first irish workshop on simulation in manufacturing, services and logistics, Limerick, Ireland, pp 83–92 Kimball GE (1988) General principles of inventory control. Journal of manufacturing and operations management 1:119-130

Kimball GE (1988) General principles of inventory control. J manufact oper manag 1:119–130

Krishnamurthy A, Suri R, Vernon M (2004) Re-examining the performance of MRP and Kanban material control strategies for multi-product flexible manufacturing systems. J Flex Manuf Syst 16:123–150

Marek RP, Elkins DA, Smith DR (2001) Understanding the fundamentals of kanban and CONWIP pull systems using simulation. In: Peters BA, Smith JS, Medeiros DJ, Rohrer MW (eds) Proceedings of the 33rd conference on winter simulation, Arlington, Virginia, vol 1. IEEE Computing Society, Washington, DC, USA, pp 921–929

Nelson BL, Swann J, Goldsman D, Song W (2001) Simple procedures for selecting the best simulated system when the number of alternatives is large. Oper Res 49:950–963

Olaitan OA, Geraghty J (2013) Evaluation of production control strategies for negligible-setup, multi-product, serial lines with consideration for robustness. J Manuf Technol Manage 24 (3):331–357

Onyeocha CE, Geraghty J (2012) A modification of the hybrid kanban-CONWIP production control strategy for multi-product manufacturing systems. In: IMC29: international manufacturing conference proceedings, University of Ulster, Belfast, UK

Onyeocha CE, Khoury J, Geraghty J (2013). A comparison of kanban-like control strategies in a multi-product manufacturing system under erratic demand. In: Pasupathy, R, Kim S-H, Tolk A, Hill R, Kuhl ME (eds) Proceedings of the 2013 conference on winter simulation, vol 1. IEEE, Washington DC, USA, pp 2730–2741

Spearman ML, Woodruff D, Hopp W (1990) CONWIP: a pull alternative to Kanban. Int J Prod Res 28:879–894

Sugimori Y, Kusunoki K, Cho F, Uchikawa S (1977) Toyota production system and Kanban system: materialization of just-in-time and respect-for-human system. Int J Prod Res 15 (6):553–564

Womack JP, Jones DT, Roos D (1990) The machine that changed the world: the story of lean production. Harper Collins Publishers, New York

Smart Decision in Industrial Site Selection: What's New in the Case of a Steel Mill in Brazil?

Alexandre de Oliveira Gomes and Carlos Alberto Nunes Cosenza

Abstract The site-selection problem on Steel industry deals with complex aspects of economy, society, engineering and environment. For a long time, there was only the logistics outlook, but until today, this is not completely overcame. The consequences of a wrong decision are adverse for most firms. In this regard, this paper proposes a hierarchical multi-criteria Fuzzy-based model and simulates a steel facility selection in Brazil to approach the problem beyond the logistics, including new factors imposed by the current business and social demands. Initially, we present the first two steel mill's locational studies in Brazil and their relationship with the classical models. Then, a hypothetical steel mill of 2.5 Mtons/year is specified to enable the assessment of four locational options over twenty factors. The results showed that environment and engineering factors gained relevance, where the former may be decisive and the later mitigates the risks for construction and operation, contrarily to the traditional view. Furthermore, transportation still plays a relevant role onto decision, but solely. Lastly, short and long terms forecasts encourage to rethink the traditional view of locating steel mills close to the resources.

Keywords Site selection · Steel industry · Multi-criteria fuzzy model · Engineering

---

A. de Oliveira Gomes (✉) · C.A.N. Cosenza
Programa de Engenharia de Produção, COPPE/UFRJ, Universidade Federal do Rio de Janeiro, Cidade Universitária—Centro de Tecnologia—Bloco F, sala 110, Ilha do Fundão, Rio de Janeiro, Brazil
e-mail: gomesk2@gmail.com

C.A.N. Cosenza
e-mail: cosenzacoppe@gmail.com

1 Introduction

The steel industry is intrinsically related to the economic development of many countries. In Brazil, the growth started in 1921 when the Belgian-Luxembourg ARBED group joined to an existing Brazilian mill (Baer 1970). In 1946, Brazil sets up the biggest steel mill in Latin America. In the 90s, the changeover: steel company's privatization due to the reflect of world globalization. The emergence of the "knowledge society" (Dziekaniak and Rover 2011) and the "Social License" (Prno and Slocombe 2012) imposed a review of the basis for site selection decision-making. The transportation costs are no longer the sole factor to be considered, not meaning that they have lost importance, but other factors emerged accordingly. This paper proposes a heuristic approach to address industrial location decision-making, in particular, steel mills in Brazil. It demonstrates the benefits of introducing engineering factors on the decision. Additionally, it discusses the country's experience on featuring a non-standard geographical distribution of its steel park.

2 Approaching the Problem

Historically, in Brazil the location of steel mill facilities has not followed the deterministic pattern of Weber's *Material Index* (1929), which rates the industries as oriented to the resources, market or labour. According to such index, steel plants were classified as resource-oriented. Weber (1929) and his followers (Predöhl, Palander, Ritschl and others), considered transportation costs the most critical locational factor (Richardson 1975).

The steel park diversification is notorious from 1960, when one-third of the national facilities were close to the sources of raw materials—at the continent, motivated by the intention of local mine owners to value their iron ore commodities. Less than one-fifth were near to the market—at the southeast coast, where existing firms were willing to produce special steel to serve the emerging automobile park located there. Furthermore, there was one single facility placed between the resources and the market (Baer 1970). The major consumption was concentrated in the southeast coast (Kafuri 1957).

This geographic diversification suggests a debate on whether the location of a facility should follow a rule like Weber's (1929) attempt or, on the contrary, it should be evaluated case-by-case, in light of the interests and restrictions involved. As Richardson (1975) cited that there is not a general locational theory, nor for steel mills, this paper approaches the locational decision beyond the logistic problem, includes new objective and subjective factors and, points out the value of engineering aspects for site selection decision-making.

3 The Rational of the First Two Locational Studies in Brazil

Kafuri (1957) and Junior (1961) reported the first two locational studies for steel plants carried out in Brazil over the 50's. The militaries ruled the country and Brazilian industrialization was just at the beginning. Influenced by Weber's (1929) study in "*The theory of location*", Lösch (1954) and Dantzig (1951 *apud* KAFURI 1957), in his study on Linear Programming, "*Application of the Simplex method to a transportation problem*", the location problem was approached using factors that defined the criteria of locating "close-to-resources", "close-to-market" or "in the middle", which had to be defined first. That differed a bit from the orthodox decision-making process where no alternative options other than "close-to-resource" were considered. The transportation cost was the most critical criterion and, in both studies, engineering was applied only to implement the decision made to a given place.

4 Model Building

The following scheme illustrates the Multi-criteria Coppe-Cosenza Fuzzy Model framework (Cosenza 1997, 2001) used with some modifications to support a decision of a steel plant location in Brazil (Fig. 1).

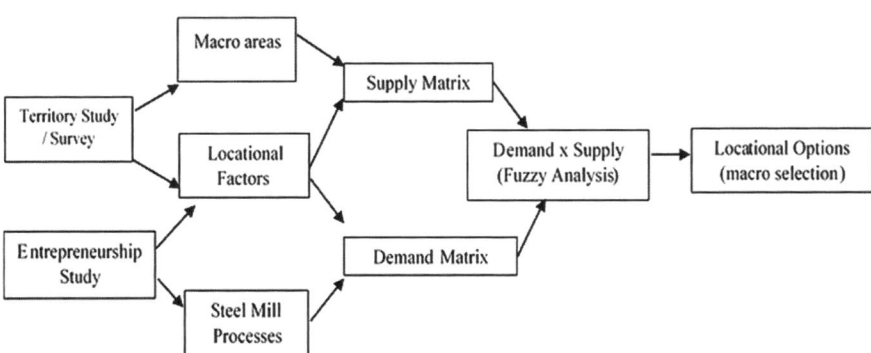

Fig. 1 Site selection algorithm

5 Steel Plant Specification

The hypothetical steel facility simulated has the following main specification:

- Integrated mill with a production capacity of 2.5 Mtons/year;
- Finished products are flat slabs and rolled coils—slabs for exportation and rolled coils for the domestic market;
- Iron ore from domestic market and metallurgical coal is imported;
- Power supply of 150 MW, generated by an own Thermoelectric Plant;
- Own Water plant and Sewage treatment station;
- 4,500 employees (2,000 hired and 2,500 third-parties);
- Construction area + expansion: 1,200 ha (12 km^2 or 12,000,000 m^2).

6 Macro Areas—Locational Options

Next Table 1 (that is at the bottom of the page) shows the selected macro areas as locational options.

Table 1 Macro areas—locational options

Macro area	Area (km^2)	Latitude	Longitude	City	State	Location
Açu	19.68	21°51′53.56″S	41°4′27.64″O	São João da Barra	RJ	BMR
Ubu	19.63	20°46′56.25″S	40°36′48.36″O	Anchieta	ES	CTM
Marabá	19.78	5°21′48.65″S	49°12′59.64″O	Marabá	PA	CTR
Suape	20.16	8°24′2.88″S	35°2′2.74″O	Cabo de Santo Agostinho	PE	CTM

Source GIS database; *BMR* Between market and resources; *CTM* Close-to-market; *CTR* Close-to-resources

7 Locational Factors

The use of environment and engineering locational factors represents the difference between the traditional and the contemporaneous approach. Among these, engineering is the newest trend on supporting site selection decision. The engineering factors attempt to mitigate the construction risks and to reduce investment and maintenance costs, e.g. construction works, piping, electrical power transmission lines and equipment size and quantity (Table 2).

Table 2 Locational factors

Production	Transportation	Environment	Engineering
Raw material (iron ore)	Port	APPs	Topography and declivity
Electrical power	Railway	Traditional community	Liquid effluent reservoir
Water	Waterway	Indigenous land	Water capture
Labour	Roads	Conservation unit	Interconnection to HV lines
		City master plan	Interconnection to PSTN
		Atmospheric emission impact on communities	Floodable area

8 Demand and Supply Matrices

8.1 Project Typology (h) and Demand Matrix ($A_{h \times n}$)

The project typology (h) is unitary because only one type of project is being considered. The Demand Matrix ($A_{1 \times 20}$) is represented in Table 3.

Table 3 Locational factors—degree of relevance

Production	Crucial (A)	Conditional (B)	Not very conditional (C)	Irrelevant (D)
1 Raw material (iron ore)		X		
2 Electrical power	X			
3 Water	X			
4 Labour			X	
5 Port	X			
6 Railway		X		
7 Waterway		X		
8 Roads		X		
9 APPs		X		
10 Traditional community			X	
11 Indigenous land			X	
12 Conservation unit		X		
13 City master plan		X		
14 Atmospheric emission impact on communities		X		
15 Topography and declivity			X	
16 Liquid effluent reservoir		X		
17 Water capture			X	
18 Interconnection to High voltage (HV) transmission lines			X	
19 Interconnection to telecommunication network (PSTN)			X	
20 Floodable area		X		

8.2 Supply Matrix ($B_{n \times m} = B_{20 \times 4}$)

$\underline{N}$ is the number of locational factors and $\underline{m}$, the locational options. The ratings A, B, C and D resulted from the comparison between actual field data and the parameter's set up by experts on steel industry. Distances, incidences and topography data were obtained through Geographic Information System (GIS) (Rushton 2001; NCGIA 2013), and the remaining via questionnaires and interview with experts or search on database (Table 4).

8.3 Relationship Between Demand and Supply Matrices

Establishes a vis-à-vis comparison. It follows the matrix product logic, but instead of figuring each $a_{ij} \times b_{ij}$ multiplication, the values are compared according to the Table 5, that is elaborated by the model's expert according to the problem.

Table 4 Supply matrix $B_{20 \times 4}$

Locational factors	Açu	Ubu	Marabá	Suape
1 Raw material (iron ore)	A	A	A	D
2 Electrical power	B	B	B	B
3 Water	A	A	A	A
4 Labour	A	A	A	A
5 Port	A	A	C	A
6 Railway	B	B	B	A
7 Waterway	A	D	D	D
8 Roads	A	A	A	A
9 APPs	B	C	B	A
10 Traditional community	A	A	A	A
11 Indigenous land	A	A	A	A
12 Conservation unit	A	A	A	A
13 City master plan	A	A	A	A
14 Atmospheric emission impact on communities	C	C	B	C
15 Topography and declivity	A	D	D	C
16 Liquid effluent reservoir	C	A	A	A
17 Water capture	B	A	A	A
18 Interconnection to high voltage (HV) transmission lines	B	B	A	B
19 Interconnection to telecommunication network (PSTN)	B	B	B	B
20 Floodable area	A	B	B	B

A Surplus; *B* Required; *C* Insufficient; *D* Absent

Table 5 Relationship between demand and supply matrices

Project demand/Macro area supply		Surplus	Required	Insufficient	Absent
		A	B	C	D
Crucial	A	1 + 4/n	1	1 − 4/n	0
Conditional	B	1 + 3/n	1	1 − 3/n	1/(n − 16)!
Not very conditional	C	1 + 2/n	1	1 − 2/n	1/(n − 17)!
Irrelevant	D	1 + 1/n	1	1 − 1/n	1/(n − 18)!

8.4 Site Selection Matrix ($C_{h \times m}$)

The result of Σ ($a_{ij} \times b_{ij}$) is [C] = {Açu (21.40); Ubu (19.46); Marabá (19.46) and Suape (19.53)}. Multiplying each element Σ (c_{ik}) above by the scalar $\underline{E}$ ($E_{h \times h} = E_{1 \times 1}$), we obtain Matrix $C_{h \times m}$ (Table 6), which lists the index of suitability for each macro area.

$$E = 1/n = 1/20 = 0.05 \tag{1}$$

9 Results

According to Table 5, the project is feasible when [A] × [B] results at least in 1 ("required" or "surplus"), thus on short-term only Açu would be eligible. The far distance from raw material (Table 4) led Suape to disqualification, otherwise, e.g. if equally assessed (rate "A"), it would rank the 2° position, with suitability index of 1.03.

9.1 Long-Term

For long-term simulation, we took "insufficient" and "absent" factors and forecasted them in a five year's horizon. Factors related to natural resources (mine, waterway or liquid effluent reservoir), geological (topography) and environment (APPs) tend

Table 6 Matrix $C_{h \times m}$: short-term ranking

Category	Açu	Suape	Marabá	Ubu
Production	1°	2°	1°	1°
Transportation	1°	2°	4°	3°
Environment	2°	1°	1°	3°
Engineering	1°	1°	2°	3°
Suitability index	1.07	0.98	0.97	0.97
Final ranking	1°	2°	3°	4°

Table 7 Matrix $C_{h \times m}$: long-term ranking

	Açu	Marabá	Suape	Ubu
Suitability index	1.07	1.05	0.93	0.93
Final ranking	1°	2°	3°	4°

not to vary, whereas ongoing construction works in Port and in existing waterway might be completed, therefore, it improves location's rates. The distance among communities and locations tend to shorten due to population's growth, and then the factor "atmospheric emission impact on communities" decreases for all locations (Table 7).

To measure only the environmental factor's impact, we repeated the assumption made in 9 (rate "A" to Suape on raw material's factor) and held the others. The outcome disqualifies Suape, its suitability index drops from 1.03 to 0.99.

10 Conclusion

The results showed that an intermediate position between resources and market is more proper to Brazil, but it does not mean another attempt to set a rule. The long-term simulation introduced Marabá as the second interesting option, which indicated the more plentiful is the supply, *coeteris paribus,* the locational option varies independently. The transportation remained important, but sole. Finally, the introducing of engineering and environmental factors divided the past and the present's approach and fits the model in the current business, economic, social and political environment. They expanded the site-selection outlook beyond the logistics. Engineering factors added asymmetry and enriched the comparison, besides contributing to mitigate the risks and to reduce the costs associated to plant's construction and operation. Environment factors proved that may be decisive. Moreover, the model delivered satisfactory results and demonstrated great performance on processing different data: numerical, linguistic, objective and subjective.

References

Baer W (1970) Siderurgia e desenvolvimento brasileiro. Zahar Editores, Rio de Janeiro

Cosenza CAN, Vieira LEV (1997) Modelo do confronto entre requerimentos e satisfação de critérios para problemas multicriteriais discretos. In: XVII ENEGEP–Encontro Nacional de Engenharia de Produção

Cosenza, CAN (2001) A evolução do modelo coppetec-cosenza de localização industrial para modelo de hierarquia fuzzy e sua aplicabilidade. In: XXI Encontro Nacional de Engenharia de Produção, Salvador, BA, Brazil, vol 1, Aug 2001, pp 30–35

Dantzig, George B (1951) Application of the simplex method to a transportation problem. In: Activity analysis of production and allocation, vol 13, p 359–373

Dziekaniak G, Rover A (2011) Sociedade do conhecimento: características, demandas e requisitos. Datagramazero—Revista da Informação 12(5)

Junior AL (1961) O projeto da Usiminas e sua justificativa no planejamento da siderurgia brasileira. Geologia e Metalurgia, Universidade do Estado de São Paulo—USP, São Paulo, n° 23

Kafuri, JF (1957) Estudo da localização de uma indústria siderúrgica. Revista Brasileira de Economia, Rio de Janeiro, n° 3, Ano 11, Setembro

Lösch A (1954) The economics of location. Yale University Press, New Haven

NCGIA—National Center of Geographic Information and Analysis (2013) GIS definition [online]. Available: http://www.ncgia.ucsb.edu. Accessed 12 April 2013]

Prno J, Slocombe D (2012) Exploring the origins of social license to operate in the mining sector: perspectives from governance and sustainability theories, vol 37. Elsevier, London, pp 346–357

Richardson HW (1975) Economia regional: teoria da localização, estrutura urbana e crescimento regional. Zahar Editores, Rio de Janeiro

Rushton G (2001) Spatial decision support systems. In: Smelser NJ, Baltes PB (eds) International encyclopedia of the social & behavioral sciences, vol 22. Pergamon, Oxford, pp 14785–14788

Weber A (1929) Theory of the location of industries. University of Chicago Press, Chicago

Dynamic Multi-objective Maximal Covering Location Problem with Gradual Coverage

Mohammad Forghani Youshanlo and Rashed Sahraeian

Abstract In this study, a dynamic multiple-objective model of maximal covering location problem has been presented in which the gradual coverage radius has been taken into account. The model has several aims, including: maximizing initial coverage, maximizing backup coverage, and minimizing the fixed and variable expenses related to the location, transportation vehicles, and variable demands in the course of programming. This model intends to find the best location for the centers of emergency services which can meet the maximum demands of present and future. In this model, the variables of backup and initial coverage have been presented fractionally in order to cover more demands. Then, an example with random numbers has been given and solved by lexicographic multi-objective linear programming (LMOLP) and fuzzy goal programming (FGP) approaches. The obtained results show the superiority of fuzzy goal programming approaches.

Keywords Maximal covering location problem (MCLP) · Fuzzy goal programming (FGP) · Multi-objective models · Dynamic location

1 Introduction

Creating systems of emergency services and establishing the related facilities have a pivotal role in the success of planning. The model of maximal covering location has a lot of applications in this field. In MCLP, the objective is to determine the location

---

M. Forghani Youshanlo
Department of Industrial Engineering, Najafabad Branch, Islamic Azad University,
Isfahan, Iran
e-mail: Mohammadind@sin.iaun.ac.ir

R. Sahraeian (✉)
Department of Industrial Engineering, College of Engineering, Shahed University,
Tehran, Iran
e-mail: sahraeian@shahed.ac.ir

© Springer International Publishing Switzerland 2015
P. Cortés et al. (eds.), *Enhancing Synergies in a Collaborative Environment*,
Lecture Notes in Management and Industrial Engineering,
DOI 10.1007/978-3-319-14078-0_5

of a limited number of facilities in order to maximize the extent of coverage based on coverage radius. In this study, another multiple-objective model of maximal covering location has been presented. Populations and the extent of demands are continuously changing throughout the time in different places. Based on these changes, the number of transportation vehicles in the centers of emergency services must change. Determining the initial coverage and backup coverage in a fractional manner are among the other features of the model. In many models, coverage radius is determined in a fixed manner, which is not consistent with reality. In the presented model, coverage radius is gradual. In addition, capacity constraint of facilities is taken into account. The potential locations have fixed and variable expenses in the installation of facilities. Up to now, different methods have been suggested for the solution of multiple-objective models, including: goal programming and weighted methods. The goal programming enables decision-makers to determine the level of expectations for each goal. Since in reality, it is difficult to find the expected figure exactly, the intended figures might not be found with assurance. In situations which the decision-makers cannot determine the objective figures, the fuzzy goal programming can help them.

2 Literature Review

Up to now, many models have been suggested in this field. All these models have developed the basic models. In the field of multiple-objective models, Araz et al. (2007) proposed a multi-objective model. The objectives considered in the model are maximization of the population covered by one vehicle, maximization of the population with backup coverage, and increasing the service level. Beraldi and Bruni (2009) suggested a probability model of the positioning of emergency transportation vehicles. This model seeks to minimize fixed and variable expenses related to the stationing of transportation vehicles. Coskun and Erol (2010) proposed a single-objective model whose focus was to minimize the expenses related to the stationing and the allocation of transportation vehicles. In the suggested models, the coverage radii are fixed. The gradual coverage is a concept which helps to bring the models closer to the realities. The suggested model of the current study is a model with gradual coverage. Berman et al. (2003) in their article proposed two coverage radii. In their model, the demands within the smaller radius were covered completely; the demands outside the bigger radius were not covered; and the demands between the two radii were covered according to a linear function. Another effective factor which increased the efficiency of the model was the consideration of demand dispersion in each zone. The majority of studies on maximal covering location are static, and the variable of time has not been taken into account in the process of positioning (Fazel Zarandi et al. 2013). Since demand dispersion changes throughout years, the dynamic positioning of emergency facilities can improve the quality of these services. Başar et al. (2011) suggested a dynamic model with double-coverage approach for medical service stations. This single-objective model optimizes the level of demands which have been covered at least twice.

3 Proposed Model

In the proposed model, the problem is considered in a discrete condition. The expenses are considered to be fixed throughout time. Each demand point can be covered by one or more centers. Then, some of the parameters and variables are introduced. r_1, r_2: Smaller and bigger radius; d_{ij}: The shortest distance from i to j; K_j: The workload capacity for a facility in j; f_j: Fixed costs; v_j: Variable costs (per-unit capacity cost); h_{it}: The population at node i in period t; M_i: The set of zones that are within a distance less than r_2 from i; U_{it}: The fraction of population in i covered by more than one vehicle in t; Y_{it}: The fraction of population in node i covered by one vehicle in t; X_{jt}: The integer number of vehicles located at potential site j in period t; W_{ijt}: The fraction of population.

The suggested model has several advantages over previous models. Firstly, it is multiple-objective model whose aims are maximizing the initial coverage, maximizing backup coverage, and minimizing expenses. Secondly, since population dispersion is not fixed throughout time, the variable of time has been taken as an important factor in order to prepare the ground for meeting the future demands. Thirdly, U_{it} and Y_{it} have been defined as continuous variables in order to make it possible that part of the demands of a zone are covered. Finally, in contrast to the previous models whose coverage radii are fixed, the coverage radius in this model is gradual. In the following parts, the suggested mathematical model is explained.

$$\text{Max } Z_1 = \sum_{t=1}^{T} \sum_{i=1}^{I} h_{it} \cdot Y_{it} \qquad (1)$$

$$\text{Max } Z_2 = \sum_{t=1}^{T} \sum_{i=1}^{I} h_{it} \cdot U_{it} \qquad (2)$$

$$\text{Max } Z_3 = \sum_{t=1}^{T} \sum_{j=1}^{J} (f_j \cdot Z_{jt} + v_j \cdot X_{jt}) \qquad (3)$$

Subject to:

$$\sum_{j=1}^{J} X_{jt} \leq P_t \quad \forall t \in T, (P_t \leq P_{t+1}) \qquad (4)$$

$$X_{jt} \leq S_j \cdot Z_{jt} \quad \forall t \in T, \forall j \in J \qquad (5)$$

$$Z_{jt} \leq X_{jt} \quad \forall t \in T, \; \forall j \in J \tag{6}$$

$$Z_{j(t-1)} \leq Z_{jt} \quad \forall (t \geq 2) \; \forall j \in J \tag{7}$$

$$U_{it} - q_{it} \leq 0 \quad \forall t \in T, \; \forall i \in I \tag{8}$$

$$Q_{it} - Y_{it} \leq 0 \quad \forall t \in T, \; \forall i \in I \tag{9}$$

$$\sum_{j \in M_i} W_{ijt} - Y_{it} - U_{it} \geq 0 \quad \forall t \in T, \; \forall i \in I \tag{10}$$

$$W_{ijt} \leq C_{ij} \quad \forall t \in T, \; \forall i \in I, \; \forall j \in M_i \tag{11}$$

$$\sum_{i=1}^{I} W_{ijt} \cdot h_{it} \leq K_j \cdot X_{jt} \quad \forall t \in T, \; \forall j \in J \tag{12}$$

$$U_{it} \leq 1 \quad \forall t \in T, \; \forall i \in I \tag{13}$$

$$Y_{it} \leq 1 \quad \forall t \in T, \; \forall i \in I \tag{14}$$

$$C_{ij} = \begin{cases} 1 & d_{ij} \leq r_1 \\ \frac{(r_2 - d_{ij})}{(r_2 - r_1)} & r_1 < d_{ij} \leq r_2 \\ 0 & d_{ij} > r_2 \end{cases} \quad \forall i \in I, \; \forall j \in J \tag{15}$$

$$X_{jt} \geq 0 \text{ integer} \quad \forall t \in T, \; \forall j \in J \tag{16}$$

$$Z_{jt} \in \{0, 1\} \quad \forall t \in T, \; \forall j \in J \tag{17}$$

$$q_{it} \in \{0, 1\} \quad \forall t \in T, \; \forall i \in I \tag{18}$$

In this model, the objective functions (1) and (2) seek to maximize the initial and backup coverage, respectively. The objective function (3) intends to minimize the fixed and variable expenses related to the stationing of transportation vehicles. The constraint (4) shows that in the period t, at most P_t facilities can be stationed in the potential locations (j). Also, the maximum number of facilities in each period should be more than or equal to the number of facilities in the previous period ($P_t \leq P_{t+1}$). The constraints (5) and (6) show that if the potential location j in the period t is selected for the facility, at least one facility and at most S_j facilities can be allocated to this station. Constraint (7) shows that if the potential location j is selected as a location for stationing facilities, it is also selected in the following periods. In other words, if a station is built for a given period, it is also used in the following periods. The constraint (8) and (9) guarantee that U_{it} can be assigned value if $Y_{it} = 1$. The constraint (10) shows the number of zones which are able to cover the demand point i. The constraint (11) guarantees that the demand points can be covered to a level equal to the coverage function C_{ij}, if the constraint (12) is met.

In constraints (13) and (14), U_{it} and Y_{it} have defined as continuous variables within [0,1]. Constraint (15) shows the gradual coverage function. In Constraint (16) X_{jt} is the integer number of vehicles. In constraint (17), the variable Z_{jt} is equal to 1 if the location j in the period t is selected as a station; otherwise zero. Based on constraint (18), the variable q_{it} is equal to 1, the backup coverage is made possible for that demand point.

4 Solution Approach

In order to solve the suggested model, four approaches have been used. In lexicographic multi-objective linear programming (LMOLP) approach which is crisp, the first objective is met completely at the beginning; then, the second and last objectives are met, respectively. In this approach, an ordinal ranking of the objectives are specified as follows:

Priority level 1: Max Z_1, Priority level 2: Max Z_2, Priority level 3: Min Z_3.

In the first FGP approach (FGP1), the objectives have no priority over each other and all objectives are wanted to be satisfied simultaneously and also no relative importance assigned to objectives. So the first FGP based on lower bounds (LB), upper bounds (UB) of objectives and α_t (the satisfaction-level of tth period) can be defined as follows:

$$\text{Max} \sum_{t=1}^{3} \alpha_t \quad (19)$$

Subject to:

$$\alpha_t \leq \frac{\sum_{i=1}^{I} h_{it} \cdot Y_{it} - LB(Z_1)_t}{UB(Z_1)_t - LB(Z_1)_t}, \alpha_t \leq \frac{\sum_{i=1}^{I} h_{it} \cdot U_{it} - LB(Z_2)_t}{UB(Z_2)_t - LB(Z_2)_t} \quad \forall t \in T \quad (20)$$

$$\alpha_t \leq \frac{UB(Z_3)_t - \sum_{j=1}^{J} (f_j \cdot Z_{jt} + v_j \cdot X_{jt})}{UB(Z_3)_t - LB(Z_3)_t} \quad \forall t \in T \quad (21)$$

and constraints (3)–(18). Also, $\alpha_t \in [0, 1]$.

In additive fuzzy goal programming (Chen and Tsai 2001)—(A-FGP-C) the first objective has priority over the second objective, and the second objective has priority over the third objective. So A-FGP-C model can be formulated as follows:

$$\text{Max} \sum_{z=1}^{3} \sum_{t=1}^{3} \alpha_{tz} \quad (22)$$

Subject to:

$$\alpha_{t1} \le \frac{\sum_{i=1}^{I} h_{it} \cdot Y_{it} - LB(Z_1)_t}{UB(Z_1)_t - LB(Z_1)_t}, \alpha_{t2} \le \frac{\sum_{i=1}^{I} h_{it} \cdot U_{it} - LB(Z_2)_t}{UB(Z_2)_t - LB(Z_2)_t} \quad \forall t \in T \quad (23)$$

$$\alpha_{t3} \le \frac{UB(Z_3)_t - \sum_{j=1}^{J} (f_j \cdot Z_{jt} + v_j \cdot X_{jt})}{UB(Z_3)_t - LB(Z_3)_t} \quad \forall t \in T \quad (24)$$

$$\alpha_{t1} \ge \alpha_{t2} \ge \alpha_{t3} \quad \forall t \in T \quad (25)$$

and constraints (3)–(18), $\alpha_{tz} \in [0, 1]$ where denotes the satisfaction-level of tth period and zth fuzzy goal (Z_1, Z_2, Z_3).

In weighted additive fuzzy goal programming (Tiwari et al. 1987) the objectives have different weights and importance which is denoted by λ_z. WA-FGP-T model can be written as follows:

$$\text{Max} \sum_{t=1}^{3} \sum_{z=1}^{3} \lambda_z \alpha_{tz} \quad (26)$$

and constraints (3)–(18), (23), (24). Also, $\alpha_{tz} \in [0, 1]$.

5 Computational Experience

In this section, an example with random numbers is presented. Then, the proposed model is solved and evaluated. The example is presented with 30 points within [0, 45] by choosing random numbers and uniform distribution. In this example, in the first period, 5 transportation vehicles are allocated, and then 3 other vehicles are added in each following period. The model of the problem has been made by three FGP approaches as well as LMOLP approach. Based on the programming, in each one of the periods, the interval must be determined for each objective. In Table 1, the value of each objective in different periods have been shown, which are determined by decision-makers. Since the periods are dependent on each other, wrong decision on the intervals leads to a situation in which the problem cannot be

Table 1 Aspiration levels of fuzzy goals

Objective	Period 1		Period 2		Period 3	
	Upper	Lower	Upper	Lower	Upper	Lower
Z_1	250,000	230,000	410,000	390,000	570,000	550,000
Z_2	600,000	400,000	90,000	70,000	120,000	100,000
Z_3	7,000	4,000	7,000	4,000	7,000	4,000

solved. Also, by changing the intervals, the coverage priority of the population in different periods can be changed. In order to obtain the results, the problem is solved by lingo 11.0 optimization software. The results obtained by the solution of suggested model and approaches have been shown in Table 2.

In LMOLP model, the initial coverage has been taken as the first objective with the highest priority, and the obtained value in the three periods is equal to 1,244,236 totally. The minimization of expenses in this method is the last priority. According

Table 2 Results of the proposed model

Solution approach	P[a]	N[b]	The zones representing vehicle locations	Covered population by one vehicle	Covered population with backup coverage	CC[c]
LMOLP	1	5	8, 9, 11, 23, 24	254,316 (81.77 %)	1,300 (0.42 %)	4,554
	2	8	6.8, 9, 11, 23, 24, 25, 26	419,521 (96.40 %)	31,392 (7.21 %)	7,059
	3	11	6, 8, 9, 11, 14, 23, 24(2), 25, 26(2)	570,399 (98.52 %)	53,733 (9.28 %)	7,688
Total population				1,244,236 (93.89 %)	86,425 (6.52 %)	
FGP1	1	5	16, 20, 21, 24, 26	239,627 (77.05 %)	49,627 (15.95 %)	4,066
	2	8	7, 8, 9, 16, 20, 21, 24, 26	391,040 (89.86 %)	71,040 (16.32 %)	6,844
	3	11	7, 8, 9, 16, 20, 21 (2), 24(2), 26(2)	550,334 (95.06 %)	100,334 (17.33 %)	6,907
Total population				1,181,001 (89.12 %)	221,001 (16.68 %)	
WA-FGP-T	1	5	8, 16, 19, 24, 25	230,445 (74.09 %)	50,153 (16.12 %)	4,042
	2	8	8, 9, 11, 16, 19(2), 24, 25	395,318 (90.84 %)	72,076 (16.56 %)	5,807
	3	11	7, 8, 9, 11, 16, 19 (2), 24(2), 25(2)	560,000 (96.73 %)	100,000 (17.27 %)	6,799
Total population				1185,763 (89.48 %)	222,229 (16.77 %)	
A-FGP-C	1	5	16, 20, 21, 24, 26	239,627 (77.05 %)	49,627 (15.95 %)	4,066
	2	8	7, 8, 9, 16, 20, 21, 24, 26	391,930 (90.06 %)	71,930 (16.53 %)	6,844
	3	11	7, 8, 9, 16, 20, 21 (2), 24(2), 26(2)	550,334 (95.06)	100,334 (17.33 %)	6,907
Total population				1,181,891 (89.19 %)	221,891 (16.74 %)	

[a] Period, [b] Number of vehicles, [c] Cumulative cost

to this solution, at the end of three periods, at least 7,688 units are expended in order to establish service centers. In order to show how initial coverage variable (which is fractional) affects coverage increase, the suggested model has been solved by LMOLP approach and the binary variable of y. In this condition, the maximum of initial coverage throughout the total of three periods is equal to 1,230,544, which shows a substantial reduction compared to the main model. So, when the initial coverage is taken fractionally, the model becomes more flexible and the maximum demand is covered. Although LMOLP is superior in initial coverage, the backup coverage throughout the total of periods is equal to 86,425, and only 6.25 % of the population has backup coverage, which is not satisfactory. Therefore, lexicographic multiple-objective linear programming which is based on absolute priority is not efficient in backup coverage.

In fuzzy approaches, the backup coverage is improved substantially, although there is a little decrease in initial coverage. According to the obtained results from FGP, the percentage of backup coverage in the total of periods is equal to 16.68 %, while the initial coverage is equal to 89.12 % totally. In other words, a 4.77 % decrease in initial coverage (from 93.89 to 89.12 %) leads to an improvement equal to 10.16 % (from 6.25 to 16.68 %). Compared to LMOLP method, the total of expenses of the three periods has decreased from 7,688 to 6,907. In WA-FGP-T approach, different weights show the relative importance of the goals. According to the obtained results, the WA-FGP-T approach leads to the best results among the different approaches. Programming for future demands can lead to the maximum coverage as well as a reduction in the expenses of stationing. It can be observed that in the first period, 25, 24, 19, 16, and 8 nodes were selected for five transportation vehicles. In the following period, two new stations have been created and other vehicle has been stationed in station 19. In the third period, only one new station has been created, and two other vehicles have been created in the previous established stations. In this way, this model can cover the maximum level of future demands based on the changes in the congestion and with the minimum amount of expenses. The A-FGP-C approach has a structure similar to LMOLP method, but the priorities are not absolute. Although this approach does not determine the priorities of the goals, it guarantees that a fuzzy goal with highest priority has been achieved with the highest level. The obtained results from this approach are similar to the obtained results from FGP1.

6 Conclusion and Future Researches

In this article, a dynamic multiple-objective fuzzy model of maximal covering location problem with gradual coverage radius has been presented. In addition to initial coverage, the backup coverage objective, fixed expenses of establishing stations and variable expenses of facility stationing have been taken into account. Taking the coverage radius as a gradual factor and taking the initial coverage as a fractional factor brings the model closer to reality. The efficiency of the model is evaluated by presenting an example with random numbers and its solution by different approaches. The obtained results show that the suggested model is closer to the reality when it is compared with other models. Also, the FGP approaches are more efficient for the solution of location problem. As the proposed model is highly complex, it is suggested that heuristic and meta-heuristic methods are developed for the solution of the problem in large scales. In addition to variable demands, other factors such as expenses can be taken as variables for future researches.

References

Araz C, Selim H, Özkarahan I (2007) A fuzzy multi-objective covering-based vehicle location model for emergency services. Comput Oper Res 34:705–726

Başar A, Çatay B, Ünlüyurt T (2011) A multi-period double coverage approach for locating the emergency medical service stations in Istanbul. J Oper Res Soc 62:627–637

Beraldi P, Bruni ME (2009) A probabilistic model applied to emergency service vehicle location. European J Oper Res 196:323–331

Berman O, Krass D, Drezner Z (2003) The gradual covering decay location problem on a network. Eur J Oper Res 151:474–480

Chen LH, Tsai FC (2001) Fuzzy goal programming with different importance and priorities. Eur J Oper Res 133:548–56

Coskun N, Erol R (2010) An optimization model for locating and sizing emergency medical service stations. J Med Syst 34:43–49

Fazel Zarandi MH, Davari S, Haddad Sisakht SA (2013) The large-scale dynamic maximal covering location problem. Math Comput Model 57:710–719

Tiwari RN, Dharmar S, Rao JR (1987) Fuzzy goal programming–an additive method. Fuzzy Sets Syst 24:27–34

Part II
Logistics, Production and Information Systems

Drivers and Stages in "Packaging Logistics": An Analysis in the Food Sector

Jesús García-Arca, José Carlos Prado-Prado and A. Trinidad González-Portela Garrido

Abstract Packaging design stands out as one of the elements that eases a sustainable and efficient supply chain management. This paper sets out to define and characterize both the drivers and the evolutionary stages for implementing "packaging logistics", through a "case study" analysis of eight Spanish food manufacturers. Likewise, one example of packaging improvement in one of these companies is supplied for illustrating the potential of its implementation.

Keywords Packaging logistics · Supply chain · Sustainability

1 Introduction

In markets increasingly turbulent and volatile, companies should increase efforts to improve competitiveness in their chains eliminating activities that do not add value and developing innovations (Christopher 2005; Andersen and Skjoett-Larsen 2009). Simultaneously, different "stakeholders" show a growing interest in the supply chain sustainability. The concept of sustainability (Brundtland Report 2000) is based on three pillars: environmental, economic and social. Authors such as Seuring and Müller (2008) indicate the potential of adopting a sustainable supply chain, as it saves resources, reduce waste and provide competitive advantages.

In this context, packaging is one of the key and transversal elements that supports the improvement of efficiency and sustainability in supply chain management (Klevas 2005). Beyond the traditional and essential protection of the product, the packaging should be designed not only to provide product differentiation capacity, but also to provide efficiency in production/logistics (direct logistics). In parallel, this sight has expanded to its reverse side (reuse/recycling/recovery), which has led

---

J. García-Arca (✉) · J.C. Prado-Prado · A.T.G.-P. Garrido
Grupo de Ingeniería de Organización (GIO), Escola de Enxeñería Industrial, Campus Lagoas-Marcosende, C/Maxwell, 36310 Vigo, Spain
e-mail: jgarca@uvigo.es

to the development of specific legislation (e.g., the European Directive 94/62 and its update, 2004/12/EC). Therefore, the packaging would eliminate waste along the supply chain (Hellström and Nilsson 2011; Dickner 2012). Authors such as Jönson (2000) or García-Arca and Prado-Prado (2008) associated with the packaging three main functions: commercial, logistics (direct) and environmental (both direct and reverse logistics). This wide vision has enabled the development of the "packaging logistics" approach.

In this context, this paper sets out to define and characterize the evolutionary stages for deploying the "packaging logistics" approach. In this sense, the research approach develops two different phases: on one hand, theory building based on literature review of drivers that can aid the packaging logistics deployment; on the other hand, theory testing in order to characterize the stages in packaging logistics implementation. For theory testing, the authors have chosen the "case study" technique (Yin 2002), analysing eight frozen product manufacturers in an exploratory way; these companies rank among the top Spanish 15 frozen product manufacturers. The methodology used in the cases analysis was the personal interview with structured questionnaire based on qualitative and quantitative scales (Likert scale). The interviewees were the logistics or supply chain managers.

2 Drivers in Packaging Logistics Deployment

As a prelude to the conceptualization of "packaging logistics" is necessary to consider the roles and requirements to be met by packaging, aspect discussed above when main functions are mentioned, and the hierarchical structure associated with them. The hierarchical structure of the packaging is organized on three interconnected levels: primary packaging (consumer packaging or sales packaging), secondary packaging (group of several primary packages for easing handling/ displaying) and tertiary packaging (a number of primary or secondary packages assembled on a pallet or roll container) (Jönson 2000).

Authors such as Saghir (2002), Hellström and Saghir (2006) and García-Arca and Prado-Prado (2008) show the different requirements that the "agents" of the supply chain require the packaging. In addition, these requirements are not distributed homogeneously in the different levels of packaging, which require an integrated view of packaging and logistics. Unfortunately, there are few reliable methods to objectively measure the efficiency and overall sustainability of the supply chain, so that almost all efforts are focused on the extent of partial efficiency of some logistics and productive processes but not the whole (Saghir and Jönson 2001; Saghir 2002; García-Arca et al. 2014). In this context, Saghir (2002) presents the concept of Packaging Logistics "…the process of planning, implementing and controlling the coordinated Packaging system of preparing goods for safe, secure, efficient and effective handling, transport, distribution, storage, retailing, consumption and recovery, reuse or disposal and related information combined with maximizing consumer value, sales and hence profit."

As a conceptual proposal, the authors synthetize previous proposals of Saghir (2002), Hellström and Saghir (2006), García-Arca and Prado-Prado (2008), Olander-Roase and Nilsson (2009), Vernuccio et al. (2010), Svanes et al. (2010), Azzi et al. (2012) and García-Arca et al. (2014) to establish the three main drivers that could promote the implementation and development of "Packaging Logistics" from a sustainable perspective:

- The definition of all design requirements based on commercial, logistics/production and environmental aspects.
- The adoption of an organizational structure in packaging design and new product development that allows the coordination of all departments/areas related to each company, both internally and externally, in the supply chain.
- The definition of a system, not only to measure the impact on a particular packaging, but also to compare different alternatives from an holistic perspective; that should mean, for example, the measurement of the total costs, the sales, the customers´ satisfaction and the environmental impact. One of the greatest problems in developing a system to evaluate alternatives is the difficulty in objectively weighing up design requirements on the same scale. One of these proposals is the "packaging scorecard" system (Dominic and Olsmats 2001) based on a mix of qualitative and quantitative scales. In practice, this should involve the combination of different evaluation methods. The authors propose a system based on objective cost measurement, but also subjectively adjusted to other measurement systems (environmental and/or commercial).

The development of these three drivers should promote the application of "best practices" in packaging design in order to eliminate "waste" throughout the supply chain. Among these best practices appear: product size redesign, packaging downsizing, palletizing changes, changes in the number of primary packaging/secondary packaging, standardization in materials and sizes, materials changes, graphic art changes, packing changes and returnable packaging implementation (García-Arca and Prado-Prado 2008).

3 Results and Discussion

The authors have analysed the three drivers in "packaging logistics" implementation (design requirements, organizational structure and measurement system) in the eight companies in order to identify the different stages of evolution of packaging logistics deployment. In relation to the first driver, companies valued the importance of the different packaging design requirements (Likert scale): differentiation capacity (commercial), production costs (packing process), purchase costs (packaging materials), logistics costs (handling, storage and transport) and environmental impact (raw material consumption and waste generation). With minor differences, the results give maximum importance to commercial requirements and minimal

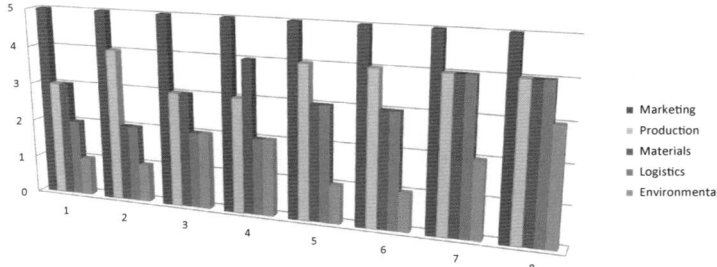

Fig. 1 Importance of requirements in packaging design in the sample of eight companies (*Likert scale 1–5*; *1* low importance; *5* high importance)

importance to environmental requirements. The other requirements are in an intermediate position (see Fig. 1) with differences among companies.

In the analysis of the second driver (see Fig. 2), focusing on the packaging design decisions (dimensions, materials, groupings and graphic art), it was found that in most companies all design decisions are almost exclusive responsibility of

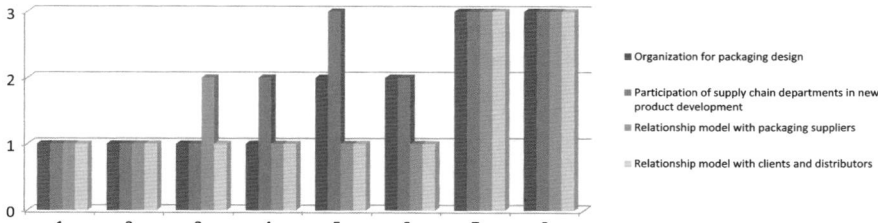

Fig. 2 Internal and external organization for packaging design in the sample of eight companies. *Internal organization for packaging design*: *1* all packaging design decisions are almost exclusive responsibility of the commercial areas. *2* packaging design decisions are mainly responsibility of the commercial areas. Purchases and production areas are partially involved in these decisions. *3* packaging design decisions are mainly responsibility of the commercial areas. Purchases, production and logistics areas are completely involved in these decisions. *Supply chain participation in new product development*: *1* No formal coordination mechanisms in the area of logistics involved in any stage of product design (conceptual development and pre-industrialization). There could be no systematic contacts. *2* There is formal coordination mechanisms with the areas of production/logistics involved in the final stages of product design (pre-industrialization). *3* There is formal coordination mechanisms with the areas of production/logistics involved in almost all stages of product design (conceptual development and pre-industrialization). *Relationships with packaging suppliers, distributors and clients*: *1* "Basic" negotiation (price is almost the one and only aspect considered and continual relationship is not ensured a priori); "Friendly" negotiation (annual contracts). *2* "Cooperation" (long-term contract; information sharing; relationships of confidence). *3* "Coordination" (long-term contract, agreements on logistic efficiency where information is shared and supported by ICTs); Collaboration (long-term relationships, multi-company hardware, joint planning, etc.)

the commercial areas, although with different level of production, purchases and logistics involvement. Simultaneously, the authors have found that the coordination between the areas related to the supply chain management and the areas related to new product design has a low/intermediate level.

On the other hand, the authors analysed the organizational model for external supply chain management (upstream and downstream; see Fig. 2). To this end, three levels have been considered: basic/friendly, cooperation and coordination/collaboration. Particularly, if the analysis focuses on packaging suppliers is observed that, mostly, the companies choose basic relationships. Obviously, all this does not help in finding packaging solutions that have a positive impact on the performance of the whole supply chain. Likewise, companies also opt mostly for implementing basic relationships with their clients and distributors. Thus, main clients and distributors mostly direct their contributions of improvements to graphic art changes (proposed almost exclusively through commercial areas).

Moreover, in relation to the availability of a cost system for measuring objectively the impact of packaging design decisions in the supply chain performance (efficiency and sustainability), we found a low level of development (see Fig. 3). The measurement of this impact focuses almost exclusively on the direct costs related to packaging purchases, and to a lesser extent, on the indirect productive costs (packing process). Curiously, among companies, there is a considerable ignorance of the costs of packaging waste (green dot) despite its direct nature.

The main problem identified with the development of these measurement systems is that the hierarchical structure of the packaging (primary, secondary and tertiary) is not usually well connected with the costs of logistics processes (where they exist). All this makes it difficult to relate both concepts. The most common variable to measure the impact on costs is the kilogram, which hinders the visibility of the impact of the proposed changes, not only in primary packaging but also in secondary and tertiary packaging (boxes and pallets), particularly with regard to the costs of handling, storage and transportation.

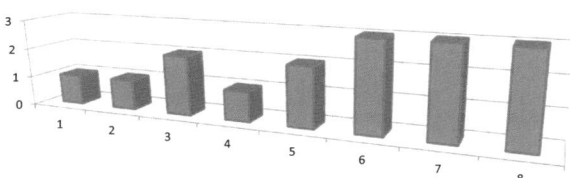

Fig. 3 Development of a system for measuring the impact of packaging design decisions in the sample of eight companies. *Scale*: *1* "No specific system for measuring detailed costs of the production/logistic processes under a packaging perspective. Materials cost availability. Global cost under a kilo perspective"; *2* "No specific system for measuring detailed costs of the logistic processes under a packaging perspective. Materials and packing cost availability under primary and secondary packaging perspective. Global cost under a kilo perspective"; *3* "Materials, packing and logistics costs availability. Global cost under a kilo perspective, but also under primary, secondary and tertiary packaging perspective"

Going off the analysis made, it has been possible to justify and characterize four theoretical stages in terms of the extent to which the 3 drivers of "packaging logistics" have been deployed (see Table 1). Each of the eight companies has been placed in this evolution process, although only companies 1, 2, 7 and 8 would present a theoretical and clear behaviour under the proposed model (see Fig. 4).

Table 1 Stages in "packaging logistics" deployment

Stage	Design requirements	Organizational structure	Measurement system
1 Basic stage	Protection and differentiation	Internally, the main and almost only responsibility of the commercial area/department. No involvement of logistics/production areas in new product development Low level of integration for managing internal supply chain. Primary relationships with packaging manufacturers and clients	No specific system for measuring the production/logistic level in relation to primary, secondary and tertiary packaging
2 Conscious stage	+ Reduction of indirect costs in packing process and packaging purchases	Main commercial responsibility with some involvement of purchasing and production departments/areas. Some involvement of production areas in new product development. Low/Medium level of integration for managing internal supply chain. Primary relationships with packaging manufacturers and clients	Measurement system to make the direct relationship visible in purchases and production costs
3 Advanced stage	+ Logistics cost reduction. + Environmental impact (raw materials consumption and waste).	Internal coordination (commercial, production, logistics and purchases areas/departments). Coordination and collaboration with packaging manufacturers and logistics operators. Logistics coordination with clients	Measurement system to make the direct and indirect relationship visible between the packaging design decisions and the total costs in supply chain
4 Systematic stage	All the above on a regular basis (updated/reviewed) and from a sustainable and holistic perspective	Systematic integrated coordination (internal and external) in new products development	Systematization in order to improve and update the measurement system mentioned above. Comparison between packaging alternatives (simulation)

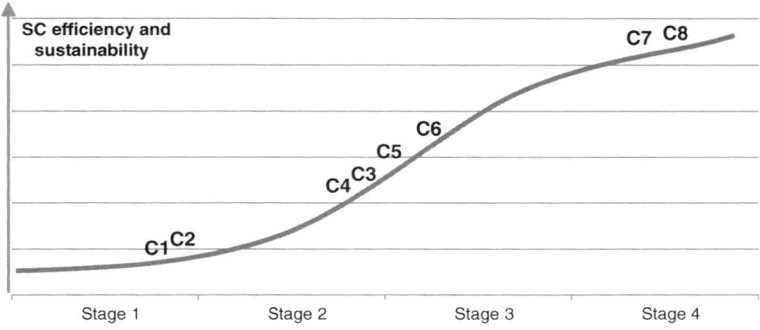

Fig. 4 Stages and level of deployment of "packaging logistics" in the sample of 8 companies

As an example, the impact of changing the packaging of hake slice in plastic tray, in which the authors have worked with the "company 8" following the approach "action research" (Coughlan and Coghlan 2002), is presented. This change comprises the increase in the number of trays per carton (3–4 trays per carton; 5.4 kilos to 7.2 kilos per carton) (see Fig. 5). That means the resizing of carton and plastic tray; the resizing of tray has been possible thanks to the change of the slices' angle in the tray, keeping the number of slices and the weight in each tray. Obviously, prior to the implementation of this change, the approval of the commercial department was necessary, but it was eased thanks to the cost reduction.

In this line, this change has supplied cost reductions in several areas: a lower consumption of cardboard and plastic per kilo, leading to savings both from the perspective of purchases (despite the investment of a new die for the new plastic tray) and from an environmental perspective to generate less waste; the improvement of a 25 % in pallet efficiency (from 240 trays/pallet to 300 trays/pallet) produces reductions in logistics costs associated with handling, storage and transportation throughout the supply chain including distributors and clients; on this last point, note that a 25 % fewer boxes is handled along the supply chain to sell the

Fig. 5 Example with initial (*left side*) and final (*right side*) packaging and palletization

same number of kilos. Overall, only in the supply chain managed by the company, savings of 52,000 euros per year (54 % logistics savings and 46 % materials savings), have been achieved with the change of this packaging.

4 Conclusions

The evolution proposed in the paper is particularly new, since in the literature the methodological treatment of the deployment of "packaging logistics" is little focus. This paper may be of interest both for researchers and professionals, because drivers and stages can aid companies in improving their vision of product and packaging design in order to increase sustainability and efficiency in the supply chain, not only in food sector, but also in other sectors. However, the preliminary results would demand a wider empirical base in order to validate the proposal.

References

Andersen M, Skjoett-Larsen T (2009) Corporate social responsibility in global supply chains. Supp Chain Manage: Int J 12(2):75–86
Azzi A, Battini D, Persona A, Sgarbossa F (2012) Packaging design: general framework and research agenda. Packag Technol Sci 25:435–456
Carter CR, Rogers DS (2008) A framework of sustainable supply chain management: moving toward new theory. Int J Phys Distrib Logistics Manage 38(5):360–387
Christopher M (2005). Logistics and supply chain management strategies for reducing cost and improving service. Financial Times Pitman Publishing, Londres
Coughlan P, Coghlan D (2002) Action research for operations management. Int J Oper Prod Manage 22(2):220–240
Dickner A (2012) Sustainable packaging. A IKEA prevension. Pro-Europe Seminar: Waste Prevention in Practice, Brussels
Dominic C, Olsmats C (2001) Packaging Scorecard—A method to evaluate packaging contribution in the supply chain. Results from Report N° 200. Packforsk, Estocolmo
European Commission (1994) Directive 94/62/EC on packaging and packaging waste, Brussels
European Commission (2004) Directive 04/12/EC on packaging and packaging waste, Brussels
García-Arca J, Prado-Prado JC (2008) Packaging design model from a supply chain approach. Supp Chain Manage: Int J 13(5):375–380
García-Arca J, Prado-Prado JC, Garrido ATG-P (2014) "Packaging logistics": promoting sustainable efficiency in supply chains. Int J Phys Distrib Logistics Manage 44(4):325–346
Hellström D, Saghir M (2006) Packaging and logistics interactions in retail supply chain. Packag Technol Sci 20:197–216
Hellström D, Nilsson F (2011) Logistics-driven packaging innovation: a case study at IKEA. Int J Retail Distrib Manage 39(9):638–657
Jönson, G. (2000). Packaging Technology for the Logistician, 2nd ed., Department of Design Sciences, Division of Packaging Logistics, Lund University, Lund
Klevas J (2005) Organization of packaging resources at a product-developing company. Int J Phys Distrib Logistics Manage 35(2):116–131

Olander-Roase M, Nilsson F (2009) Competitive advantages through packaging design-prepositions for supply chain effectiveness and efficiency. In: International conference on engineering design, ICED 2009, Stanford University, USA, pp 279–290

Saghir M, Jönson G (2001) Packaging handling evaluation methods in the grocery retail industry. Packag Technol Sci 14:21–29

Saghir M (2002) Packaging logistics evaluation in the Swedish retail supply chain, Universidad de Lund, Lund

Svanes E, Vold M, Møller H, Pettersen MK, Larsen H, Hanssen OJ (2010) Sustainable packaging design: a holistic methodology for packaging design. Packag Technol Sci 23:161–175

Vernuccio M, Cozzolino A, Michelini L (2010) An exploratory study of marketing, logistics, and ethics in packaging innovation. Eur J Innov Manage 13(3):333–354

Yin RK (2002) Applications of case study research. Sage Publications, Thousands Oaks

A Tabu Search Approach for Production and Sustainable Routing Planning Decisions for Inbound Logistics in an Automotive Supply Chain

David Peidro, Manuel Díaz-Madroñero, Josefa Mula and Abraham Navalón

Abstract In this paper, a mixed-integer mathematical programming model is proposed to address a production and routing problem related to inbound logistics processes in supply chains environments. This model is also enriched with sustainable issues related to routing decisions by introducing additional fuel consumption and pollutants emissions calculations into the objective function. For the solution methodology, a two-phase decoupled solution procedure based on exact algorithms for the production model and a tabu search algorithm for the routing model is adopted. Results of computational experiments performed with a real-world automotive supply chain confirm the efficiency of the proposed solution method in terms of total cost, fuel consumptions and CPU time.

Keywords Material requirements planning · Vehicle routing · Green logistics · Tabu search

1 Introduction

In recent years, with the growing globalization of supply chains in high competitive business scenarios, industrial firms can not consider production planning and materials procurement as independent processes from transportation planning because their possible suboptimality from an economic perspective or their infeasibility due to capacity constraints. Moreover, apart from the economic cost viewpoint, these plans would include ecological and social issues, in order to meet needs of firms' stakeholders, and hence, improve their position with respect to their competitors. In this

D. Peidro · M. Díaz-Madroñero (✉) · J. Mula · A. Navalón
Research Centre on Production Management and Engineering (CIGIP),
Universitat Politècnica de València, Escuela Politécnica Superior de Alcoy,
Plaza Ferrándiz y Carbonell, Alcoy, Spain
e-mail: fcodiama@cigip.upv.es

context, considering environmental criteria in the transport processes could have a tangible impact on responsible supply chain management because a better use of natural resources and a reduction of greenhouse gases can be achieved.

The literature in production and routing problems (PRP), which integrate a lot-sizing problem to determine production amounts and a vehicle routing problem (VRP) to determine delivery routes to distribute final products to customers, is scarce because the PRP is a recent research area focused on efficient solution procedures. In this sense, metaheuristics such as tabu search (TS) are considered a proper solution method for obtaining good quality solutions in reduced computational times (Bard and Nananukul 2009; Armentano et al. 2011). However, all these proposed solution methods are evaluated by performing computational experiments with randomly generated instances and only evaluate plans from only an economic cost viewpoint, neglecting an ecological and social perspective. However, different green logistics practices and calculation methods have been incorporated recently into transport planning decisions in order to reducing fuel consumption (Jabali et al. 2012; Ubeda et al. 2011). Readers are referred to Demir et al. (2014) for a recent review of factors affecting fuel consumption in VRP optimization models.

After a review process, we highlight the following issues related to the PRP problem: (1) there is a need for PRP focused on inbound logistics processes; (2) PRP are validated, generally, with artificially generated instances, and hence, real-world applications are scarce; (3) new constraints and/or objective functions components can enrich current PRP models to tackle environmental requirements. In this sense, factors such as vehicle speed, transported load and gradient road factor can be considered significant. In this paper, we propose a mathematical programming planning for production and vehicle routing planning related to the procurement of parts and raw materials taking into account environmental issues such as fuel consumption and pollutants emissions, which is solved by a sequential approach based on a TS procedure. The paper is arranged as follows. Section 2 describes the fuel consumption calculation methodology considered in this study. Section 3 presents the problem and the corresponding formulation. Section 4 describes the solution methodology. Next, Sect. 5 evaluates the behavior of the proposed model in a real-world automobile firm. Finally, Sect. 6 provides conclusions and directions for further research.

2 Fuel Consumption Calculation Method

In this study, MEET methodology has been adopted as a fuel consumption and pollutants calculation method and included in the proposed PRP mathematical programming model. Based on real-life experiments and diverse on-road measurements, Hickman et al. (1999) proposed a method for energy consumption calculations by using a "carbon-balance" based expression for pollutants such as CO_2, CO, unburned hydrocarbons and particulate matter emissions. The total emissions of each exhaust pollutant (i) correspond to the cold and hot emissions of each one, and can be calculated according to Eqs. (1)–(4). Cold emissions ($E_{i,cold}$) correspond to gases generated during each engine cold start, and can be defined according to Eq. (2), where $\varepsilon_{i,cold}$ is the cold emission factor for pollutant i (in g/cold start) and N is the number of cold starts in each route. Hot emissions are proportional to distance travelled in each vehicle route d_{route}, and depend on the hot emission factor $\varepsilon_{i,hot}$ (in g/km) which can be calculated by using Eq. (4) as a function of the average vehicle speed v. However, emissions factors of MEET refer to standard conditions (e.g. flat road gradient, empty vehicle, etc.) have to be corrected in order to take into account road gradient and load effects. The road gradient and load correction factors shown in Eqs. (5) and (6), respectively, are used to take into account the effect of road slope and payload. Parameters k_1, a, b, c, d, e, f, A_6, A_5, A_4, A_3, A_2, A_1, A_0, k_2, r, s, t, u and $\varepsilon_{i,cold}$ can be extracted from Hickman et al. (1999) for each type of road and vehicle, and z is the percentage of total load in each vehicle. Finally, the total value of hot emissions is given in Eq. (7).

$$E_i = \sum_{\text{vehicle route}} \left(E_{i,cold} + E_{i,hot} \right) \quad (1)$$

$$E_{i,cold} = \varepsilon_{i,cold} \cdot N \quad (2)$$

$$E_{i,hot} = \varepsilon_{i,hot} \cdot d_{route} \quad (3)$$

$$\varepsilon_{i,hot} = k_1 + av + bv^2 + cv^3 + \frac{d}{v} + \frac{e}{v^2} + \frac{f}{v^3} \quad (4)$$

$$GC(v) = A_6 v^6 + A_5 v^5 + A_4 v^4 + A_3 v^3 + A_2 v^2 + A_1 v + A_0 \quad (5)$$

$$LC(v) = \left(k_2 + rv + sv^2 + tv^3 + \frac{u}{v} - 1 \right) z + 1 \quad (6)$$

$$E_{i,hot} = \varepsilon_{i,hot} \cdot GC(v) \cdot LC(v) \cdot d_{route} \quad (7)$$

3 Problem Description and Model Formulation

The PRP addressed in this paper can be defined in a network $G = (N, A)$, where N represents the set of nodes comprising production plant and suppliers, and A represents the set of arcs connecting the nodes, where $A = \{(s, v) : s, v \in N, s \neq v\}$. Nodes are indexed by $s \in \{0, ..., n\}$, where node 0 corresponds to the production plant, while suppliers are represented by $s \in \{1, ..., n\}$ or the set $Ns = N\setminus\{0\}$. In a finite planning horizon, composed of a set of equal planning periods $t = \{1, ..., T\}$, the production plant manufactures, in a set of productive resources r with a production capacity CAP_r, a set of items represented by i. Moreover, extra production time (Tex_{rt}) and idle time (Toc_{rt}) are also included. Manufactured products p, are considered finished goods composed of parts and raw materials according to the corresponding

BOM. Finished goods and parts and raw materials can be stored in dedicated warehouses with capacity $CAPI_i$. Transportation of raw materials and parts from suppliers is done by a set of identical vehicles $K = \{1, \ldots, k\}$ with capacity $CAPT$ over the set of arcs A, which have an associated cost c_{sv} to travel from node s to node v. Each supplier can only be visited once per period. Transport routes are completed during one period and can start at any supplier and always finish at production plant. Moreover, travel times between nodes, waiting, loading and unloading times are not considered. Moreover, due to legal and labour restrictions transport routes can not exceed a maximum distance MD. The nomenclature defines the sets of indices, parameters and decision variables (Table 1).

Objective function

$$\text{Min } z = \sum_i \sum_t cp_i \times P_{it} + \sum_i \sum_t \sum_k \sum_s cp_i \cdot Q_{itks} + \sum_i \sum_t ci_i \cdot Inv_{it} + \sum_i \sum_t crd_i \cdot Rd_{it}$$
$$+ \sum_r \sum_t ctoc_{rt} \cdot Toc_{rt} + \sum_r \sum_t ctex_{rt} \cdot Tex_{rt} + \sum_t \sum_k \sum_s \sum_v cdis_{sv} \cdot Y_{ksvt}$$

(8)

Table 1 Nomenclature

Sets			
T	Set of time periods	S, V	Set of suppliers
I	Set of products	R	Set of resources
J	Set of parent products in BOM	K	Set of vehicles
Parameters			
d_{it}	Demand of i in t	ci_i	Unitary holding cost of i
Inv_{i0}	Initial inventory of i	crd_i	Unitary backorder cost of i
Rd_{i0}	Initial backorders of i	$ctoc_r$	Idle time cost in r
RP_{it}	Programmed receptions of i in t	$ctex_r$	Overtime cost in r
A_{ij}	Required quantity of i to produce a unit of j	$cdis_{sv}$	Travel cost between s and v
		dis_{sv}	Travel distance between s and v
ar_i	Unitary processing time of i	cp_i	Unitary production or purchasing cost of i
CAP_r	Production capacity of r	$CAPT$	Vehicle capacity
$CAPI_i$	Warehouse capacity for i	MD	Maximum distance per route
Decision variables			
P_{it}	Production amount of i in t	Inv_{it}	Inventory of i in t
Q_{itks}	Purchasing amount of i in t transported in k from s	Rd_{it}	Backorders of i in t
Toc_{rt}	Idle time in r in t	Tex_{rt}	Overtime in r in t
Y_{ksvt}	1 if k travels from s to v in t	X_{kst}	1 if k visits s in t

Subject to

$$INVT_{i,t-1} + P_{it} + RP_{it} + \sum_k \sum_s Q_{itks} - INVT_{it} - Rd_{i,t-1}$$
$$+ Rd_{it} - \sum_{j=1}^{I} a_{ij} P_{jt} = d_{it} \quad \forall i \ \forall t \tag{9}$$

$$\sum_{i=1}^{I} AR_{ir} P_{it} + Toc_{rt} - Tex_{rt} = CAP_{rt} \quad \forall r, \ \forall t \tag{10}$$

$$INV_{it} \leq CAPI_i \quad \forall i, \ \forall t \tag{11}$$

$$\sum_s \sum_v dist_{sv} \cdot Y_{ksvt} \leq MD \quad \forall k, \ \forall t \tag{12}$$

$$\text{Routing constraints} \tag{13}$$

$$P_{it}, Q_{itks}, Inv_{it}, Rd_{it} \geq 0 \text{ and integer} \quad \forall i, \ \forall k, \forall t \tag{14}$$

$$Toc_{rt}, Tex_{rt} \geq 0 \quad \forall r, \ \forall t \tag{15}$$

$$Y_{ksvt}, X_{kst} \in \{0,1\} \quad \forall k, \ \forall s, \ \forall v, \ \forall t \tag{16}$$

The objective function (8) corresponds to the minimization of total costs relating to production, inventories, backlogs, overtime and undertime production costs, and routing costs over the planning horizon. Constraints (9) represent the inventory flow balance at the plant. Constraint (10) establishes the available capacity for normal, overtime production and idle time. Constraint (11) establishes the inventory limits of each product at manufacturing plant. Constraint (12) establishes the distance limitation for each route and constraint (13) deals with the typical constraints for capacitated VRP. Constraints (14–16) define the variable decision domains. In order to include environmental and sustainable issues in the previous PRP model, the objective function (8) is modified to take into account the minimization of fuel consumption according to the MEET methodology.

4 Solution Approach

A two-phase approach based on, firstly, the elaboration of a production and a material requirement plan and, secondly, the generation of a vehicle routing plan. According to Boudia et al. (2008), this uncoupled heuristic is quite standard in the industry. Production and material requirement plans are optimally solved by using a standard MIP solver which returns for each period the amounts to manufacture for each finished good, the amounts of parts and raw materials to order to each supplier, inventory and backorder levels, as well as idle and overtime times. The resulting planning is imposed on the inbound vehicle routing planning phase and cannot be modified.

4.1 Tabu Search

Due to the combinatorial nature of VRP problem and in order to solve it in a reasonable computational time, a TS procedure (Glover and McMillan 1986) has been developed. The algorithm works as follows:

1. An initial feasible solution s is achieved by using a modified version of the Clarke and Wright (1964) algorithm to address asymmetric VRP problems according to Vigo (1996).
2. The TS algorithm tries to find the best neighbourhood solution from the current solution by moving the positions of the nodes between the routes. This neighbour solution will be the next one in the solution space. This solution procedure removes a node from its current route and inserts it at a minimum cost into another route, based on the following steps: (1) for each node, it is looked for the best solution of moving it to other route without exceeding the capacity constraint; and (2) after all solutions have been calculated, the next neighbour solution is the best obtained that is not tabu or achieves the aspiration condition. Moreover, in this algorithm when a node is inserted into a route the position chosen is the one that minimizes the distance on this route.
3. The best solution obtained s' that it is not tabu or complies with the aspiration criterion will be the next feasible solution $s = s'$ and the algorithm TS is restarted. This process is repeated until the stop criterion is found.

The main feature of TS is the tabu list that implements a diversification strategy. When a new neighbour solution is obtained, the pair node-route that has been moved is added to the tabu list and is set it with a tabu duration θ, called tabu memory. This means that in the next θ iterations it is forbidden to re-inserting that node into that route. The value of the tabu duration θ is the key parameter that allows the success operation of the algorithm.

5 Application to an Automotive Sector Manufacturer

The proposed model and solution approach have been validated as a tool for decision-making related to the production planning and green transport planning for procurement parts in an European automotive sector manufacturer. This firm manufactures three finished goods in a single production plant in which parts and raw materials are procured from 21 different suppliers. Their facilities are close to the production plant, so that the duration of transport routes is minor than one day

Table 2 Vehicle routing planning model results comparison

Number of suppliers	CPLEX			Tabu search		
	Solution cost (€)	Solution CPU time (s)	Fuel consumption (l)	Solution cost (€)	Solution CPU time (s)	Fuel consumption (l)
10	1091.44	0.34	292.11	1091.44	0.240	292.11
21	1890.42	10888	532.92	1869.40	0.607	528.80

with a maximum of 635 km. per each one. Moreover, it is assumed that there will always be enough vehicles available for the procurement of parts and raw materials with a total capacity of 13.6 linear meters per truck, therefore an only single time period has been considered. For emissions and fuel consumption calculations, a maximum speed average of 70 km/h has been considered. Gradient factors have been obtained as a difference between elevations of start and end nodes in each arc.

The production and vehicle routing problems were developed with the MPL modeling language and solved by the CPLEX 12.1.0 solver in an Intel Core i5, at 2.80 GHz, with 6 GB of RAM. The proposed TS procedure to solve the vehicle routing problem was implemented in Java. A preliminary analysis was conducted to fine-tune the TS attributes. Two instances, with 10 and 21 suppliers, respectively, without considering fuel consumption minimization, were solved by using MPL and CPLEX solver with its defaults settings and compared to those results obtained by the TS algorithm. Purchasing amounts of parts and raw materials obtained from production and material requirement planning model are considered in order to determine collecting routes from suppliers. The results of the experiments related to the VRP problem by using CPLEX solver and TS procedure are presented in Table 2 for instances with 10 and 21 suppliers. For the instance problem with 10 suppliers, the results obtained by CPLEX solver and TS algorithm are identical in terms of solution cost and fuel consumption and similar in terms of CPU time. However, the TS procedure outperforms the results obtained by CPLEX solver for the instance problem with 21 suppliers. In this case, the TS algorithm provides a lower fuel consumption level and a better solution cost than CPLEX solver in a significantly shorter CPU time. In this sense, these better results are obtained in 0.607 s in comparison to more than 3 h needed by CPLEX. Moreover, a better solution cost implies a lower travelled distance and a lower fuel consumption, and hence a lower level of pollutants and greenhouse emissions.

6 Conclusions

This paper has proposed a mathematical programming model for production and sustainable routing planning decisions for inbound logistics processes in supply chain environments. A two-phase solution methodology based on a production and material requirement plan solved by a MIP solver and a VRP with fuel consumption minimization solved by a TS procedure has been adopted. The advantages of this proposal are related to: (1) the modelling of a new PRP for inbound logistics processes which includes sustainability issues such as fuel consumption and pollutants emissions; (2) the development of a highly effective TS procedure in finding good-quality solutions in short CPU times. With respect to the limitations of this proposal, we have described them through further research proposals: (1) consideration of travel times between nodes, waiting, loading and unloading times; (2) extend the set of nodes to include those suppliers located more than a day away; (3) allow multiple visits per period at suppliers for collecting parts and raw materials; (4) more accurately determination of gradient factors, by considering intermediate nodes in each arc and their corresponding elevations; (5) to develop solution procedures to tackle the proposed PRP in an integrated way, by considering the synchronization of production and inbound routing plans.

Acknowledgments This work has been funded by the Universitat Politècnica de València projects: 'Material Requirement Planning Fourth Generation (MRPIV)' (Ref. PAID-05-12) and 'Quantitative Models for the Design of Socially Responsible Supply Chains under Uncertainty Conditions. Application of Solution Strategies based on Hybrid Metaheuristics' (PAID-06-12).

References

Armentano VA, Shiguemoto AL, Løkketangen A (2011) Tabu search with path relinking for an integrated production distribution problem. Comput Oper Res 38(8):1199–1209

Bard JF, Nananukul N (2009) The integrated production-inventory-distribution-routing problem. J Sched 12(3):257–280

Boudia M, Louly MAO, Prins C (2008) Fast heuristics for a combined production planning and vehicle routing problem. Prod Plan Control 19(2):85–96

Clarke G, Wright JW (1964) Scheduling of vehicles from a central depot to a number of delivery points. Oper Res 12(4):568–581

Demir E, Bektaş T, Laporte G (2014) A review of recent research on green road freight transportation. Eur J Oper Res 237(3):775–793

Glover F, McMillan C (1986) The general employee scheduling problem. An integration of MS and AI. Comput Oper Res 13(5):563–573

Hickman J et al (1999) Methodology for calculating transport emissions and energy consumption. Office for Official Publications of the European Communities, Luxembourg

Jabali O, Van Woensel T, de Kok AG (2012) Analysis of travel times and CO_2 emissions in time-dependent vehicle routing. Prod Oper Manag 21(6):1060–1074

Ubeda S, Arcelus FJ, Faulin J (2011) Green logistics at Eroski: A case study. Int J Prod Econ 131 (1):44–51

Vigo D (1996) A heuristic algorithm for the asymmetric capacitated vehicle routing problem. Eur J Oper Res 89:108–126

Clothing Industry: Main Challenges in the Supply Chain Management of Value Brand Retailers

Sandra Martínez, Ander Errasti, Martin Rudberg and Miguel Mediavilla

Abstract The clothing sector is currently characterized by frequent assortment rotation in stores, a quick response product development and a focus on minimizing end-of-campaign stock levels. This paper is based on the literature review and a case study carried out at a value brand retailer, with a purpose to identify the main challenges that value brand retailers deal in the supply chain management and how they could reconfigure their internal and external supply chain in order to can stay competitive in a market requiring more responsiveness. The research shows that the garment's categorization, the development of a global purchasing strategy and a global and local supplier network, the introduction of mini-collections, the implementation of concurrent engineering and the redesign business processes and the supply chain leads to a sales increase and purchasing savings.

Keywords Clothing industry · Quick response supply chain · "Pronto Moda" · Global purchasing

---

S. Martínez (✉) · M. Mediavilla
GLOBOPE Research and Consulting, c/Idiaquez 4-3, 200004 San Sebastián, Spain
e-mail: sandra.martinez@globope.es

M. Mediavilla
e-mail: miguel.mediavilla@globope.es

A. Errasti
Natra Group, Carretera de Aránzazu, 20560 Oñate, Guipuzcoa, Spain
e-mail: ander.errasti@natra.com

M. Rudberg
Department of Science and Technology, Linköping University, 601 74 Norrköping, Sweden
e-mail: martin.rudberg@liu.se

© Springer International Publishing Switzerland 2015
P. Cortés et al. (eds.), *Enhancing Synergies in a Collaborative Environment*,
Lecture Notes in Management and Industrial Engineering,
DOI 10.1007/978-3-319-14078-0_8

1 Introduction: Hints About the Clothing Industry

The clothing industry is one of the most dynamic retailer-driven and global economic sectors (European Monitoring Centre on Change 2008). This sector is also organizationally complex (Forman and Joregnsen 2004) and supply chains are very long with many different parties involved taking different roles. Globalization trends have made supply chains broader and more international (De Brito et al. 2008) and the clothing industry has seen the outsourcing of most of its production activities to suppliers in developing countries (Bergvall-Forsberg and Towers 2007). Moreover, this industry has been characterized by price sensitive customers, short product life cycles, a wide product range, as well as volatile and unpredictable demand (Sen 2008).

Despite this, the traditional configuration is a forecast driven supply chain distributing two campaigns; Spring-Summer and Fall-Winter. To facilitate this, the major retailers and brand owners lead and coordinate different agents and activities, such as (MacCarthy and Jayarathne 2011a; Navaretti et al. 2001), franchised stores, own design departments, product category manufacturing suppliers, garment manufacturing and yarn procurement in low cost countries. Besides these actors also service providers, like quality auditors, forwarders and third party logistics providers, have to be managed and coordinated. The supply chain complexity is not only a function of the number of links, nodes and items in the network. It is also affected by the level of inter-relationship between the organizational units (Jonsson et al. 2013). Hence, all these agents and activities have to be integrated in order to be effective in two key value chain processes; new collection (product) development and order fulfilment. The effectiveness of this approach has however been questioned because of: (1) Difficulties to get collections on time, in full and error free, leading to loss of sales in stores and loss of orders at sales conventions with customer retailers; (2) overstock after the campaigns due to forecast driven manufacturing and procurement with difficulties to have an accurate demand forecast and (3) out of stock for garments with better than forecasted sales for the campaign or season, due to the lack of manufacturing responsiveness hindering replenishment orders within the campaign.

During the last decade, this traditional configuration has been employed, to a greater or lesser extent, by different types of clothing retailers, serving different market segment, such as (MacCarthy and Jayarathne 2011b); *leading brand retailers* (e.g. Benetton, H&M, Mango), *value brand retailers* (e.g. Desigual, Peak Performance, Ternua) and *hypermarket brands* for supermarket retailers (e.g. Walmart, Tesco, Marks and Spencer). However, the Spanish clothier Zara broke this paradigm; developing a super-responsive supply chain for leading brands (see e.g. Ferdows et al. 2004) known as "Pronto Moda" or "Rapid-Fire Fulfilment".

The purpose of this paper is therefore to investigate which are the major issues that value brand retailers confront and how *value brand retailers* can develop a supply chain structure and operations strategy to stay competitive in a market requiring more responsiveness. In so doing, we address the following research questions: (1) Which are the main challenges that value brand retailers cope in the supply chain management? (2) How should the internal and external supply chain be reconfigured to value brand retailers stay competitive in the market?

The paper is organized as follows. After this introduction follows a brief literature about the supply chain management in the clothing industry and the identification of the main problems that value brand retailers deal with. Finally, we present the discussion and the concluding remarks of the paper.

2 Supply Chain Management in the Clothing Industry

Supply chain management in the clothing industry has for long been focused on the two-campaign paradigm, developing collections for a Spring-Summer and a Fall-Winter campaign respectively. Nevertheless, the Spanish clothier Zara broke this paradigm, developing a super-responsive and quick supply chain for leading brands, which has been known as "Pronto Moda" or "Rapid-Fire Fulfilment" (Ferdows et al. 2004; Tokatli 2008; Gallaugher 2010). Zara's supply chain management has of course affected the industry and has also forced other retailer types to rethink their supply chain setting. Considering the three retailer types that MacCarthy and Jayarathne (2011a) identify, Fig. 1 compares *hypermarket brands*, *value brands* and *leading brands* in terms of supply chain configurations related to; the design departments, the range of collections in terms of width (gender, product lines, product families, garments number) and depth (colours, size, etc.), the degree of vertical integration in manufacturing and the global spread of production (affecting the lead-time to market), the degree of supply chain integration (i.e. the fabric and manufacturing process), and whether ownership of the stores are integrated in the value chain or not.

Referring to Table 1, for example, the leading brand retailer Zara has developed a competitive advantage through an unique distribution channel with vertically integrated sell points and stores and a tightly integrated and coordinated supply

Table 1 Supply chain configurations for the tree major retailer types in the clothing industry

Retailer type	Design	Range		Manufacturing		Supply chain coordination (fabric and processing)	Ownership of stores and brands
		Width	Depth	Degree of vertical integration	Location relative major markets		
Hypermarket Brand	Outsourced	Limited	Limited	Low	Far away	Low level of integration	Yes, shared between brands
Value Brand	In-house	Large	Medium	Low	Far away	Tight for quick-response, loose for low-cost	Seldom
Leading Brand	In-house	Very large	Deep	High	Both far away and near	Tight for quick-response, loose for low-cost	Yes, to a very large extent

chain. This, combined with the option of using both low-cost manufacturing and medium-cost manufacturing (closer to the market with shorter replenishment times), facilitates the use of customer driven replenishment with quick response product development and frequent assortment rotation in stores. These characteristics allow leading brands to have responsiveness in variety, time, and quantity (Holweg 2005). These characteristics are on the other hand also the reasons why value brand retailers are struggling to survive facing a market demanding high variety, but being stuck with a traditional supply chain configuration.

Therefore, the value brand retailers face many difficulties to leave the two-campaign paradigm while keeping the profit margins at reasonable levels. Some of the difficulties they face are (Bruce et al. 2004): (1) The complexity of multichannel distribution (own stores, franchised stores, wholesalers) with different requirements in product assortment and quality service, (2) being able to balance assortment rotation improvements and growth of sales with adequate margins and (3) lack of supply chain responsiveness due to, e.g., difficulties to integrate design requirements with supply chain agents and the lack of own manufacturing facilities and responsive suppliers.

Besides the difficulties mentioned above, the value brand retailers also face a volatile and cost-conscious environment, with requirements for high level of responsiveness in both product mix and volume (Holweg and Pil 2004). The competitive environment also leads to a combined pressure on price and quick response forcing retailers to be able to switch between, or combine, multi-channel manufacturing focusing on both quick replenishment and cost efficient production (Ferdows et al. 2004). In order to achieve this, it is necessary to build different supply chains combining agile and lean attributes (Christopher and Towill 2001) and different supply strategies. A successful approach to combine lean and agile methods allow companies to become competitive and fulfil the objective of customer satisfaction, but also to satisfy supply chain performance objectives concerning time, costs and working capital. In this context, the position of the customer order decoupling point (CODP) becomes important (see e.g. Wikner and Rudberg 2005). The CODP concept enables an analysis of the lead time gap between the production lead time (how long it takes to plan, source, manufacture and deliver a product) and the delivery lead time (how long customers are willing to wait for the order to be completed), which is a key element of the supply chain to obtain effectiveness and efficiency (Simchi Levi et al. 2000).

2.1 Problems with the Usual Supply Chain Management

To value brand retailers, the major factor triggering the change process is the poor performance obtained in their campaigns. This forces the companies to rethink both their business strategy and their operations strategy. The main inefficiencies are: (1) The offered range and variety of garments per collection that is not usually perceived necessary by the channel distributors and wholesalers; (2) overstock when finishing the campaign due to minimum order quantities in production, meaning that the quantities to procure and manufacture for some garments were higher than the actual demand; (3) overstock due to a forecast driven supply chain. The orders to manufacturers have to be settled before the sales convention, due to long lead times of fabric and trims procurement and (4) no possibility to procure and manufacture a replenishment (second) order for the campaign in garments that were selling in higher quantities than forecasted.

These deficiencies lead the companies to develop a new business strategy based on the following two objectives; accomplishing profitable campaigns and increasing sales by introducing more campaigns, what we called "mini-collections", between the two major ones.

3 Discussion

Nowadays, fashion companies do not have to face just the challenges posed by demand unpredictability (Priest 2005), but have also to adapt to a new competitive environment. The increased time-based competition (Jacobs 2006), the new consumers' sensibility to environmental issues (Caniato et al. 2011), the growing relevance of BRIC (Brazil, Russia, India, China) markets (Abecassis-Moedas 2007) and the rising of labor costs in emerging countries have driven fashion companies to find a new balance between local and global sourcing and production in their supply networks (Abesassis-Moedas 2007). By sourcing and producing globally, fashion companies can reduce production costs (Puig et al. 2009), but may not be agile enough to meet consumer's needs on a timely basis. To optimize the cost/agility trade off, several companies are now blending global and local sourcing and production activities (Jin 2004). As an example, Jin (2004) states that the greater the demand uncertainty the higher portion of domestic sourcing required in a mixture of global and domestic sourcing strategies.

Besides the balancing of global and local production, the choice between lean and agile manufacturing and the supply chain strategy are key aspects for the clothing industry. Furthermore, it is clear that the need to increase the new product development frequency far beyond Spring-Summer and Fall-Winter campaigns implies that it is necessary to reconfigure the supply chain to allow for quick response also for value brand retailers.

Concerning the reconfiguration of the supply chain and manufacturing operations, the value brand retailers used to employ a two-campaign low cost strategy. However, the future regarding product segmentation and categories called for a more responsive supply chain. The matrix presented by Fisher (1997), which links product types with supply chains, can be used to explain the new strategy for the value brand retailer (see Fig. 1).

The horizontal axis defines the product type, differentiating between functional and innovative products, whereas the vertical axis divides the supply chain structure into efficient or responsive supply chains. Therefore, according Fisher's (1997) matrix:

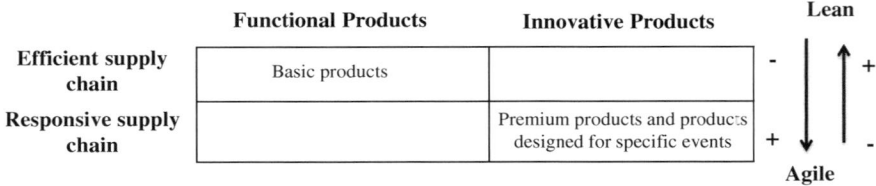

Fig. 1 Product categories classification in an adaption of Fisher's matrix (1997)

- The basic product categories could belong to the quadrant of functional products matched with an efficient supply chain because these products do not change much over time. They have stable, predictable demand and long life cycles, but their stability invites competition, which often leads to low profit margins. The main purpose of these products is therefore to supply predictable demand efficiently at the lowest possible cost. For this reason, the selection of the suppliers should be based on cost and the companies choose suppliers in low-cost countries focusing on high efficiency.
- The premium product categories and products for specific events would belong to the quadrant of innovative products matched with a responsive supply chain. These products are characterized by short life cycles, unpredictable demand and higher profit margins. Then, the primary goal is to respond quickly to the demand in order to minimize stockouts, forced markdowns and obsolete inventory. For this reason, the selection of the suppliers should be based on speed, flexibility and quality and the companies focus on suppliers in medium- or even high-cost countries to handle the demand for responsiveness.

4 Conclusions

Even though the literature review reveals a lot of research and case studies on leading retailers and some research on hypermarket clothing retailers, it has not still found a research about the main challenges that value brand retailers have to tackle due to the "Pronto Moda" paradigm.

Hence, the purpose of this paper was to expose the main challenges that value brand retailers have to face in order to manage their supply chain and how the new internal and external supply chain should be reconfigured in the future without losing efficiency. As such, we have used the literature review and a case study to investigate how one value brand retailer has addressed the transition from a strategy focusing on low cost and two campaigns, to a more responsive strategy also offering unique products with higher assortment rotations in stores. Therefore, this paper reveals the following concluding remarks:

- Campaigns could be more profitable if the collections are segmented taking into account the demand characteristics (continuity, season, etc.) and prices, establishing the most adequate supply strategy for each respective segment.
- Introducing a two-dimensional CODP and a product platform strategy facilitate concurrent engineering (designers, purchasers, manufacturers, forwarders), which in turn allows for reducing product development and order fulfilment lead times.
- Improving assortment rotations requires that the new product development and order fulfilment processes are redesigned. The latter typically leading to reconfiguration of the supply chain and manufacturing operations.

On the other hand, some of the actions that can be taken to mitigate the deficiencies (above mentioned) and to change the supply chain structure in a profitable way are the following: (1) Applying lean principles (Womack and Jones 2003), the companies managed to relax design resources and eliminate waste in the new collection development within a campaign. (2) Employing concurrent engineering among designers, purchasers, manufacturers and forwarders to reduce product development lead time. This also included the establishment of a critical chain and multi-project environment in order to avoid resource design constraints and resource allocation constraints for the manufacturing suppliers. (3) Categorizing garments according to demand behavior (continuity, season) and prices. (4) The development of a global purchasing strategy and a global and local supplier network taking into account price position, assortment rotation and demand pattern.

References

Abecassis-Moedas C (2007) Globalisation and regionalisation in the clothing industry: survival strategies for UK firms. Int J Entrepreneurship Small Bus 4(3):291–304
Bergvall-Forsberg J, Towers N (2007) Creating agile supply networks in the fashion industry: a pilot study of the European textile and clothing industry. J Text Inst 98:377–385
Bruce M, Daly L, Towers N (2004) Lean or agile. A solution for supply chain management in the textiles and clothing industry? Int J Oper Prod Manag 24(2):151–170
Caniato F, Caridi M, Crippa L, Moretto A (2011) Environmental sustainability in fashion supply chains: an exploratory case based research. Int J Prod Econ 135(2):659–670
Christopher M, Towill D (2001) An integrated model for the design of agile supply chains. Int J Phys Distrib Logist 31(4):235–246
De Brito MP, Carbone V, Blanquart CM (2008) Towards a sustainable fashion retail supply chain in Europe: organisation and performance. Int J Prod Econ 114:534–553
European Monitoring Centre on Change (2008) Trends and drivers of change in the European textiles and clothing sector: mapping report. http://www.eurofound.europa.eu/pubdocs/2008/15/en/1/ef0815en.pdf
Ferdows K, Lewis MA, Machuca JAD (2004) Rapid-fire fulfillment. Harv Bus Rev 82(11):104–117
Fisher ML (1997) What is the right supply chain for your product? Harv Bus Rev 75:105–117
Forman M, Jorgensen MS (2004) Organising environmental supply chain management. Greener Manag Int 45:43–62

Gallaugher J (2010) Information systems: a manager's guide to harnessing technology. Flat World Knowledge, L.L.C, Washington, p 324

Holweg M (2005) The three dimensions of responsiveness. Int J Oper Prod Manag 25(7):603–622

Holweg M, Pil F (2004) The second century: moving beyond mass and lean production in the auto industry. MIT Press, Cambridge

Jacobs D (2006) The promise of demand chain management in fashion. J Fash Mark Manag 10(1):84–96

Jin B (2004) Achieving an optimal global versus domestic sourcing balance under demand uncertainty. Int J Oper Prod Manag 24(12):1292–1305

Jonsson P, Rudberg M, Holmberg S (2013) Centralised supply chain planning at IKEA. Supply Chain Manag Int J 18(3):337–350

MacCarthy B, Jayarathne A (2011a) Global supply networks and responsiveness in the international clothing industry: differences across different retailer types. Paper presented at Euroma Conference proceedings, Cambridge, UK

MacCarthy BL, Jayarathne PGSA (2011b) Sustainable collaborative supply networks in the international clothing industry: a comparative analysis of two retailers. Prod Plan Control 1(17)

Navaretti GB, Falzoni A, Turrini A (2001) The decision to invest in a low-wage country: evidence from Italian textiles and clothing multinationals. J Int Trade Econ Dev 10(4):451–470

Priest A (2005) Uniformity and differentiation in fashion. Int J Cloth Sci Technol 17:253–263

Puig F, Marques H, Ghauri P (2009) Globalization and its impact on operational decisions: the role of industrial districts in the textile industry. Int J Oper Prod Manag 29(7):692–719

Sen A (2008) The US fashion industry: a supply chain review. Int J Prod Econ 114:571–593

Simchi Levi D, Kaminsky P, Simchi Levi E (2000) Designing and managing the supply chain: concepts, strategies, and cases. Mac Graw Hill, USA

Tokatli N (2008) Global sourcing: insights from the global clothing industry-the case of Zara, a fast fashion retailer. J Econ Geogr 8:21–38

Wikner J, Rudberg M (2005) Introducing a customer order decoupling zone in logistics decision-making. Int J Logist Res Appl 8(3):211–224

Womack J, Jones D (2003) Lean thinking. Free press, New York

Structural Equation Modeling for Analyzing the Barriers and Benefits of E-procurement

Peral Toktaş-Palut, Ecem Baylav, Seyhan Teoman and Mustafa Altunbey

Abstract We perform an empirical study to analyze whether it would be beneficial for a company to switch from traditional procurement system to e-procurement system. For this purpose, we determine the main barriers and benefits of e-procurement systems, and using Structural Equation Modeling, we analyze the effects of the barriers and benefits on the e-procurement adoption decision. The results denote that barriers (benefits) of e-procurement systems have negative (positive) effect on the adoption decision. We also find that the effect of the benefits is higher than that of the barriers, indicating that it would be beneficial for the company to adopt the e-procurement system.

Keywords Barrier · Benefit · E-procurement · Structural equation modeling

1 Introduction

Supply chains can be managed more effectively through the developments in the information and communication technology. One of the developments in supply chains is the e-procurement system, which helps the integration of the procurement process throughout the supply chain. Morris et al. (2000) define e-procurement as a system that utilizes Internet technologies and services to automate and streamline an organization's processes—from requisition to payment.

P. Toktaş-Palut (✉)
Department of Industrial Engineering, Doğuş University, 34722 Istanbul, Turkey
e-mail: ppalut@dogus.edu.tr

E. Baylav · M. Altunbey
Logistics and Supply Chain Management Doctorate Program, Doğuş University, 34722 Istanbul, Turkey

S. Teoman
Logistics and Supply Chain Management Doctorate Program, Maltepe University, 34857 Istanbul, Turkey

E-procurement, which has become one of the fundamental elements of a supply chain, is still in the development phase and there are several studies in the literature on this subject. Gunasekaran and Ngai (2008) state that without the adoption of e-procurement systems, the supply chain of a company cannot be integrated successfully. The authors conduct a questionnaire-based survey in order to understand the adoption process of e-procurement in Hong Kong. Panayiotou et al. (2004) work on a case study about the Greek purchasing process and indicate the problems which may occur. The authors also study e-procurement system design. Gunasekaran et al. (2009) analyze the current state of e-procurement in SMEs located in the Southcoast of Massachusetts, and they also examine the factors that affect the e-procurement adoption.

In the light of the previous studies on e-procurement systems, this study aims at investigating the effects of the barriers and benefits of e-procurement on its adoption decision. For this purpose, by using Structural Equation Modeling (SEM), we perform an empirical analysis for a retail store chain which operates in book and stationery sector in Turkey.

SEM can be used for multivariate data as it is appropriate for illustrating the relations between exogenous and endogenous latent variables in one model (Kline 1998). SEM is a widely used tool especially for psychology, sociology, and econometrics and there are several studies applying SEM in the literature. Thus, we only mention the studies in a related field. Madeja and Schoder (2003) examine the impacts of e-procurement adoption process on e-business success. Lee and Quaddus (2006) conduct a study in Singapore about the impacts of buyer-supplier relationship on e-purchasing adoption decision. Vaidyanathan and Devaraj (2008) study the impacts of the quality of information flow process and the quality of logistics fulfillment on the satisfaction of e-procurement performance. Finally, Devaraj et al. (2012) investigate the effects of mixing flexibilities and purchase volumes on e-procurement performance.

The rest of the paper is organized as follows. The theoretical background, research hypotheses, and data collection are given in Sect. 2. The data analysis and results are presented in Sect. 3. Finally, concluding remarks are given in Sect. 4.

2 Research Background and Data Collection

In this study, we consider a retail store chain which operates in book and stationery sector in Turkey. The company has an annual procurement volume of approximately 100 million USD, which covers its demand for more than 20,000 different stock keeping units provided by 185 suppliers. Although the company uses traditional procurement methods, the senior management wants to decide whether to adopt an e-procurement system or not based on its barriers and benefits.

First, 20 barriers and 20 benefits of e-procurement systems are determined through a detailed literature review.

In the second step, eight top managers from the company have rated these barriers and benefits on a five-point scale. By taking the geometric mean of their responses and after a brainstorming session, we get the final set of 14 barriers and 15 benefits that are rated above three.

In the third step, Interpretive Structural Modeling (ISM) technique (Warfield 1973) is applied to determine the barriers and benefits that have high driving power and the capability to influence the other drivers. Based on the results of the ISM model, the numbers of barriers and benefits are decreased to 3 and 7, respectively.

Finally, these barriers and benefits are entered into the SEM model. In this step, another five-point scale questionnaire is prepared. Here, the aim is to analyze the effects of these barriers and benefits on the e-procurement adoption decision and the necessity of the e-procurement system for the company. The questionnaire is uploaded to the Intranet of the company. Among 916 people working in a related field, 277 respondents answer the questionnaire, resulting in a response rate of 30.24 %. Among these responses, 21 cases contain missing values, giving a percentage of 7.58 %. Since sample size is especially important for SEM, Bayesian imputation is used to estimate the missing values.

The barriers of e-procurement systems taken into consideration in this study are given below (Eadie et al. 2007; Eei et al. 2012):

Lack of e-procurement knowledge/skilled personnel (R1). Mainly related to personnel issues such as older generations that have not kept up to the advances in IT related fields, but relying heavily on traditional forms and means of procurement.

Lack of adequate technical/IT infrastructure (R2). Need of adequate IT infrastructure to carry out e-procurement processes and/or technology to operate IT.

Inadequate IT infrastructure of suppliers/business partners (R3). The external parties of the supply chain do not have adequate IT infrastructure compatible with the e-procurement system.

On the other hand, the benefits of e-procurement systems considered in this study are as follows (Morris et al. 2000; Eadie et al. 2007; Eei et al. 2012):

Easier access to market data and enhanced intelligence (N1). Enables to monitor and scan external sources of data and intelligence easily and to share information with others pro-actively.

Quicker response to problems through real-time information (N2). Increased speed in transactions, tracking and reporting through real-time information helps quicker problem solving and reactive decisions.

On-line and real-time reporting (N3). Real-time reporting system that enables management to have a fast and reliable way to compare the spending with budget, allowing quick reaction to problems.

Improved supply chain transparency (N4). Transparency of product specifications, prices, contract details, such as the contractual conditions, time, terms of orders, etc., making these visible to relevant parties both internally and externally.

Simplified and streamlined purchasing process (N5). E-procurement solutions simplify the purchasing process by bringing all suppliers together, accessible from a single e-platform and eliminating the need for a paper form, while providing streamlined procedures to expedite order to payment processing.

Integrated information sharing (N6). Accelerates the flow of important information between internal and external business partners; also provides real-time information sharing within a broader structure.

Improved communication and collaboration in supply chain (N7). E-procurement allows sections of electronic documentation to flow through the supply chain. As it improves the speed of returns and makes it easy to communicate requirements in a quicker and more accessible manner, it will result in a better understanding of requirements and due compliance.

This study uses SEM to analyze the effects of the barriers and benefits of e-procurement on its adoption decision and we have two main research hypotheses given below:

Hypothesis 1 (H1). Barriers of e-procurement systems have negative impact on the adoption decision.

Hypothesis 2 (H2). Benefits of e-procurement systems have positive impact on the adoption decision.

In addition to these hypotheses, we are also interested in understanding whether the effects of the barriers or benefits are higher on the e-procurement adoption decision. Accordingly, we may recommend the company to adopt the e-procurement system or not.

3 Data Analysis and Results

We use Anderson and Gerbing's (1988) two-step approach in this study, where the measurement model is estimated prior to the structural model. The AMOS 16.0 software is used to test the measurement and structural models based on the maximum likelihood estimation method.

The measurement model has two latent variables, i.e., Barriers and Benefits, with three and seven indicators, respectively. We test the measurement model using the confirmatory factor analysis (CFA) method as described in Hair et al. (2010). Here, we present the results for the final measurement model. We calculate several goodness-of-fit measures based on the recommendations of Jöreskog and Sörbom (1993) and Hu and Bentler (1998): Comparing the magnitude of χ^2 with the number of degrees of freedom, i.e., χ^2/df; standardized root mean square (SRMR); goodness-of-fit index (GFI); adjusted goodness-of-fit index (AGFI); normed fit index (NFI); comparative fit index (CFI); and root mean square error of approximation (RMSEA). The recommended values for these fit indices and the results for the measurement model are given in Table 1. As one can see from the table, all values

Table 1 Goodness-of-fit measures for the measurement model

Goodness-of-fit measure	Recommended value (Schermelleh-Engel et al. 2003)	Result
χ^2/df	≤3.00	2.118
SRMR	≤0.10	0.052
GFI	≥0.90	0.957
AGFI	≥0.85	0.923
NFI	≥0.90	0.957
CFI	≥0.95	0.977
RMSEA	≤0.08	0.064

are within the recommended ranges indicating that the measurement model has a good fit.

After evaluating the fit of the measurement model, we assess the construct validity that consists of convergent validity and discriminant validity.

We test the convergent validity by examining the factor loadings first. All factor loadings are significant at the 0.001 level and except two of them all are above 0.70. Only R1 and R3 have factor loadings of 0.65 and 0.67, respectively; but since these are above 0.50, they are also acceptable. Second, the average variance extracted (AVE) is calculated for each latent construct. AVE values are calculated as 0.49 and 0.62 for Barriers and Benefits, respectively. Since an AVE of 0.50 or higher suggests adequate convergent validity, one can see that the AVE for Barriers is just below this critical level. Third, we calculate the construct reliability (CR), which is also an indicator of convergent validity. CR values are calculated as 0.74 and 0.92 for Barriers and Benefits, respectively. Since they both exceed the threshold of 0.70, this also validates convergent validity. The results for the measurement model can be found in Table 2.

Table 2 Results for the measurement model

Latent variable	Indicator	Factor loading	AVE	CR
Barriers			0.490	0.742
	R1	0.654*		
	R2	0.774*		
	R3	0.665*		
Benefits			0.624	0.924
	N1	0.743*		
	N2	0.790*		
	N3	0.748*		
	N4	0.809*		
	N5	0.865*		
	N6	0.756*		
	N7	0.810*		

* Significant at the 0.001 level (two-tailed)

To test the discriminant validity, the AVE values for the constructs are compared with the square of the correlation estimate between these two constructs. The correlation estimate between Barriers and Benefits is calculated as 0.30. Since AVE values for both constructs are greater than the squared correlation estimate, this result provides good evidence of discriminant validity.

After assessing the convergent and discriminant validities, we examine the standardized residual covariances and modification indices to identify the problems in the measurement model. Since all standardized residual covariances are less than |4.00|, we conclude that the standardized residual covariances indicate no problem in the measurement model. On the other hand, we also examine the modification indices. High modification indices suggest that the fit could be improved significantly by freeing the corresponding path to be estimated. Since the modification indices of the measurement model are reasonable, this result also provides good evidence of model validation.

After testing the measurement model, we test our structural model and our main focus is to test the hypothesized relationships. The structural model consists of two exogenous variables (Barriers and Benefits) and one endogenous variable (Adoption). Recall that the exogenous variables Barriers and Benefits have three and seven indicators, respectively. On the other hand, the endogenous variable Adoption has a single indicator, which is measured through the question that investigates the necessity of the e-procurement system for the company.

The goodness-of-fit of the structural model is evaluated with the same measures used to test the measurement model and the results are given in Table 3. The results denote that the structural model fits the data well.

Once we evaluate the goodness-of-fit measures for the structural model, we analyze the path coefficients and loading estimates, which are given in Fig. 1. It can be seen that the path coefficient estimate between Barriers and Adoption is −0.15 and it is significant at the 0.05 level. Thus, our first hypothesis is supported, i.e., barriers of e-procurement systems have negative effect on the adoption decision. On the other hand, the path coefficient estimate between Benefits and Adoption is 0.48 and it is significant at the 0.001 level. Thus, our second hypothesis is also supported, i.e., benefits of e-procurement systems have positive effect on the adoption decision. As stated before, besides these hypotheses, another point of interest is to

Table 3 Goodness-of-fit measures for the structural model

Goodness-of-fit measure	Recommended value (Schermelleh-Engel et al. 2003)	Result
χ^2/df	≤3.00	1.836
SRMR	≤0.10	0.050
GFI	≥0.90	0.956
AGFI	≥0.85	0.926
NFI	≥0.90	0.955
CFI	≥0.95	0.979
RMSEA	≤0.08	0.055

Fig. 1 Results of the path analysis

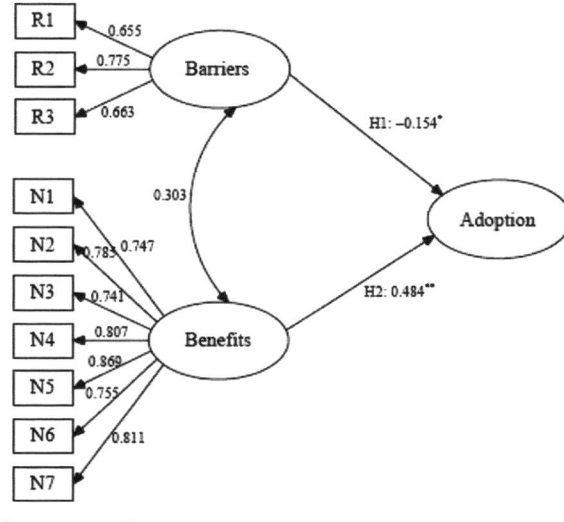

examine whether the barriers or benefits of e-procurement systems have higher impact on the adoption decision. Since the path coefficient estimate of Benefits is higher than that of the Barriers, we conclude that benefits of an e-procurement system overweigh its barriers. Thus, it would be beneficial for the company to adopt the e-procurement system. However, it is also noteworthy that the squared multiple correlations (i.e., R^2) of Adoption is calculated as 0.21. Therefore, adding other latent constructs that may affect the e-procurement adoption decision would increase the variance explained.

4 Conclusion

The aim of this study is to analyze the effects of the barriers and benefits of e-procurement systems on the e-procurement adoption decision. An empirical analysis has been performed for a retail store chain which operates in book and stationery sector in Turkey.

Using the ISM technique, the barriers of e-procurement that have high driving power and the capability to influence the other drivers are found to be lack of e-procurement knowledge/skilled personnel; lack of adequate technical/IT infrastructure; and inadequate IT infrastructure of suppliers/business partners. On the other hand, the important benefits are easier access to market data and enhanced intelligence; quicker response to problems through real-time information; on-line and real-time reporting; improved supply chain transparency; simplified and streamlined purchasing process; integrated information sharing; and improved communication and collaboration in supply chain.

Once we determine the main barriers and benefits of e-procurement systems, they are integrated into the structural equation model. The results denote that barriers (benefits) of e-procurement systems have negative (positive) effect on the e-procurement adoption decision and the effect of benefits on the adoption decision is higher than that of the barriers. Thus, based on this analysis, it would be beneficial for the company to adopt the e-procurement system. As a further study, besides the barriers and benefits of e-procurement systems, other factors could also be included into the models.

References

Anderson JC, Gerbing DW (1988) Structural equation modeling in practice: a review and recommended two-step approach. Psychol Bull 103(3):411–423
Devaraj S, Vaidyanathan G, Mishra AN (2012) Effect of purchase volume flexibility and purchase mix flexibility on e-procurement performance: an analysis of two perspectives. J Oper Manag 30(7–8):509–520
Eadie R, Perera S, Heaney G, Carlisle J (2007) Drivers and barriers to public sector e-procurement within Northern Ireland's construction industry. Electr J Inf Technol Constr 12:103–120
Eei KS, Husain W, Mustaffa N (2012) Survey on benefits and barriers of e-procurement: Malaysian SMEs perspective. Int J Adv Sci Eng Inf Technol 2(6):14–19
Gunasekaran A, McGaughey RE, Ngai EWT, Rai BK (2009) E-procurement adoption in the Southcoast SMEs. Int J Prod Econ 122(1):161–175
Gunasekaran A, Ngai EWT (2008) Adoption of e-procurement in Hong Kong: an empirical research. Int J Prod Econ 113(1):159–175
Hair JF, Black WC, Babin BJ, Anderson RE (2010) Multivariate data analysis. Pearson Prentice Hall, New Jersey
Hu L, Bentler PM (1998) Fit indices in covariance structure modeling: Sensitivity to unparameterized model misspecification. Psychol Methods 3(4):424–453
Jöreskog KG, Sörbom D (1993) Structural equation modeling with the SIMPLIS command language. Scientific Software, Chicago
Kline RB (1998) Principles and practice of structural equation modeling. The Guilford Press, New York
Lee W, Quaddus M (2006) Buyer-supplier relationship in the adoption of e-purchasing in the small and medium printing industries in Singapore: an empirical test using structural equation modelling. In: Proceedings of the 10th pacific Asia conference on information systems, pp 337–345
Madeja N, Schoder D (2003) The adoption of e-procurement and its impact on corporate success in electronic business. In: The 3rd international conference on electronic business, Singapore
Morris A, Stahl A, Herbert R (2000) E-procurement: streamlining processes to maximize effectiveness. Luminant Worldwide Corporation, USA
Panayiotou NA, Gayialis SP, Tatsiopoulos IP (2004) An e-procurement system for governmental purchasing. Int J Prod Econ 90(1):79–102
Schermelleh-Engel K, Moosbrugger H, Müller H (2003) Evaluating the fit of structural equation models: tests of significance and descriptive goodness-of-fit measures. Methods Psychol Res Online 8(2):23–74
Vaidyanathan G, Devaraj S (2008) The role of quality in e-procurement performance: an empirical analysis. J Oper Manag 26(3):407–425
Warfield JN (1973) Binary matrices in system modeling. IEEE Trans Syst Man Cybern SMC 3(5):441–449

Ergonomic Risk Minimisation in Assembly Line Balancing

Joaquín Bautista, Cristina Batalla-García and Rocío Alfaro-Pozo

Abstract In this work it is presented a variant for the assembly line balancing that considers simultaneously production conditions and labour conditions. Specifically, a new mathematical model is formulated to solve the assembly line balancing problem. The objective of this model is minimising the maximum ergonomic risk, given a specific cycle time and a number of workstations. This affords the evaluation of the ergonomic risk impact on the number of workstations of the line through an application linked to the Nissan plant of engines, located in Barcelona.

Keywords Manufacturing systems · Assembly line balancing · Ergonomic risk · Mathematical model

1 Introduction

Production and assembly lines present a lot of factors that can harm the safety and welfare of the workers assigned to workstations. Indeed, a task can suppose ergonomic risk due to psychic factors, physical factors, or both.

This work is supported by the Spanish Ministerio de Economía y Competitividad under Project DPI2010-16759 (PROTHIUS-III) including EDRF fundings.

J. Bautista (✉) · C. Batalla-García · R. Alfaro-Pozo
Research Group OPE-PROTHIUS, Dpto. de Organización de Empresas,
Universitat Politècnica de Catalunya, Avda Diagonal, 647,
7th Floor, 08028 Barcelona, Spain
e-mail: joaquin.bautista@upc.edu

C. Batalla-García
e-mail: cristina.batalla@upc.edu

R. Alfaro-Pozo
e-mail: rocio.alfaro@upc.edu

© Springer International Publishing Switzerland 2015
P. Cortés et al. (eds.), *Enhancing Synergies in a Collaborative Environment*,
Lecture Notes in Management and Industrial Engineering,
DOI 10.1007/978-3-319-14078-0_10

In terms of physical load, there are many factors that can affect the health of workers (Landau et al. 2008). In fact, there are different methods of evaluation which only analyse a specific factor (postural loads, repetitive movements, or manual handling). Among them are the *RULA* method (Rapid Upper Limb Assessment) to analyse postural loads, the *NIOSH* method (National Institute for Occupational Safety and Health) to evaluate the manual handlings due to different actions as raise, move, push, grasp, and transport objects, and the *OCRA Check List* (Occupational Repetitive Action) for repetitive movements.

However, the Ergonomics is not the only problem presented in actual assembly lines. Currently, assembly lines should be able to treat several product types that, although similar, they can suppose operations with different processing times, different use of resources and different component consumptions. This variety together the precedence restrictions of the operations make necessary the assembly line balancing. Thus, a classic problem of literature appears, the *ALBP* (Assembly Line Balancing Problem) (Salvenson 1955; Becker and Scholl 2006; Battaïa and Dolgui 2012).

Given a set of elemental tasks or operations, J, and a set of workstations, K, the assembly line balancing problem consist of establishing task assignments to the set of workstations satisfying the set of technological conditions that defines the order in which the operations must be realised (precedence constraints).

In addition to the precedence constraints, other variants of the assembly line balancing include more line attributes, such as the available time for the workstations (at normal activity), i.e., the cycle time, and the available lineal area for materials and tools on workstations. This type of problems corresponds to the family *TSALBP* (Time and Space Constrained Assembly Line Balancing Problems) (Chica et al. 2010).

Moreover, some recent works incorporate ergonomic attributes into the assembly line balancing problems. In particular, Rajabalipour et al. (2012) consider the monotonous corporal postures that the operators suffer during their workday. To do this, they use a simplification of the *OWAS* method (Ovako Working Analysis System) that only considers three corporal areas: back, arms and legs.

On the other hand, Otto and Scholl (2011) incorporate the ergonomic risk into the *SALBP-m* (Simple Assembly Line Balancing Problem minimising the number of workstations). They propose two ways to consider the ergonomic risk: (1) constraints that limit the maximum allowed ergonomic risk $Erg\,(F(S_k) \leq Erg, \forall k)$; (2) a new objective function that minimises the number of workstations and the global ergonomic risk of the line using a weighting coefficient $(\min K'(x) = K(x) + \omega \cdot \xi(F(S_k)))$. In both proposals, they evaluate the ergonomic risk with three methods: the *NIOSH*, the *OCRA* and the *EAWS* (European Assembly Worksheet) method which was created for assembly production systems.

Similarly, Bautista et al. (2013a), expand the family of problems *TSALBP* through the incorporation of constraints that limit the minimum and maximum ergonomic risk to which workers may be exposed. Specifically, they consider psychological and physical factors. In other work, Bautista et al. (2013b) analyse the impact of the ergonomic risk reduction on the number of workstations of the line. In this work they add the elemental tasks in blocks of tasks to realise an experience computational.

Therefore, following the trend of satisfying both the productive and the labour conditions in this work a new model for assembly line balancing is presented. In particular this model minimises the maximum risk of line given, previously, the number of workstations, the cycle time (at normal activity) and the available maximum area per workstation.

2 Ergonomic Risk Minimisation into the *TSALBP*

The determination of the ergonomic risk associated to each operation is a preliminary step in the formulation of the mathematical model for the assembly line balancing problem. Therefore, considering the set of ergonomic risk factors Φ ($\phi = 1, \ldots, |\Phi|$), all the tasks will be evaluated classifying them by risk categories. Thus, the task j ($j = 1, \ldots, |J|$) will have the category $\chi_{\phi,j}$ associated with the ergonomic risk factor ϕ ($\phi = 1, \ldots, |\Phi|$).

In such conditions, the following model is proposed for the line balancing problems whose parameters and variables are:

Parameters	
J	Set of elemental tasks ($j = 1, \ldots, |J|$)
K	Set of workstations ($k = 1, \ldots, |K|$)
Φ	Set of ergonomic risk factors ($\phi = 1, \ldots, |\Phi|$)
t_j	Processing time at normal activity required by the elemental task j ($j = 1, \ldots, |J|$)
a_j	Linear area required by the elemental task j ($j = 1, \ldots, |J|$)
$\chi_{\phi,j}$	Category of the task j ($j = 1, \ldots, |J|$) associated to the risk factor ϕ ($\phi = 1, \ldots, |\Phi|$). Here it is a non-negative integer value between 1 and 4
$R_{\phi,j}$	Ergonomic risk of task j ($j = 1, \ldots, |J|$) associated to the risk factor ϕ ($\phi = 1, \ldots, |\Phi|$). Here $R_{\phi,j} = t_j \cdot \chi_{\phi,j}$
P_j	Set of direct precedent tasks of the task j ($j = 1, \ldots, |J|$)
c	Cycle time. Standard time assigned to each workstation ($k = 1, \ldots, |K|$) to process its workload (S_k)
m	Number of workstations. In this case $m = |K|$
A	Available space or linear area assigned to each workstation
Variables	
$x_{j,k}$	Binary variable equal to 1 if the elemental task j ($j = 1, \ldots, |J|$) is assigned to the workstation k ($k = 1, \ldots, |K|$), and to 0 otherwise
R_ϕ	Maximum ergonomic risk associated to the risk factor ϕ ($\phi = 1, \ldots, |\Phi|$), and allowed to each workstation

TSALBP-R_erg:

$$\min \bar{R}(\Phi) = \frac{1}{|\Phi|} \sum_{\phi=1}^{|\Phi|} R_\phi \qquad (1)$$

Subject to:

$$\sum_{\forall k \in K} x_{j,k} = 1 \quad (j = 1, \ldots, |J|) \tag{2}$$

$$\sum_{\forall j \in J} t_j \cdot x_{j,k} \leq c \quad (k = 1, \ldots, |K|) \tag{3}$$

$$\sum_{\forall j \in J} a_j \cdot x_{j,k} \leq A \quad (k = 1, \ldots, |K|) \tag{4}$$

$$\sum_{\forall j \in J} R_{\phi,j} \cdot x_{j,k} \leq R_\phi \quad (k = 1, \ldots, |K|) \wedge (\phi = 1, \ldots, |\Phi|) \tag{5}$$

$$\sum_{\forall k \in K} k \left(x_{i,k} - x_{j,k} \right) \leq 0 \quad \left(1 \leq i, j \leq |J| : i \in P_j \right) \tag{6}$$

$$\sum_{\forall k \in K} k \cdot x_{j,k} \leq m \quad (j = 1, \ldots, |J|) \tag{7}$$

$$\sum_{\forall j \in J} x_{j,k} \geq 1 \quad (k = 1, \ldots, |K|) \tag{8}$$

$$x_{j,k} \in \{0, 1\} \quad (j = 1, \ldots, |J|) \wedge (k = 1, \ldots, |K|) \tag{9}$$

In the model, the objective function (1) expresses the minimisation of the average of maximum ergonomic risks resulting of the aggregation of the set of risk factors Φ. Constraints (2) indicate that each task can only be assigned to one workstation. Constraints (3) and (4) impose the maximum limitation of the workload time and the maximum linear area allowed by the workload of each workstation. Constraints (5) determine the maximun ergonomic risk associated to the factor $\phi \in \Phi$ allowed at each workstation. Constraints (6) correspond to the precedence task bindings. Constraints (7) and (8) limit the number of workstations and force that there is no empty workstation, respectively. Finally, constraints (9) require the assignment variables be binary.

3 Computational Experience

A production plan linked to a case study of Nissan engine plant in Barcelona (NMISA: Nissan Motor Ibérica SA) is used to evaluate the proposed model, the *TSALBP-R_erg*. In this way the impact of the number of workstations on the maximum ergonomic risk is evaluated. In particular the production plan corresponds to a total demand of $T = 270$ engines which is characteristic of a workday with two shifts. This demand is divided, equally, into nine types of engine: three for 4 × 4 vehicles (p_1, p_2 and p_3), two for vans (p_4 and p_5) and four for trucks (p_6, p_7, p_8 and p_9). Despite their differences, the assembly of the three engine classes requires 378 elementary tasks (including rapid testing). These tasks have been grouped into 140 operations (Bautista and Pereira 2007; Chica et al. 2010). To facilitate this aggregation of tasks into different workstations of the line at the time of balancing the appropriate precedence rules have been taken into account.

To implement the experiment a Mathematical Programming Solver (the Solver CPLEX v11.0) has been used on a MacPro computer with an Intel Xeon 3.0 GHz CPU and 2 GB RAM using Windows XP with a CPU time limit of 7200 s.

Moreover, the cycle time has been set in $c = 180$ s (see labour conditions in Llovera et al. 2014); the maximum allowed area has been considered as infinite, $A \rightarrow \infty$, in order to avoid its impact; several values for the number of workstations have been also considered, these values range from 19 to 23 workstations; and finally, only the physical ergonomic risk factor has been taken into account, considering postural loads, repetitive movements and manual handlings at once. The corresponding methods used for the evaluation of those ergonomic risk factors have been the *RULA*, *OCRA* and *NIOSH* method, respectively, by modifying the levels of risk of each method on a common scale, ranging from 1 (acceptable risk) to 4 (unacceptable risk). Thus, a global physical risk category for each task has been obtained.

Figure 1 shows the effect of increasing the number of workstations of the line on the maximum ergonomic risk. Specifically, when the line has 19 workstations the maximum risk is 350 e-s (ergo-seconds) which is equivalent to a risk category of 1.94. In this case, a risk analysis should be necessary in order to carry on corrective actions for the improvement of the line in the future. However, this ergonomic risk decreases with the increased number of workstations. Indeed, when the line is composed by 20 workstations the maximum ergonomic risk is 315 e-s; when there are 21 or 22 workstations, the risks obtained correspond to 300 e-s and 285 e-s, respectively; finally, the maximum risk is 280 e-s when the line has 23 workstations.

On the other hand, if we analyse the obtained results according to the temporal, spatial and ergonomic attributes (see Table 1), we see that the workload times of

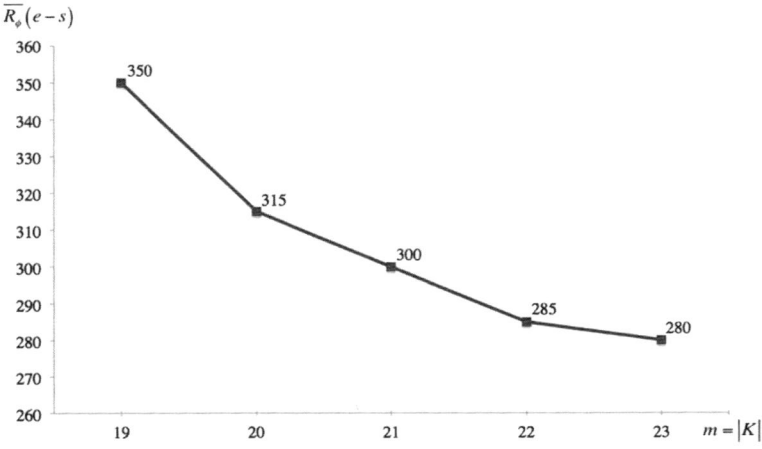

Fig. 1 Maximum ergonomic risk (e-s) depending on the number of workstations of the line

Table 1 Processing times, $t(k)$, required linear area, $a(k)$, and ergonomic risk, associated to the physical factors $R_\phi(k)$, corresponding to each workstation ($k \in K$)

k	$|K|=19$			$|K|=20$			$|K|=21$			$|K|=22$			$|K|=23$		
	$t(k)$	$a(k)$	$R_\phi(k)$	$t(k)$	$a(k)$	$R_\phi(k)$	$t(k)$	$a(k)$	$R_\phi(k)$	$t(k)$	$a(k)$	$R_\phi(k)$	$t(k)$	$a(k)$	$R_\phi(k)$
1	180	7.5	265	180	6	300	180	6.5	285	171	8	267	173	7.5	271
2	176	5	332	173	4.5	311	161	5	282	180	3	260	160	2.5	280
3	179	5	298	180	6	280	175	4	290	142	4.5	284	168	5.5	276
4	175	5	320	177	6.5	299	174	6.5	298	172	5.5	284	169	6	278
5	125	7.5	350	105	6.5	315	105	5.5	290	105	4	275	110	3.5	280
6	115	1.5	345	105	1.5	315	100	2.5	300	100	3.5	285	90	3.5	270
7	115	1.5	345	105	2	315	100	2	300	95	4	285	80	3.5	240
8	120	1.5	340	105	0	315	100	1	300	95	1.5	285	90	1	270
9	170	4.5	340	140	3.5	315	110	2	290	95	0	285	90	1	270
10	175	5.5	350	155	4.5	310	150	4	300	130	3	285	115	2	280
11	175	6.5	350	155	6	310	150	5	300	140	4	280	130	3.5	260
12	130	2.75	335	130	3.75	315	150	5.5	300	135	4.5	270	130	4.5	260
13	140	1.5	340	120	2.5	315	110	2.25	285	140	5.5	280	140	5.5	280
14	155	3	350	130	1	290	95	1.5	285	120	2.25	285	110	2.25	275
15	170	2	340	155	2.5	315	150	2	300	95	2.5	285	75	1.5	225
16	180	4	345	170	3	315	145	3	295	125	1	285	120	0.5	270
17	170	5	285	175	4.5	315	140	1.5	280	140	3	285	130	2.5	265
18	160	4.25	265	180	4.5	315	175	4	295	140	1.5	280	140	3	280

(continued)

Table 1 (continued)

k	$\|K\|=19$			$\|K\|=20$			$\|K\|=21$			$\|K\|=22$			$\|K\|=23$		
	$t(k)$	$a(k)$	$R_\phi(k)$	$t(k)$	$a(k)$	$R_\phi(k)$	$t(k)$	$a(k)$	$R_\phi(k)$	$t(k)$	$a(k)$	$R_\phi(k)$	$t(k)$	$a(k)$	$R_\phi(k)$
19	180	2	250	170	3.75	295	165	4	295	140	3	285	160	2.5	280
20				180	3	285	175	4.75	290	170	4.5	280	150	3.5	275
21							180	3	285	180	4.25	280	165	5	280
22										180	2.5	255	115	2.75	225
23													180	2.5	255
Avg	157	4.0	323	149	3.8	307	142	3.6	293	136	3.4	279	130	3.3	267
Min	115	1.5	250	105	0	280	95	1	280	95	0	255	75	0.5	225
Max	180	7.5	350	180	6.5	315	180	6.5	300	180	8	285	180	7.5	280

workstations ($t(k)$ $\forall k \in K$) are more variable when there are more workstations in the line. Indeed, when the line has 19 workstations the minimum workload time is 115 s, whereas with 23 workstations the minimum workload is 75 s. However, this trend is not repeated with the linear area required by the workstations ($a(k)$ $\forall k \in K$), because regardless of the number of workstations, the maximum linear area of workstations is more than 6 m in all cases, reaching even 8 m when there are 22 workstations.

Regarding the ergonomic risk, the best situation in terms of maximum, minimum and average risk occurs when the line has 23 workstations. In this case, the values obtained are 280, 225 and 267.2 e-s, respectively. However, the more balanced situation between workstations occurs when the line has 21 workstations, situation in which the range of values (20 e-s) and the standard deviation (6.9) are smaller.

Finally, the obtained results shown that the five line designs (from 19 to 23 workstations) present a maximum ergonomic risk category lower than 2 (from 1.94 to 1.56). Because of this, we can conclude that all the line configurations have an acceptable and moderate ergonomic risk. In fact, the average risk category of the five lines is between 1.48 and 1.80 for 23 and 19 workstations, respectively.

4 Conclusions

Ergonomics is fundamental in the design of work places in any production system. When these jobs correspond to modules or workstations of production lines it is necessary to take into account not only the safety of workers (measured through the risk) but also a set of constraints that affect the number of workstations (m), the repetitive cycle of the workload (c) and the available area at each workstation (A).

In this work, taking as a starting point the family *TSALBP*, we have proposed a line balancing model with the aim of minimising ergonomic risk and compliance with the temporary and spatial restrictions.

Through a case study, from the plant of engines from Nissan in Barcelona, we have proposed a progressive reduction of the risk category of the line (which is translated into operators with fewer injuries) through the creation of new jobs without changing the production capacity of the plant.

In future works we will measure the impact of the limitation of the available area at the workstations and we will study if the savings due to the risk reduction (fewer injuries) are economically compensated with the costs due to the creation of new jobs.

References

Battaïa O, Dolgui A (2012) Reduction approaches for a generalized line balancing problem. Comput Oper Res 39(10):2337–2345

Bautista J, Batalla C, Alfaro R (2013a) Incorporating ergonomics factors into the TSALBP. Advances in Production Management Systems. Competitive manufacturing for innovative products and services IFIP advances in information and communication technology. Springer, Holanda. ISBN 978-3-642-40351-4 (Print) 978-3-642-40352-1 (Online), 397:413–420

Bautista J, Batalla C, Alfaro R et al. (2013b) Impact of ergonomic risk reduction in the TSALBP-1. Industrial Engineering and Complexity Management. Book of Proceedings of the 7th International Conference on Industrial Engineering and Industrial Management—XVII Congreso de Ingeniería de Organización. ISBN 978-84-616-5410-9, pp 436–444

Bautista J, Pereira J (2007) Ant algorithms for a time and space constrained assembly line balancing problem. Eur J Oper Res 177(3):2016–2032

Becker C, Scholl A (2006) A survey on problems and methods in generalized assembly line balancing. Eur J Oper Res 168:694–715

Chica M, Cordón O, Damas S et al (2010) Multiobjective constructive heuristics for the 1/3 variant of the time and space assembly line balancing problem: ACO and random greedy search. Inf Sci 180(18):3465–3487

Landau K, Rademacher H, Meschke H et al (2008) Musculoskeletal disorders in assembly jobs in the automotive industry with special reference to age management aspects. Ind Ergon 38:561–576

Llovera S, Bautista J, Llovera J, Alfaro R (2014) Tiempo efectivo de trabajo: Un análisis normativo de la Jornada Laboral en el Sector de Automoción. DOI: 10.13140/2.1.1946.3680, Report number: OPE-WP.2014/06 (http://hdl.handle.net/2117/24508), Universitat Politècnica de Catalunya

Otto A, Scholl A (2011) Incorporating ergonomic risks into assembly line balancing. Eur J Oper Res 212(2):277–286

Rajabalipour CH, Haron H, Kazemipour F et al (2012) Accumulated risk of body postures in assembly line balancing problem and modeling through a multi-criteria fuzzy-genetic algorithm. Comput Ind Eng 63:503–512

Salveson ME (1955) The assembly line balancing problem. J Ind Eng 6:18–25

Reverse Logistics Barriers: An Analysis Using Interpretive Structural Modeling

Marina Bouzon, Kannan Govindan and Carlos Manuel Taboada Rodriguez

Abstract The objective of this research is to identify and analyze the interactions of the barriers that impede reverse logistics (RL) development from a Brazilian perspective. Firstly, international peer-reviewed publications were considered to select the barriers and classify them into categories. Secondly, an empirical research was conducted using Interpretive Structural Modeling (ISM), in order to evaluate the relationship between the barrier categories for the Brazilian RL context.

Keywords Reverse logistics · Barrier analysis · ISM · Brazil

1 Introduction

Reverse logistics (RL) has emerged as an important element for organizations to build their strategic advantage (Govindan et al. 2013), but little research has been conducted on the subject (Van Der Wiel et al. 2012). In the Green Supply Chain Management (GSCM) domain, RL is generally considered as the most difficult initiative to implement (Hsu et al. 2013). According to Abdulrahman et al. (2014), concerning the prior research on the barriers for RL implementation, little attention has been paid to emerging economies. As a BRIC country, Brazil is the largest Latin America economy and the seventh largest world economy. RL is gaining

---

M. Bouzon (✉) · C.M.T. Rodriguez
Department of Production and Systems Engineering, Federal University of Santa Catarina, Campus Universitário, Caixa Postal 476, Trindade, Florianópolis, Brazil
e-mail: marinabouzon@gmail.com

C.M.T. Rodriguez
e-mail: taboada@deps.ufsc.br

M. Bouzon · K. Govindan
Department of Business and Economics, University of Southern Denmark, Campusvej 55, 5230 Odense, Denmark
e-mail: gov@sam.sdu.dk

importance in Brazil, but, at the same time, Brazilian organizations encounter the challenge of a poor logistics infrastructure (Arkader and Ferreira 2004; da Rocha and Dib 2002). Based on this, further research is needed to understand, analyze and overcome these barriers. In this matter, this article aims at identifying RL barriers in an international panorama and analyze their interaction in the Brazilian context. The proposed research framework includes a systematic literature review process and Interpretive Structural Modeling (ISM) for the interdependency analysis of barriers.

2 Theoretical Background

RL is the process of moving goods from their typical final destination for the purpose of capturing value or proper disposal. RL comprises all the activities involved in processing, managing, reducing, and disposing of hazardous or non-hazardous waste from production, packaging, and use of products (Govindan et al. 2013; Rogers and Tibben-Lembke 1999). Although companies are increasingly being pressured to engage green initiatives, alongside, there are many barriers to the development of RL that limit their implementation (Kapetanopoulou and Tagaras 2011). In general, RL is considered by firms as an undervalued part of the supply chain (SC) due to a variety of reasons (Abdulrahman et al. 2014). The profitability of RL is multidimensional because it depends on several aspects, such as: market price of the materials, technological innovation that could contribute to lowering the price of the recovery process, and the quantities of recovered materials (Van Der Wiel et al. 2012). In the domain of GSCM, RL is considered as the most difficult initiative to implement when compared to design for environment and green purchasing (Hsu et al. 2013). Based on a systematic literature review process (for details, please refer to Sect. 3), 25 barriers have been identified and categorized based on their meaning and similarities (Table 1).

Although products gradually are being recycled and reused in developed countries, the most common practices in emerging economies continue to send used products to landfills, causing considerable costs and harm to the environment (Hsu et al. 2013). RL in developing countries seems to be an immature practice in most industry sectors (Lau and Wang 2009). To date, most existing research on the drivers and barriers of RL implementation is focused on developed countries (Abdulrahman et al. 2014; Lau and Wang 2009). Thus, more research is needed on the barriers for RL adoption in developing countries such as Brazil. As already mentioned, Brazil is seventh largest world economy, but the country's magnitude has its drawbacks for the environment. In 2011, Brazil's population generated almost 62 million tons of solid waste (de Sousa Jabbour et al. 2013a, b). In addition, companies consider RL an undervalued part of the SC (Abdulrahman et al. 2014). RL is recently gaining importance in this country due to some reasons: the implementation of the National Policy on Solid Waste (NPSW), economic issues as the recovery of the value of used products, green marketing, and improving social conditions. Nevertheless, Brazilian companies might face the challenge of a deficient logistics infrastructure to cope

Table 1 Reverse logistics barriers classification

Category	Barriers
Technology and infrastructure related issues (T&I)	T&I-1. Lack of personnel technical skills
	T&I-2. Lack of IT systems standards
	T&I-3. Lack of latest technologies
	T&I-4. Lack of in-house facilities (infrastructure)
	T&I-5. Technology and the R&D issues related to product recovery
Governance and supply chain process related issues (G&SC)	G&SC-1. Difficulties with supply chain members (poor coordination)
	G&SC-2. Limited forecasting and planning
	G&SC-3. Inconsistent quality
Economic related issues (E)	E-1. Lack of initial capital
	E-2. Lack of financial support for investments in return monitoring system/storage and handling
	E-3. Uncertainty related to economic issues
	E-4. Lack of economy of scale
Knowledge related issues (K)	K-1. Lack of knowledge on RL practices
	K-2. Lack of information on take back channels
	K-3. Lack of awareness concerning reverse logistics and its benefits
	K-4. Lack of taxation knowledge on returned products
Policy related issues	P-1. Lack of specific laws
	P-2. Lack of waste management practices
	P-3. Lack of inter-ministerial communication
	P-4. Lack of motivation laws
Market and competitors related issues (M&C)	M&C-1. Perception of a poorer quality product
	M&C-2. Undeveloped recovery marketplaces
	M&C-3. Little recognition of RL competitive advantage
Management related issues	M-1. Low importance of RL relative to other issues
	M-2. Low involvement of top management and strategic planning

with the NPSW. This circumstance encouraged us to analyze the issues related to RL implementation, specifically to identify the interaction among the 25 RL barriers from Table 1 for the Brazilian context.

3 Solution Methodology and Results

First, international peer-reviewed publications were considered to select the barriers and classify them into categories. Second, an empirical research was conducted using ISM, involving a Brazilian professional, in order to evaluate the relationship between the barrier categories for the Brazilian RL context. Interpretive Structural Modeling (ISM) is used to understand the mutual effects among the barrier categories, in order to identify, for example, the driving barrier categories, which can aggravate few more barrier categories.

ISM was developed for dealing with complex situations as a communication tool (Diabat and Govindan 2011; Mathiyazhagan et al. 2013). The main idea is to use experts' knowledge and experience to frame a complicated system into several subsystems and construct a multilevel structural model. ISM methodology helps create order and direction on the complexity of relationships among elements of a system (Sage 1977; Warfield 1974). The solution methodology is showed in Fig. 1. The following topics corresponds the steps of the ISM methodology.

- *Literature review* (*Step* 1)

  In order to better establish the research gap, international peer-reviewed publications on RL were investigated to identify the RL barriers listed and classified in Table 1. The bibliographic databases searched include: Springer, Science Direct, Emerald, Taylor & Francis, Wiley, ISI Web of Science, Scopus, and Google Scholar.

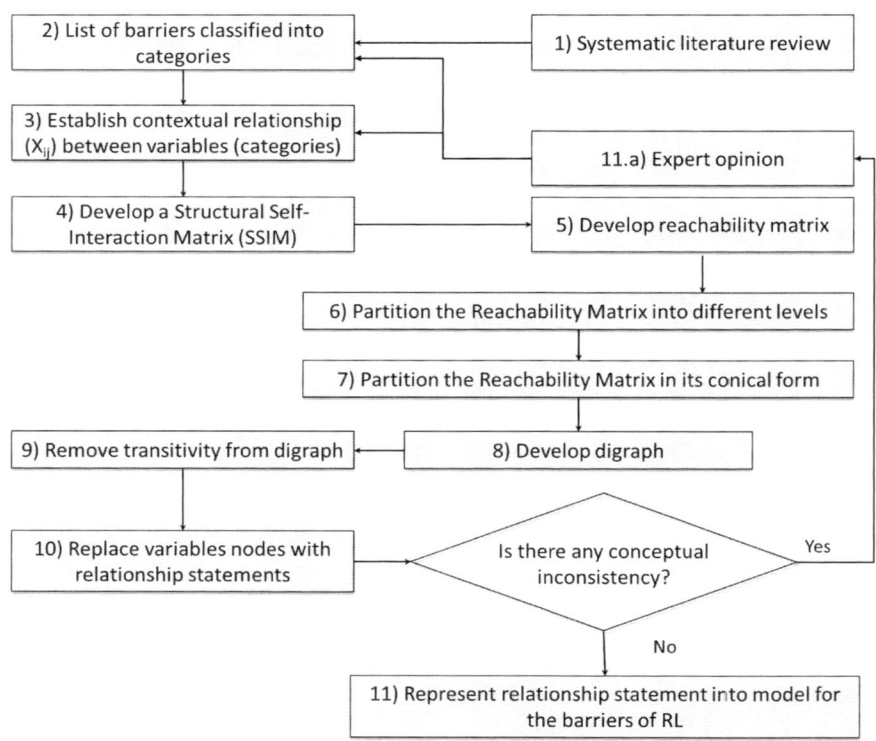

Fig. 1 Flow chart for ISM (modified from Diabat and Govindan 2011)

- *List of barriers and categories* (*Step* 2)
  The 25 identified barriers were classified into seven categories (please refer to Table 1). The categories were used as the variables for the contextual relationship in the next steps of ISM methodology.
- *Establish contextual relationship between variables* (*Step* 3) *and develop SSIM* (*Step* 4)
  From the variables identified in Step 2, a contextual relationship is established among the categories in order to identify which pairs of categories should be examined (Step 3). A structural self-interaction matrix (SSIM) is developed for variables (categories), indicating pairwise relationships among variables of the system. For analyzing the barrier categories, a contextual relationship of "has effect on" type is chosen. This means that one category influences another category. Thus, contextual relationship between the variables is developed according to the classification in Mathiyazhagan et al. (2013).
- *Reachability matrix* (*Step* 5)
  Reachability matrix (Table 2) is derived from SSIM developed in the previous step, using the following rules: (i) If $X_{i,j}$ entry in SSIM is V, then $X_{i,j}$ is set to 1, and $X_{j,i}$ is set as 0; (ii) If $X_{i,j}$ entry in SSIM is A, then $X_{i,j}$ is set to 0, and $X_{j,i}$ is set as 1; (iii) If $X_{i,j}$ entry in SSIM is X, both $X_{i,j}$ and $X_{j,i}$ is set as 1; (iv) If $X_{i,j}$ entry in SSIM is O, both $X_{i,j}$ and $X_{j,i}$ is set as 0. Table 2 also presents the driving power and dependence of each category. The driving power of a particular category is the total number of categories (including itself) which it may has an effect on. The dependence is the total number of categories which may have an effect on it.
- *Level partitions* (*Step* 6)
  The reachability matrix obtained above was partitioned into levels. The reachability and antecedent sets for each category (Warfield 1974) were found from the values on Table 2. The reachability set for an individual category consists of itself and the other categories which may have effect on. The antecedent set of an individual category is the list of the categories themselves which may have effect on it. The intersection of these sets was also derived for all categories.

Table 2 Final reachability matrix

		1	2	3	4	5	6	7	Driving power
1	T&I	1	0	0	0	0	0	0	1
2	G&SC	1	1	1	0	0	0	1	4
3	E	1	0	1	0	0	0	0	2
4	K	1	1	1	1	0	1	1	6
5	P	1	1	1	1	1	1	1	7
6	M&C	1	1	1	1	0	1	1	6
7	M	1	0	1	0	0	0	1	3
Dependence power		7	4	6	3	1	3	5	

A category is considered to be in level I (top position in ISM) if the reachability set and the intersection set for a given category are the same (Diabat and Govindan 2011).

In this case, as can be seen in Table 3, the barrier category Technology and infrastructure related issues (T&I) is found at level I. Consequently, it would be positioned at the top of the ISM model. After the first iteration, the category forming level I is discarded and the above mentioned process is continued with the remaining categories until the levels of each category has been found. The identified levels aid in building the digraph and the model of ISM.

- *Conical matrix form (Step 7) and Formation of ISM model (Steps 8, 9, 10 and 11)*
  Based on the level partition presented in Table 3, a model with the barrier categories for RL was developed. An initial digraph including transitivity links is derived from the conical form of the reachability matrix. The conical matrix is achieved from the partitioned reachability matrix by rearranging the elements according to their level, which means all the elements having the same level are pooled (Pfohl et al. 2011). The digraph is finally converted into the ISM model when the transitivities are removed, as described in the ISM methodology. The resulting digraph is shown in Fig. 2.

Table 3 Level partition for categories—iteration 1–6

		Reachability set	Antecedent set	Intersection	Level
1	T&I	1	1,2,3,4,5,6,7	1	I
2	G&SC	1,2,3,7	2,4,5,6	2	IV
3	E	1,3	2,3,4,5,6,7	3	II
4	K	1,2,3,4,6,7	4,5,6	4,6	V
5	P	1,2,3,4,5,6,7	5	5	VI
6	M&C	1,2,3,4,6,7	4,5,6	4,6	V
7	M	1,3,7	2,4,5,6,7	7	III

Fig. 2 ISM model for the barrier category affecting RL in Brazil

4 Discussion and Final Remarks

By analyzing the barrier categories using ISM, crucial barriers that hinder RL activities can be extracted. From Fig. 2, it is evident that "P" is the barrier category with greater influence on all remaining categories. This means that the lack of specific laws for product take back, waste management, or motivational legislation are significant barriers to RL practice. Legal issues are critical when implementing RL and green practices in a SC (Abdulrahman et al. 2014; Govindan et al. 2014). Sarkis et al. (2011) and Zhang et al. (2011) have stated that RL is a mandatory component of the SC in developed countries but it is still in a state of infancy in emerging economies. The implementation of National Policy on Solid Waste (NPSW) in Brazil might change the current status quo, influencing companies to cope with product end-of-life practices. "K" related issues barrier category comes next. Some authors (Abdulrahman et al. 2014; Mudgal et al. 2010; Van Der Wiel et al. 2012) have already stated that the lack of information about RL practices and return channels are major impediments for the implementation of RL. This category is influenced by barriers from "P" and "M&C" categories. In other words, the lack of knowledge on RL practices, take back channels and awareness of RL benefits is a consequence of the lack of pressure from specific laws, as well as lack of well-established recovery marketplaces, which could boost RL implementation. Previous studies (Abraham 2011; Rahimifard et al. 2009) have claimed that there is a difficulty on establishing reuse markets for recovered products or recycled materials.

"G&SC" category is influenced by P, K and M&C categories. The lack of SC coordination and integration is a major obstacle for RL practice (Abdulrahman et al. 2014; Bernon et al. 2013). This category has an effect on the "M" barrier category, impacting barriers such as low importance of RL comparing to other issues and low involvement of top management. This lack of priority for RL impacts the "E" category. In this sense, some authors (Alvarez-Gil et al. 2007; Kapetanopoulou and Tagaras 2011) posit that RL is not an economically justifiable investment.

In conclusion, RL is gaining attention recently due to many different reasons, such as environmental, social, economic and legislative issues. The identification and analysis of key barriers for RL implementation is complex, especially considering its numerous characteristics. The barriers involved in the implementation of RL place considerable challenges for organizations worldwide. Considering this problem and its complexity, this paper studied the interaction of RL barrier categories from a Brazilian perspective by means of ISM methodology. Barriers gathered from literature were classified into categories, and the latter were used as elements for comparison. Results from this study have significant contributions for RL implementation, from a practical and theoretical perspective. Knowing the barriers and understanding their relationship is one of the first steps for RL implementation. In this sense, this work might help industrial managers who occasionally have to improvise in the underdeveloped areas of RL. From theoretical lenses, this work is built on a systematic literature review, compiling 25 barriers for RL development. Further research can address the 25 barriers analysis using ISM, as well as perform the category analysis from another country perspective.

References

Abdulrahman MD, Gunasekaran A, Subramanian N (2014) Critical barriers in implementing reverse logistics in the Chinese manufacturing sectors. Int J Prod Econ B 147:460–471

Abraham N (2011) The apparel aftermarket in India—a case study focusing on reverse logistics. J Fashion Mark Manage 15(2):211–227

Alvarez-Gil MJ, Berrone P, Husillos FJ, Lado N (2007) Reverse logistics, stakeholders' influence, organizational slack, and managers' posture. J Bus Res 60(5):463–473

Arkader R, Ferreira CF (2004) Category management initiatives from the retailer perspective: a study in the Brazilian grocery industry. J Purchasing Supply Manage 10(1):41–51

Bernon M, Upperton J, Bastl M, Cullen J (2013) An exploration of supply chain integration in the retail product returns process. Int J Phys Distrib Logistics Manage 43(7):586–608

da Rocha A, Dib LA (2002) The entry of Wal-Mart in Brazil and the competitive responses of multinational and domestic firms. IJ Retail Distrib Manage 30(1):61–73

de Sousa Jabbour ABL, de Souza Azevedo F, Arantes AF, Jabbour CJC (2013a) Green supply chain management in local and multinational high-tech companies located in Brazil. Int J Adv Manuf Technol 68:807–815

de Sousa Jabbour ABL, Jabbour CJC, Sarkis J, Govindan K (2013b) Brazil's new national policy on solid waste: challenges and opportunities. Clean Technol Environ Policy 16:1–3

Diabat A, Govindan K (2011) An analysis of the drivers affecting the implementation of green supply chain management. Resour Conserv Recycl 55(6):659–667

Govindan K, Sarkis J, Palaniappan M (2013) An analytic network process-based multicriteria decision making model for a reverse supply chain. Int J Adv Manuf Technol 68:1–18

Govindan K, Kaliyan M, Kannan D, Haq A (2014) Barriers analysis for green supply chain management implementation in Indian industries using analytic hierarchy process. Int J Prod Econ 147:555–568

Hsu C-C, Tan KC, Zailani S, Jayaraman V (2013) SC drivers that foster the development of green initiatives in an emerging economy. IJ Oper Prod Manage 33(6):656–688

Kapetanopoulou P, Tagaras G (2011) Drivers and obstacles of product recovery activities in the Greek industry. Int J Oper Prod Manage 31(2):148–166

Lau KH, Wang Y (2009) Reverse logistics in the electronic industry of China: a case study. Supply Chain Manage Int J 14(6):447–465

Mathiyazhagan K, Govindan K, NoorulHaq A, Geng Y (2013) An ISM approach for the barrier analysis in implementing green SCM. J Clean Prod 47:283–297

Mudgal RK, Shankar R, Talib P, Raj T (2010) Modelling the barriers of green supply chain practices: an Indian perspective. IJ Logistics Syst Manage 7(1):81–107

Pfohl H-C, Gallus P, Thomas D (2011) Interpretive structural modeling of supply chain risks. Int J Phys Distrib Logistics Manage 41(9):839–859

Rahimifard S, Coates G, Staikos T, Edwards C, Abu-Bakar M (2009) Barriers, drivers and challenges for sustainable product recovery and recycling. Int J Sustain Eng 2(2):80–90

Rogers DS, Tibben-Lembke RS (1999) Going backwards: reverse logistics trends and practices. Reverse Logistics Executive Council, Reno

Sage A (1977) Interpretive structural modelling: methodology for large-scale systems. McGraw-Hill, New York

Sarkis J, Zhu Q, Lai KH (2011) An organizational theoretic review of green supply chain management literature. Int J Prod Econ 130(1):1–15

Van Der Wiel A, Bossink B, Masurel E (2012) Reverse logistics for waste reduction in cradle-to-cradle-oriented firms: Waste management strategies in the Dutch metal industry. Int J Technol Manage 60(1–2):96–113

Warfield J (1974) Developing interconnected matrices in structural modelling. IEEE Transcript Syst Men Cybern 4(1):51–81

Zhang T, Chu J, Wang X, Liu X, Cui P (2011) Development pattern and enhancing system of automotive components remanufacturing industry in China. Resour Conserv Recycl 55(6):613–622

Mixed Trips in the School Bus Routing Problem with Public Subsidy

Pablo Aparicio Ruiz, Jesús Muñuzuri Sanz and José Guadix Martín

Abstract In some regions, school buses have a subsidy associated with the number of school buses needed in a day, in relation to the number of schools or students. This paper presents the application of a Tabu Search (TS) metaheuristic to generate an information system that will allow searching for the best selection of routes and schedules with the minimum number of buses. The complexity and magnitude of the problem is important, especially at economic level for Andalusia government, because there are many buses and roads to go to school. The approach is more relevant for the necessary austerity in public services, considering that the quality is guaranteed by a set of restrictions. In the mixed trip problem, students of different schools can share the same bus at the same time. We use a mixed load algorithm for the school bus routing problem (SBRP) and measure its effects on the number of required vehicles. The solution could reduce the required number of vehicles compared with the current practice.

Keywords Vehicle routing problem · School bus routing · Meta-heuristics · Optimization · Tabu search

---

P. Aparicio Ruiz (✉) · J. Muñuzuri Sanz · J. Guadix Martín
Dpto. de Organización Industrial y Gestión de Empresas II, Escuela Superior de Ingeniería, Universidad de Sevilla, C/Camino de los descubrimientos, S/N, Sevilla 41092, Spain
e-mail: pabloaparicio@us.es

J. Muñuzuri Sanz
e-mail: munuzuri@us.es

J. Guadix Martín
e-mail: guadix@us.es

1 Introduction

Vehicle routing problems constitute one of the most widespread topics in scientific literature. In comparison with other problems, the school stands out as a service in which the demand does not usually change because it is projected for an entire school year, and demand tends to have little variation over the year. Therefore, this problem can be reliably optimized.

In the mixed trip problem, a trip consists of a sequence of bus stops and their designated schools, where students of different schools can get on the same bus at the same time. Buses make their movements through a route network. These routes are based on non-fixed points, in contrast to the typical problem of the industry, where the starting point is fixed (the depot). In this problem, the time from depot to first stop and from last stop to depot has not been taken into account, because the cost is evaluated by the necessary number of buses. In the transport network, each section has an associated cost or travel time. This value may depend on many factors; for example, the characteristics of fleet of vehicles or the time period during which the road is travelled.

School buses are a special case of vehicle routing problem. For the described problem, a comprehensive review of the literature is presented in Park and Kim (2010), Kim et al. (2012) and Simchi-Levi et al. (2014).

This paper describes an application where each student must be picked up at the bus stop and must be dropped off at school. Students can only be delayed because the bus has to pick up and leave other children in their schools. In the same way, the waiting time at school before classes start is limited, as travel time that students can spend on the bus. In this case, a fleet of buses is used to transport students from their bus stops to their schools. The objective is to minimize the number of buses, regarding to a set of quality criteria for public service that are related to: total time during which student is traveling, scheduled class start time (punctuality), bus occupancy rates, etc. These criteria are called restriction or feasibility rules to be met by vehicle routes.

In the routing problem family, different objectives to minimize can be found (Li and Fu 2002): total number of buses required, total travel time spent by pupils at all points, total bus travel time, and balance the loads and travel times between buses. Although the objective is minimizing the total number of buses required, by means of restriction, these factors can be combined.

The remainder of the paper is organized as follows: Sect. 2 presents a review of the school buses problem. Section 3 describes the classification scheme of this problem. Section 4 shows the randomized Location Based Heuristic (LBH). Section 5 presents the Tabu search in Routing School Buses (TRSB) algorithm. In Sect. 6, the LBH (Sect. 4) is compared to the TRSB algorithm (Sect. 5). Finally, a few conclusions are presented in Sect. 6.

2 Problem Description

There are different approaches in the literature that differ in the decomposition problem (Desrosiers et al. 1981). Two different approaches can be found: a school-based approach and a home-based approach (Spada et al. 2005). The first one does not allow mixing students from different schools at the same time, therefore the problem is greatly simplified. However, the second one aims at solving the problem for each child at the same time, the solution is more complicated when you include a new child in this problem, for this reason, exists few studies based on the home approach, as Braca et al. (1997). On the buses, mixed loads are allowed, in other words, students from different schools can travel on the same bus at the same time. This paper studies the scheduling and route generation of adjustment at the start of the school day. The resolution of the problem at the end of the school day is similar.

The problem can be classified based on their characteristics (Park and Kim 2010) as: multi-school, mostly urban, developed specially for mornings but it can be applied for evening with modifications, with mixed loads, homogenous fleet but could be designed to pre order heterogeneous fleet if there are no restrictions on its use on the road. According to the classification, the goal is to minimize the number of buses, and the restrictions are the vehicle capacity, maximum travel time, school time window and minimum number of students to create a route.

This paper optimizes the service level provided by the bus operator. Moreover, TS is applied as in Spada et al. (2005), but with a different approach because the model proposed by Spada et al. (2005) is based on an algorithm where schools are considered in increasing order of their starting times and descending distance. The routes for each school are built by using a greedy method. Thereafter a solution with the maximum number of buses is generated, then the buses are merged, and finally the TS is applied to generate students exchanges from one bus to another.

3 Location Based Heuristic

In this section, the algorithm presented in Braca et al. (1997) and named *Randomized Location Based Heuristic* (*LBH*) is presented. Because of its mixed loads approach, it clearly allows to optimize the number of buses. The Braca problem is considered as the construction of a bus route between its school and bus stop selected randomly. Then the pair established by a bus stop and its school is inserted into the bus route (if the school is not already on that bus route). This procedure is greedy. First, it selects a pair that minimizes the total length, verifying that all restrictions are met. The solution is not revised by the LBH, unless the buses contain very few children. *Randomized LBH* algorithm can be described the following way:

```
Let U be the set of indices of all unvisited pick-up points,
U = {1,2,3,…,n}
Let m = 0
WHILE (U ≠ ∅) DO{
   Pick a pick-up point at random from U. Call it j.
   Let U ← U\{j}
   Let the current route be R_m = {j → school[j]}
   REPEAT{
      For each I ∈ U, calculate ci = routelength(I,R_m)
      Let c_k = min_{i∈U} {c_i}
      IF c_k < +∞ then{
         Let R_m = buildroute(k,R_m)
         Let U ← U\{k}
      }
   } UNTIL c_k = +∞.
   If (no. of students on R_m < 11) then
   Movestops(R_1, R_2, …, R_m)
   m ← m+1
}
M ← m
The heuristic solution is {R_1, R_2,… , R_m}
```

This algorithm chooses a random bus stop, and then it builds a bus route with a bus stop and its associated school. After inserting the bus stops, it is checked on the bus route if the restrictions are fulfilled and the route cost is minimal. The function Routelength(I, R) is defined, which determines the approximate cost of inserting the stop I into route R for each unvisited index; as a consequence, the approach is not efficient for route finding.

The function finds the best insertion point for the bus stop I, so it is found that the stop can be inserted between each consecutive pair of points along the route. Finally, if the school bus stop is not on the route, then it finds the best insertion point of the school from the insertion point of the bus stop. If a feasible insertion point cannot be found, then the value is $+\infty$. If an insertion is found that results in a feasible route, then the Routelength(I, R) value is made to be exactly the additional distance travelled.

Braca considered only capacity constraints. It only checks if a point can be added to a route and if the maximum load is less than the capacity of the vehicle.

The function Buildroute(k, R) creates the route that results from the insertion of the stop k into route R. Again, k is simply inserted between two consecutive points that make the shortest route. This route is guaranteed to be feasible because the cost of k is $c_k < +\infty$.

When the limitation of minimum number of students on a bus is not carried, the algorithm backtracks and restarts the cycle with a different seed point [movestops $(R_1, R_2, …, R_m)$].

LBH algorithm has great difficulty in finding the optimum after choose the first stop randomly. The algorithm chooses the lowest cost stop, namely, it calculates the route with minimum length, but not minimize the number of buses because the bus is filled with nearby bus stops. With the aim of minimize the number of buses, these may have to make a route with a longer distance to remove the stops which cannot find the optimal solution.

The algorithm can be used as a heuristic initialization, but the iterated application of this is wrongly focused on. However, it can be applied on an algorithm TS as in the work (Spada et al. 2005), where the students are exchanged from one bus to another, with the idea of minimizing the number of buses.

In this paper an alternative design is presented. This is a direct application of TS heuristics, which reduces the number of operations because Routelength is not calculated on the entire set of stops.

4 Tabu Search in Routing School Buses

The Tabu search in Routing School Bus (TRSB) algorithm is based on a preprocessing previous preparation of the data. The bus stops, the students and the corresponding location have been calculated. Before applying the algorithm, a procedure is performed to generate the complete graph of costs between all bus stops and schools. Finally, the algorithm applied for the TS as follows:

```
Let k=0;
iticialSolution();
iticialTabuList();
while (!isEnd(k) && isPosibleToContinue()){
    neighborhood=newNeighbors(currentSolution,tabuList)
    CurrentSolution = chooseBestSolution(neighborhood)
    if(isBestSolution(currentSolution))
       setBestSolution(currentSolution);
    changeTabuList();
    k=k+1;
}
```

The function initialSituation() generates an initial solution of the problem. The first solution may be found by student insertion and bus saturation, provided the restrictions. Another possibility is to apply a randomizedLBH modification, where the bus stop is inserted in the bus route when the cost $c_k \neq +\infty$, without comparing all possibilities.

The function isEnd(k) determines that the specified number of interactions makes it possible to continue the algorithm, whereas isPosibleToContinue() checks that the Tabu allows to continue the algorithm.

The function newNeighbors() generates a possible set of solutions to consider, from a general list where the bus stop position is exchanged which are not included in the Tabu list.

The function chooseBestSolution() generates a list of possible solutions related to bus stops, after applying the exchanges of stops nearby (not Tabu), and it returns the lower cost solution regarding the number of buses.

The evaluation cost is made by inserting stops in the bus in order until the conditions are not met. When this happens it creates a new bus, and so on. In other words, if the bus is full or it is the first stop and there are stops which are not allocated, the first item is assigned to the stops list and it is removed. When the bus exists, the elements are included in different positions on the bus route, always before the school stop. Otherwise, while there are unassigned stops, the algorithm looks for the best position for each stop. But if at the end of the route construction, the restrictions are not met, the route is undone. The conditions to be met are: the occupation and the time that students use on their way to school, and the time window allowed.

At the end of the algorithm, the capacity condition may not be complied with minimum number of students; in this case the algorithm is forced to stop.

Other functions are general, for example changeTabuList() updates the tabu list, or isBestSolution(currentSolution) checks the lowest cost of the current solution, or setBestSolution(currentSolution) saves the solution.

5 The Comparison Tests

This study has addressed only the comparison of both algorithms with synthetic data generated to analyze the relative efficiency. In the test experiments, the location of the bus stops and the school were selected randomly, as well as the nodes and the network. The algorithm Randomized LBH was chosen to compare with TRSB because Braca is the author who has studied the greatest number of constraints (Park and Kim 2010).

As it is shown in Table 1, the two algorithms were compared, in terms of the number of iterations Parameter (P), the average number of Buses (B), the average Time Bus Route (TBR), the Average Buses Occupancy (ABO) and the time consumed by the algorithm in seconds (TA).

In Fig. 1, the bus stops are represented by squares. The first problem contains two separate schools in a network for a 5–7u. of distance (red ellipses in Fig. 1 left), while the maximum distance between nodes is 11u. In the second problem, the same network is used and both schools are at a closer distance of 5–7u. (green circles in Fig. 1 left). As it is shown in Table 1, LBH did not find the optimal solution in the first problem, although it did in the second problem. This is due to the fact that LBH optimizes the distance between stops in its cost function, and not the number of buses. Nevertheless, the TRSB achieves the goal although the average driving time grows (this increased between 15 and 42 %).

In the third problem, the schools are placed closer, 5u. in one direction and 7u. on the other one (green circles in Fig. 1 right). In the fourth problem, the schools are placed further than in the third problem, with a distance of 13u. (red circles Fig. 1 right) in one direction and 17u. in the other one. Clearly, as it is show in Table 1, the tendency to take the shortest route in the randomized LBH algorithm does not allow to achieve the optimum solution when the schools are too far apart. Nevertheless, the solutions are optimal using the TRSB algorithm.

Table 1 Results of the experiments

Alg.	Problem P	Problem 1[a] B	TBR	ABO	TA	Problem 2[a] B	TBR	ABO	TA	Problem 3[b] B	TBR	ABO	TA	Problem 4[b] B	TBR	ABO	TA
LBH	1	3.43	12.6	26.8	0.3	2.71	13.5	34.4	0.47	7.90	13.1	26.7	0.30	12.87	13.7	16.4	0.30
	10	3.00	13.7	30.0	0.3	2.01	17.6	44.9	1.41	7.00	13.6	30.1	2.35	12.00	14.1	17.1	2.80
	100	3.00	13.8	30.0	5.3	2.00	17.6	45.0	6.44	6.72	14.3	31.4	27	12.00	14.1	17.1	25
	1,000	3.00	13.9	30.0	51	2.00	17.4	45.0	68	6.04	15.5	34.8	274	12.00	14.1	17.1	247
	10,000	3.00	13.8	30.0	525	2.00	17.1	45.0	698	6.00	15.6	35.0	277	12.00	14.1	17.1	2,528
TRSB iTb = 4	1	2.82	15.5	32.7	1	2.00	17.4	45.0	1	11.83	15.7	17.8	18	11.83	15.7	17.8	18
	10	2.74	16.0	33.9	3	2.00	17.3	45.0	3	11.00	17.1	19.1	105	11.00	17.1	19.1	105
	100	2.85	16.0	33.8	16	2.00	17.1	44.9	18	11.00	17.1	19.1	937	11.00	17.1	19.1	937
	1,000	2.71	16.3	34.4	148	2.00	17.2	45.0	163	11.00	17.1	19.1	9,310	11.00	17.1	19.1	9,310
	10,000	2.80	15.7	30.0	1,692	2.00	17.8	45.0	1,472	11.00	17.4	19.1	93,753	11.00	17.4	19.1	93,753
TRSB iTb = 3	1	2.83	15.2	32.6	1	2.00	17.3	45.0	1	6.93	16.1	30.5	22	11.72	15.8	18.0	19
	10	2.77	16.4	35.0	4	2.00	17.3	45.0	4	6.5	17.3	32.5	147	11.0	17.2	19.1	135
	100	2.52	17.2	37.2	39	2.00	17.2	45.0	36	6.42	17.5	32.9	1,387	11.0	17.1	19.1	1,305
	1,000	2.69	16.9	36.2	376	2.00	17.2	45.0	395	6.4	17.7	33.1	13,854	11.0	17.1	19.1	13,077
TRSB iTb = 2	1	2.97	15.0	32.0	1	2.00	16.8	45.0	1	7.0	16.1	30.1	22	11.88	15.6	17.7	18
	10	2.63	16.6	35.6	7	2.00	17.3	45.0	6	6.51	17.4	32.5	172	11.0	17.2	19.1	159
	100	2.64	16.6	35.4	70	2.00	17.3	45.0	67	6.62	16.9	31.9	1,682	11.0	17.2	19.1	1,608
	1,000	2.64	16.5	35.4	700	2.00	17.1	45.0	670	6.44	17.2	32.8	17,007	11.0	17.2	19.1	16,103
TRSB iTb = 1	1	2.86	15.9	33.6	1	2.01	17.5	44.9	1	6.93	16.3	30.5	21	11.8	15.8	17.9	24
	10	2.74	15.7	33.9	11	2.01	17.1	44.9	10	6.51	16.9	32.5	215	11.0	17.1	19.1	206
	100	2.85	15.8	33.8	111	2.03	17.4	44.6	105	6.54	16.8	32.3	2,192	11.0	17.2	19.1	2,138
	1,000	2.81	15.3	32.8	1,139	2.00	17.4	45.0	1,095	6.5	17.4	32.5	22,583	11.0	17.2	19.1	21,768

[a] Nodes 36, Edges 61, Schools 2, Stops 9
[b] Nodes 56, Edges 78, Schools 2, Stops 21

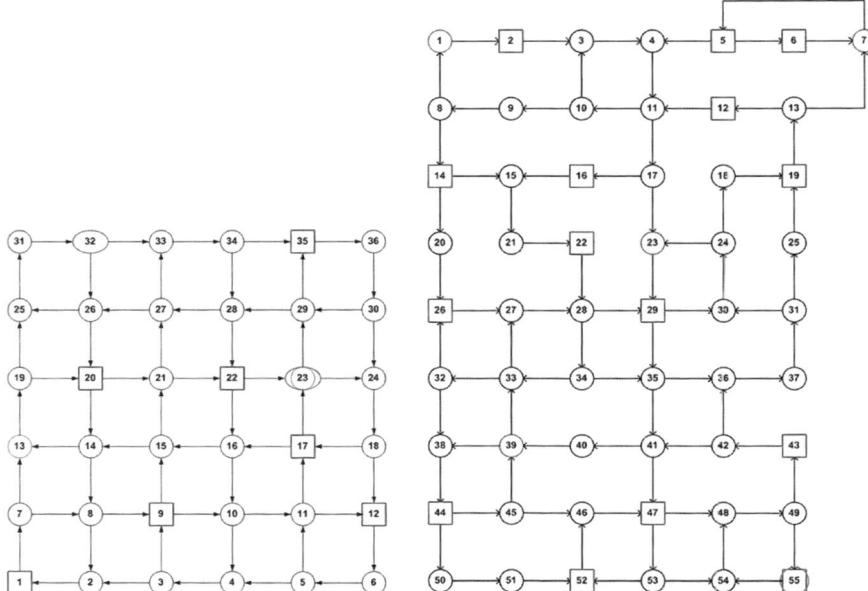

Fig. 1 (*Left*) First network to problem 1 and 2. (*Right*) Second network to problem 3 and 4

The TRSB algorithm could be improved. In future studies, the best solution could be selected by compliance with two conditions: first of all, by minimal number of buses, and then minimum route time.

6 Conclusions

The school bus planning is an important area where services costs can be extremely reduced in many countries. The process carried out to generate a scholar route is an arduous task which encourages the development of mechanisms to facilitate the process itself.

The development of this application can be accomplished by implementing a Geographical Information System (GIS). For this, a database can be used to store all the relevant information taken from the students about their stops or destinations, just as all school directions are saved.

The number of buses should be the primary objective in the transport system within the restrictions of quality of service, because austerity measures are necessary in public spending. The TS application is proposed in the school bus routing model. This strategy is compared with the randomized LBH algorithm to solve the problem. Obviously, the size of the data is too small to permit any generalizations.

Despite its preliminary character, the research seems to indicate that TRSB improves efficiency.

The TRSB algorithm seems to be more consistent with large real applications. This heuristic may be effective not only for assessment but also to find better solutions and test the effects that could cause the changes in quality restrictions on the cost of the number of buses needed for all the education system.

References

Braca J, Bramel J, Posner B, Simchi-Levi D (1997) A computerized approach to the New York City school bus routing problem. IIE Trans 29:693–702
Desrosiers J, Ferland JA, Rousseau JM, Lapalme G, Chapleau L (1981) An overview of a school busing system. Scientific Management of Transport Systems. pp 235–243
Kim B, Kim S, Park J (2012) A school bus scheduling problem. Eur J Oper Res 218:577–585
Li L, Fu (2002) The school bus routing problem: a case study. J Oper Res Soc 53:552–558
Park J, Kim B (2010) The school bus routing problem: a review. Eur J Oper Res 202:311–319
Spada M, Bierlaire M, Liebling ThM (2005) Decision-aiding methodology for the school bus routing and scheduling problem. Transp Sci 39:477–490
Simchi-Levi D, Chen X, Bramel J (2014) The logic of logistics: theory, algorithms, and applications for logistics management. Springer, Berlin

Analysis of the Criteria Used by Organizations in Supplier Selection

Joan Ignasi Moliné, Anna Maria Coves Moreno and Anna Rubio

Abstract In order to analyse the criteria used by organizations in their selection of suppliers and to compare them with those provided by the literature, a survey has been designed with which a field study has been carried out. The article shows some of the most relevant conclusions of the field study, after a statistical analysis (descriptive and principal components) of the responses. In this work the methodology used for the creation of the survey is also detailed.

Keywords Supplier selection · Supplier survey criteria

1 Introduction

The incidence of acquired products and services in the cost structures of manufacturing companies is quite significant, reaching up to 80 % of the total cost. Many strategic-type parameters are requested of suppliers. This means that the choice of suppliers is a strategic stage in purchasing management, as the ability of companies to satisfy their clients, and their own continuity, depends largely on their suppliers.

The selection of suppliers requires a list of criteria with which to evaluate them, therefore, the identification of these criteria is crucial.

This task is not obvious; there is no consensus between the different authors, and so this article contrasts supplier selection criteria—as dealt with in the published

---

J.I. Moliné · A.M.C. Moreno (✉) · A. Rubio
Department of Management, Institute of Industrial and Control Engineering,
Universitat Politècnica de Catalunya, Av. Diagonal 647, 08028 Barcelona, Spain
e-mail: anna.maria.coves@upc.edu

J.I. Moliné
e-mail: jignasi.moline@gmail.com

A. Rubio
e-mail: anna.rubio.pelegrin@gmail.com

literature—with those actually used by companies, designing a questionnaire with which a field study is carried out.

This article is organized in such a way that Sect. 2 reviews the literature, Sect. 3 defines the objectives to achieved, Sect. 4 develops the field study, Sect. 5 presents the most important results and Sect. 6 provides the conclusions.

2 Literature and Theoretical Background

Although there are articles that study the criteria used in supplier selection, there is very little literature dealing specifically with them; most of the existing literature integrates them in the procedures for selecting suppliers and/or the allocation of orders, without making a specific analysis of the criteria.

One of the conclusions of the study conducted by Dickson (1966) shows doubts about the development of a universal system for the analysis of suppliers, showing that only one or two criteria can be considered routine, the rest vary depending on the type of purchase. A review of the literature published by Moliné and Coves (2013) concludes that there is no standardization in the criteria used, suggesting that the definition of each criterion depends on the experience of the authors.

There are authors who argue that the criteria may be common among different organizations and who have presented proposals for what they considered to be standard criteria in selecting suppliers (Beamon 1999; Gunasekaran 2007; Huang 2007). On the contrary, there are authors who argue that the criteria may vary; Cho et al. (2012) considered that the differences between the industrial and service sectors justify a specific analysis of the latter, others consider that the criteria vary according to the type of product purchased (Kannan and Haq 2007; Cervera and Coves 2009). Webber et al. (1991) review the literature of the criteria used to select suppliers, with special emphasis on the specific operating environment, the Just In Time environment, in which Aksoy and Öztürk (2011) also make a proposal.

Araz and Ozkarahah (2007) take into account aspects such as the establishment of long-term relationships, innovative capacity and the willingness to cooperate. According to Chen and Chao (2012) organizations may have different "cultural backgrounds" and the criteria should be chosen according to these needs. Shaw et al. (2012) introduce criteria of environmental sustainability.

All the above makes it necessary to determine if there are any lists of standard criteria.

3 Definition of the Objective

This study aims to analyse the criteria actually used by organizations in the selection of suppliers and to contrast them with those provided by the literature.

4 Development of the Field Study

The main tool for achieving the proposed objective consists in carrying out a field study. This section outlines the development of the design of the questionnaire with which the field study and dissemination are carried out.

To achieve the proposed objectives, a hypothesis and four sub-hypotheses are presented, inspired by the review of the literature mentioned above. The studied hypothesis is:

H: There is no standard used by organizations, nor any proposed in the literature.
Being the sub-hypotheses:

- H1: The criteria are different due to the type of product purchased.
- H2: The differences are due to the sector to which the organization belongs.
- H3: The differences are due to the overall framework of the environment.
- H4: The differences are due to the resources of the Purchasing Department.

The instrument used for the field study is a questionnaire designed according to the hypothesis and the sub-hypotheses raised. Its structure is based on the literature, with the authors who have contributed most being: Dickson (1966), Beamon (1999), Gunasekaran (2007), Huang (2007), Moliné and Coves (2013) and the doctoral theses of Mendoça (2008) and Bernardo (2009). The questionnaire it is divided into four sections:

1. Organization Profile, covering general issues of the organization.
2. Purchase Management, encompassing aspects of supplier selection.
3. Description of the environment, collecting information on aspects of general policy.
4. Criteria and indicators, providing a list of criteria which can potentially be used for choosing suppliers.

Its implementation and execution is "on-line", using Google Forms, which allows the creation of a questionnaire, ensuring confidentiality and the export of the responses.

The questionnaire must be delivered to purchasing professionals so that they may carry out good management practices in their organizations. It is considered that professionals associated with AERCE (Spanish Association of Purchasing, Contracting and Supply Professionals) have the maximum interest in both the issue that we are discussing and in working optimally. After signing a partnership agreement with AERCE, two mailings were made to members of their database.

From two mailings, 50 responses have been obtained; it was found necessary to increase the responses in some types of organization, therefore a third mailing focused on them. At the time of writing this mailing is being completed.

The treatment of responses will use the statistical tools that enable the objective of this work to be achieved; descriptive analysis and Principal Components Analysis, which enables the grouping of criteria in such a way that each group has the maximum homogeneity. The 50 available responses were dealt with.

5 Results

This section presents the results of the descriptive analysis and, in a second part, contrasts the hypotheses and sub-hypotheses presented by means of Principal Components Analysis.

5.1 Descriptive Analysis

The majority of organizations (46 %) made durable consumer goods, non-durable goods organizations constitute 36 % and services make up 18 %.

The average of the approximate value of total purchases with respect to total income is 56.62 %, emphasizing that in non-durable consumer goods producers it is 59.09 and 47.63 % in services. In 39 organizations (81.25 %) this value is very significant, being greater than or equal to 40 %.

Organizations have methods both for the selection of suppliers (66 %) and for follow-up (78 %). It can also be seen how some of the management tools (II.7, II.4 and II.9 in Table 1) are greater than 66 %, indicating that companies that lack formal methods do exercise some kind of management.

Regarding the level of standardization, criteria are grouped into three blocks; the first includes those used by more than 60 % of organizations, the second consists of those considered by 35 and 60 % of organizations and the third are those considered by less than 35 % (Table 2).

All of the organizations consider price as a criterion for the selection of suppliers, while criteria linked to quality do not appear until position 13, prioritizing ability and speed in solving problems (positions 3 and 5) ahead of prevention (position 13). The criteria included in the first block are mostly linked to the "tranquillity" of the client organization. Organizations use an average of 17.4 criteria for the selection of suppliers.

Table 1 Supplier selection management

		S (%)	N (%)
II.1	Formal method for selection	66.00	34.00
II.3	Formal method for tracking	78.00	22.00
II.4	List of criteria for selection	80.00	20.00
II.5	*Are they prioritized?*	82.50	17.50
II.6	*Do they vary according to article?*	75.00	25.00
II.7	QA systems required from the supplier	72.00	28.00
II.8	Audits	64.00	36.00
II.9	Recognition	82.00	18.00

Table 2 Ranking of criteria for the selection of suppliers and their identification number (ID)

ID	Criteria	Total	% Organiz.
1	Price	50	100.00
5	On time delivery (%)	36	72.00
21	Problem solving capability	36	72.00
9	Flexibility (deliveries and quantities)	35	70.00
37	Speed of problem solving	34	68.00
27	Technical capacity for performing the purchased article /service	33	66.00
22	Supplier financial stability	32	64.00
12	Compliance or likehood to work according with your procedures	31	62.00
28	Production capacity	31	62.00
4	Price reduction capability	30	60.00
24	Communication capability	30	60.00
7	Order fulfillment lead time	29	58.00
10	Quality management practices and systems	28	56.00
35	After sales service offered by the supplier	28	56.00
15	Supplier geographical location	25	50.00
30	Personal impression/Attitude of each vendor toward your organizat.	25	50.00
11	Position in industry (including reputation)	24	48.00
19	Capacity to reflect on the item/service requirements of your organiz.	23	46.00
6	Supply quantity /Required quantity	22	44.00
16	Implementation of environmental regulations	22	44.00
32	Flexibility of technology	21	42.00
34	Supplier employment stability	21	42.00
8	Number of correct parts /Number of parts received	20	40.00
2	Transportation cost	19	38.00
14	Amount of past business that has been done with each supplier	18	36.00
23	Warranty incidents	18	36.00
26	Type of products/services	18	36.00
31	Capability to design and develop goods/services	17	34.00
36	Risk management (existence of contingency plan, …)	16	32.00
33	Size of the supplier	15	30.00
13	Product/service variety	14	28.00
29	The supplier has enterprise management systems (ERP, …)	12	24.00
17	Sustainability (Green supplier, …)	11	22.00
25	Stockout probability	11	22.00
3	Presence of discounts	10	20.00
18	Supplier ROI (Return On Investment)	9	18.00
38	Supplier monitoring of its own performance	9	18.00
20	Type of supplier organization (multinational, …)	6	12.00
39	Other	1	2.00

5.2 Principal Components Analysis

Principal Components Analysis (PCA) is a visual and an interactive technique that helps to identify hidden trends and data correlations in multidimensional data sets. A part from PCA, Multiple Correspondences Analysis (MCA) is being used for the data analysis; it allows analysing the pattern of relationships of several categorical dependent variables. Both, PCA and MCA are applied to the four sub-hypotheses and the software used is Minitab. At the time of writing this MCA is been applied.

This article shows results obtained for H1 sub-hypothesis. For the other three sub-hypotheses we will proceed in the same way.

H1: *The criteria are different due to the type of product purchased.*

Nominal variable comprises several levels, and each of them is coded as a binary variable. The survey respondent must answer yes or no. If the answer is yes, it is codified as 1, if not, as a 0. In case of no reply, it is not codified. Procedure is repeated for the other four sub questions.

PCA is applied and not until the 15th component more than 80 % of the system variability is described. Figure 1 shows the variables in the axis that correspond to their values in the two main components.

Relations between the type of product that the organization purchase and the closest criteria associated are summarized in Table 3.

All the other criteria do not cited in Table 3 are independent from the type of product purchased.

We can see in Fig. 1, by means of Biplot obtained of PCA results, the relationship between type of product and criteria.

Table 3 Relation between variables related to H1 and the closest criteria associated

Variable (type of product purchased)	Closest criteria associated
I.9.a—It is an undifferentiated product (commodity)	Cr13 Product/service variety
I.9.b—Technical complexity in their development	Cr3 Presence of discounts
I.9.c—Article characteristics must strictly comply with the specifications required	Cr28 Production capacity
	Cr10 Quality management practices and systems
	Cr4 Price reduction capability
	Cr27 Technical capacity for performing the purchased article/service
I.9.d—The article is subject to regulations under standards (food, medicine,…)	Cr5 On time delivery (%)
I.9.e—Item is customized to the requirements of your organization	Cr3 Presence of discounts
	Cr5 On time delivery (%)

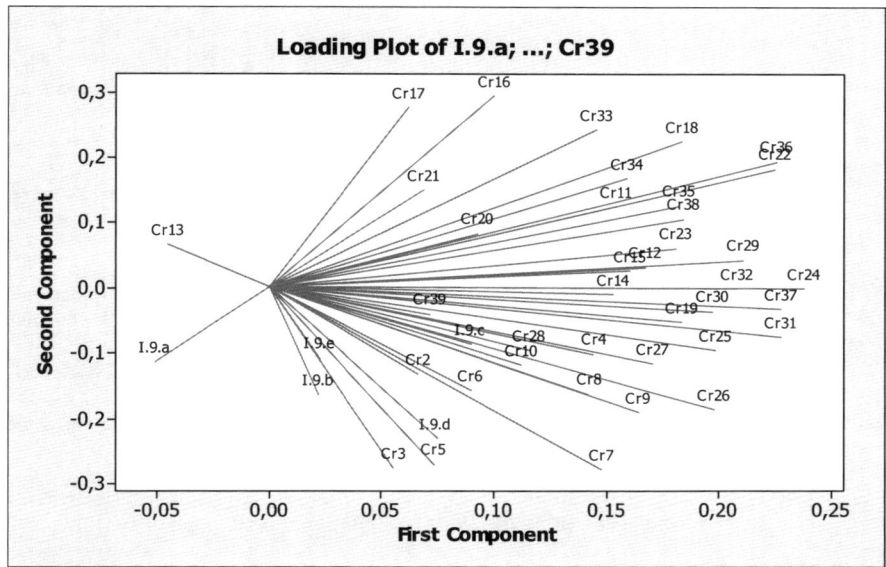

Fig. 1 Variables in the axis that correspond to their values in the two main components

6 Conclusions

Review of the literature shows that there is no standardised list of criteria for selecting providers, with some authors making proposals that are considered standard. Nevertheless, analysis shows that they do not coincide.

In order to contrast the criteria used by organizations in supplier selection with the literature, a questionnaire was designed which makes it possible to analyse the possible existence of patterns, depending on several variables.

The results show that the average value of the total purchases regarding total income is 56.62 %, a significant value which justifies the importance of purchasing and, therefore, its management. The descriptive analysis also shows how consensus on the criteria to be considered is low, where 9 criteria (out of 39) are considered by more than 60 % of companies.

For the first sub-hypotheses it is shown that for each one type of product there is some criteria associated it (PCA results).

Currently, with the aim of achieving a greater number of responses, the survey remains open and in a second stage the same statistical analysis will be carried out on the new sample, the next statistical technique, MCA, will be added.

References

Aksoy A, Öztürk N (2011) Supplier selection and performance evaluation in just-in-time production environments. Expert Syst Appl 38(5):6351–6359

Araz C, Ozkarahan I (2007) Supplier evaluation and management system for strategic sourcing based on a new multicriteria sorting procedure. Int J Prod Econ 106(2):585–606

Beamon BM (1999) Measuring supply chain performance. Int J Oper Prod Manage 19(3):275–292

Bernardo M (2009) Integració de sistemes estandarditzats de gestió: anàlisi empírica. Girona: Universitat de Girona. http://hdl.handle.net/10803/7962

Cervera D, Coves AM (2009) Analyze and designing of allocation models in multi-supplier environments, based on MILP industrial engineering: a way for a sustainable development. Book if full papers. In: 3rd international conference on industrial engineering and industrial management. ISBN 978-84-7653-388-8

Chen YH, Chao RJ (2012) Supplier selection using fuzzy preference relations. Expert Syst Appl 39(3):3233–3240

Cho DW, Lee YH, Ahn SH, Hwang MK (2012) A framework for measuring the performance of service supply chain management. Comput Ind Eng 62(3):801–818

Dickson GW (1966) An analysis of vendor selection systems and decisions. J Purchasing 2:5–17

Gunasekaran A, Kobu B (2007) Performance measures and metrics in logistics and supply chain management: a review of recent literature (1995–2004) for research and applications. Int J Prod Res 45(12):2819–2840

Huang SH, Keskar H (2007) Comprehensive and configurable metrics for supplier selection. Int J Prod Econ 105(2):510–523

Kannan G, Haq AN (2007) Analysis of interactions of criteria and sub-criteria for the selection of supplier in the built-in-order supply chain environment. Int J Prod Res 45(17):3831–3852

Mendoça E (2008) Uncertainty, integration and supply flexibility. Universitat Pompeu Fabra, Barcelona http://hdl.handle.net/10803/7386

Moliné JI, Coves AM, (2013) Supplier evaluation and selection: a review from literature from 2007. Managing complexity: challenges for industrial engineering and operations management. Series: lecture notes in management and industrial engineering vol 2 (ebook). ISBN 978-3-319-04704-1

Shaw K, Shankar R, Yadav SS, Thakur LS (2012) Supplier selection using fuzzy AHP and fuzzy multi-objective linear programming for developing low carbon supply chain. Expert Syst Appl 39(9):8182–8192

Weber CA, Current JR, Benton WC (1991) Vendor selection criteria and methods. Eur J Oper Res 50(1):2–18

Event Monitoring Software Application for Production Planning Systems

Andrés Boza, Beatriz Cortes, Maria del Mar Eva Alemany and Eduardo Vicens

Abstract In a company, unexpected events may affect the planning production system. These events should be identified and analysed with the aim of evaluate their impact and make (or not) decision about the on-going production plan. This paper submits an event monitoring software application based in expert systems for planning production systems (1) to identify and describe events that generate incidences in the system (2) to define alert criteria and (3) to check periodically the production planning system to identify possible incidences.

Keywords Event management · Decision support system · Production planning system · Expert system

1 Introduction

Planning is the process of determining enterprise objectives and selecting a future course of action to accomplish them. According with Bedeian (1986) this process includes: (a) to establish business objectives; (b) to develop premises about the environment where they must be developed; (c) to take decisions about the course

---

A. Boza (✉) · B. Cortes · M.M.E. Alemany · E. Vicens
CICIP. Centro de Investigación Gestión e Ingeniería de la Producción (Research Centre on Production Management and Engineering), Universitat Politècnica de València, Cno. de Vera s/n, 46022 Valencia, Spain
e-mail: aboza@cigip.upv.es

B. Cortes
e-mail: beacorsa@fiv.upv.es

M.M.E. Alemany
e-mail: mareva@cigip.upv.es

E. Vicens
e-mail: evicens@cigip.upv.es

of action; (d) to initiate the needed activities to translate plans into actions; (e) to plan on the fly to correct existing deficiencies.

This decision-making process is complex, and decision-makers have to manage this complex process as an important part of their management responsibilities (Duan et al. 2012). The Decision Support System (DSS) field arose as a tool to help decision-makers to make decisions useful for a period of time. Cohen and Asín (2000) defines decision support system as a set of programs and tools to obtain the necessary information that is required during the process of decision-making that is developed in an uncertainty environment. Thus, computer system could be developed to deal with the structured portion of a DSS problem, but the judgment of the decision-maker is brought to bear on the unstructured part, hence constituting a human machine, problem-solving system (Shim et al. 2002).

Decision Support Systems use information obtained from the real situation of the company, but also use projections that incorporate a certain level of uncertainty. It is not necessary to reanalyse anything when a projected scenario match with the actual situation, but, it makes sense to consider reanalyse the decision when it does not match with the planned situation in the middle of a decision period (Boza et al. 2014). An event not expected by the company may have a negative impact in the company. So, it is necessary developing methods to act on its resolution quickly and efficiently (Valero et al. 2013). Different events can produce these deviations, and an early detection of these events helps to evaluate these new situation and their impact in the current decisions. This is not an easy task and potential benefits are lost on multiple occasions because organizations do not know how to respond appropriately to unexpected events. Decision Support Systems are usually rigid systems and are not prepared for these proposals (Van Wezel et al. 2006).

The conception and implementation of appropriate information and communication systems is a basic condition for the identification of critical incidents in operational process executions, such as, machine breakdowns, lack of workers, order changes, cancellation of orders, etc. So, the main objective of event monitoring is to sense the production and logistics information on a real-time environment and detect events (Shamsuzzoha et al. 2013).

Thus, an event monitoring system in a production planning system could identify quickly events that impact in the current decision of the production planning systems.

2 Event Monitoring Software Application for a Production Planning System

An automated event-detection system can offer a real-time view to the events (Shamsuzzoha et al. 2013). But an event monitoring is more than an event-detection system. Boza et al. (2014) indicate that an event monitoring system is a part of an event management system, which should interact with the DSS used in the production planning system to manage events that might affect their decisions. Thus, a production planning system driven by DSS should interact with an event monitoring system, which should act as a supra-system to identify events that impact in the on-going decision of the planning production system.

Expert knowledge is necessary to identify and to classify potential events. Expert systems to support decision making help in this purpose. Expert systems have the capability to manage problems that normally need the specialised human intervention for their resolution (García Martínez and Britos 2004). According with Cohen and Asín (2000), experts and knowledge engineers create the expert system collecting information and creating heuristic rules to build the knowledge base (Fig. 1). Benefits generated by the use of these systems include (1) Reduction on dependence of key personal, (2) Facilitates staff training, (3) Improves the quality and efficiency of decision-making and (4) Transfer the ability of making decisions.

Due to the advantages presented for the Expert System, we propose an Event Monitoring Software Application based in Expert Systems for planning production systems to identify and classify potential events that impact in the on-going decision of the planning production system.

Unlike the conventional expert systems, we looked for a manageable system that could be used directly for the expert. Thus, experts could create their own expert system as an aid to their decision making process (Fig. 2).

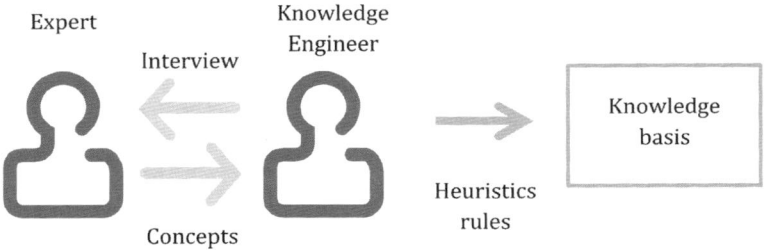

Fig. 1 Process of creating an expert system (adapted from Cohen and Asín 2000)

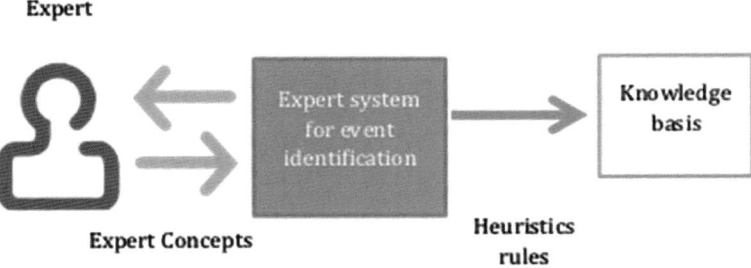

Fig. 2 Creation of an expert system directly by the expert

3 Design of the Event Monitoring Software Application

In order to contextualise our proposal, we deal with a planning system driven by mathematical-model DSS. This set of decision support systems is the context where our production Event Monitoring Software Application has been developed.

Events can impact over decision made with these DSS and the goal of the Event Monitoring Software Application is to identify and describe events that generate incidences in the system and to define alert criteria to be checked periodically (Fig. 3).

From a general point of view, the user creates their own alert criteria from the information that has been used in these DSS and these alert criteria are checked in order to identify possible incidences.

The system can be divided into three phases: (1) model and attributes selection, (2) criteria creation and visualization and (3) execution. Functional phases of the application are showed in Fig. 4:

3.1 Model and Attributes Selection

A mathematical model file, which is used in a DSS in the production planning system, is selected by the user in order to identify potential events to be checked. This proposal use Mathematical Programming Language (MPL) files as mathematical model files.

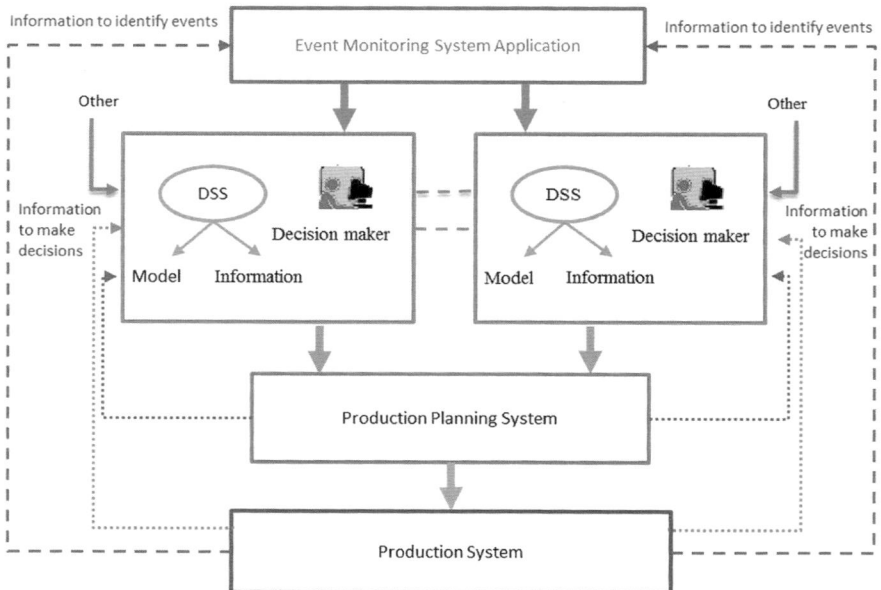

Fig. 3 Environment of the event monitoring system application

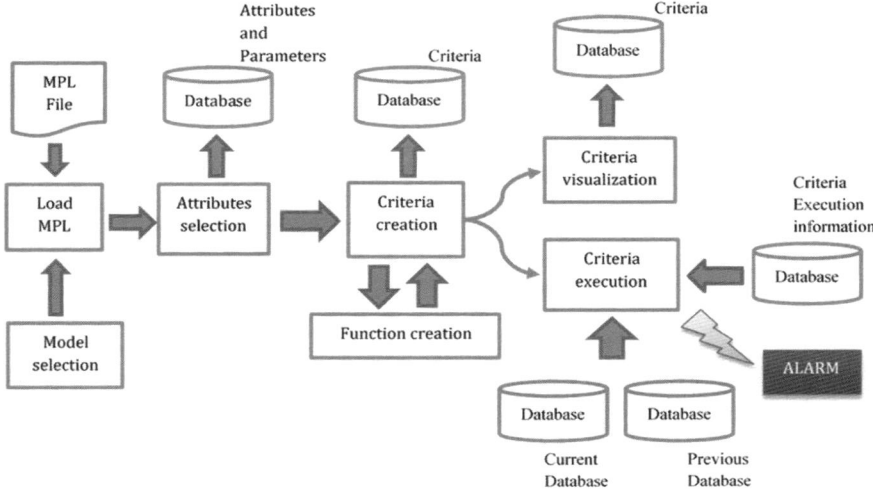

Fig. 4 Phases for the event monitoring application

Once the user has selected the MPL file, the application read the file to load the attributes (parameters and decision variables), which are potential events to be checked. Then, the user selects (of the previous set) their subset of attributes that will be finally checked.

3.2 Criteria Creation and Visualization

Once the attributes have been selected, experts can define alert criteria. Operands of each rule to build alert criteria can be attributes, constants or a function. The attributes used for these rules can be analysed using:

1. Current values and previous values of these attributes (for example the value of this attribute when the decision was made).
2. General evaluation or particular evaluation (for example the average will be a general evaluation and the evolution of a specific element (instance) of an attribute will be a particular evaluation).

For each criterion, experts have to define the name, the operands of the criteria, the logic operation to be performed with these operands, and optionally a description.

The criteria visualization allows users to review, modify or delete the alert criteria previously included in the system.

3.3 Execution (Monitoring)

Once all the alert criteria have been stored in the knowledge database, they can be checked. This execution can be launched by the user or in time intervals previously chosen (hourly, daily, weekly...). And, the source of data to be analysed are selected from the user. This monitoring process analyse the information in order to identify events that active the alert criteria. Thus, alarms can appear (or not) informing the user of an event occurrence.

4 Implementation of the Application

The application have been build using the following elements:

- MPL files with the mathematical models of the decision processes.
- Access databases. One database includes information of the current situation about a production planning. The other includes data obtained in a previous stage of the enterprise, when the decision about the production plan was taken. In this way, variations in current values from previous, or vice versa, can be detected.
- The internal database of the system includes the knowledge database (attributes, variables, functions, alert criteria,...).
- The interface has been developed using java libraries to be easily used by experts (Fig. 5).

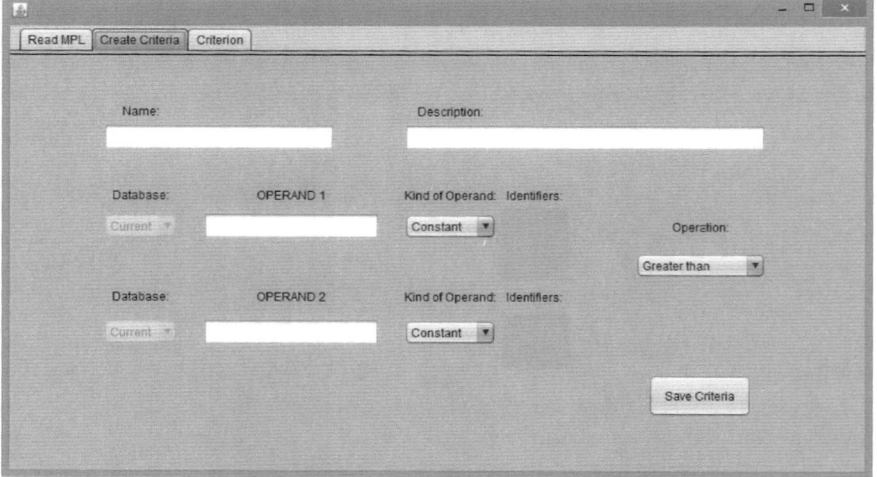

Fig. 5 Criteria creation window

5 Conclusions

Made decisions can be affected by unexpected events and it could be necessary to review the decision previously made. Decision Support Systems participate in decision processes and this research proposes an Event Monitoring Software Application based in expert systems in order to identify and classify potential events that can impact in a production planning system.

The design of the Event Monitoring Software Application has included three phases, model and attributes selection, criteria creation and visualization, and execution.

An event monitoring software application has been implemented with the goal that it could be used directly by the decision-makers. Thus, an easy interface has been developed to be used, in a simple way, by experts to create a knowledge base about alert criteria.

These alert criteria can be checked in a production planning system to identify unexpected events and to alert to the user about this situation. Thus, the system offers the automation in the monitoring of unexpected events.

The main advantages about using this application are: First, the own decision-maker can create their alert criteria to detect incidents easily without the needed of a knowledge expert to make the expert system. Second, he can execute them using different instants of the information (current production planning situation and a previous situation). Third, experts can identify general or particular events over a specific element. And finally, the monitoring system can be automatized to warn periodically about unexpected events.

Acknowledgment This research has been carried out in the framework of the project PAID-06-21Universitat Politècnica de València and GV/2014/010 Generalitat Valenciana.

References

Bedeian AG (1986) Management. The Dryden Press, Chicago

Boza A, Alemany MME, Vicens E, Cuenca L (2014) Event management in decision-making processes with decision support systems. In: Paper accepted for 5th international conference on computers communications and control, ICCCC 2014

Cohen D, Asín E (2000) Sistema de Información para los negocios. Un enfoque de toma de decisiones. McGrawHill, Mexico

Duan Y, Ong VK, Xu M, Mathews B (2012) Supporting decision making process with "ideal" software agents—What do business executives want? Expert Syst Appl 39(5):5534–5547

García Martínez R, Britos P (2004) Ingeniería de Sistemas Expertos. Editorial Nueva Librería, Buenos Aires

Shamsuzzoha AHM, Rintala S, Cunha PF, Ferreira PS, Kankaanpaa T, Carneiro LM (2013) Event monitoring and management process in a non-hierarchical business network, in book: intelligent non-hierarchical manufacturing networks (online). doi:10.1002/9781118607077.ch16

Shim JP, Warkentin M, Courtney JF, Power DJ, Sharda R, Carlsson C (2002) Past, present, and future of decision support technology. Decis Support Syst 33:111–126

Valero R, Boza A, Vicens E (2013) Sistemas de Ayuda a la Toma de Decisiones para la gestión de Incidencias. In: Proceedings of the 7th international conference on industrial engineering and industrial management, XVII Congreso de Ingeniería de Organización

Van Wezel W, Donk DPV, Gaalman G (2006) The planning flexibility bottleneck in food processing industries. J Oper Manage 24

Optimization of Public Transport Services in Small and Medium Size Towns. A Case Study on Spain

Elvira Maeso-González and Juan Carlos Carrasco-Giménez

Abstract Optimizing transit services become a challenge when economic resources are limited. Different from big cities which can afford expensive consultancy firms to support the establishment of their transportation plans, small and medium size towns have to face this problem under economic constraints. This paper aims to design a methodology to provide a pragmatic approach to transport planning in small and medium size towns, speeding up this process. The empirical investigation examined a sample of citizens based on stratification sampling methods to identify their transportation needs and preferences with the aim of enhancing the transit service performance and to increase its market share versus others modes of transport. Results show that there is a possibility in these type of towns to attract private car users to public transport when planning is effective.

Keywords Traffic management · Traffic data collection · Transport optimization · Public transport

1 Introduction

Spanish legislation requires towns with over 50,000 citizens to provide public transportation services to their inhabitants, which means that Town Councils must establish transportation plans and concern themselves with the effective implementation of the plans. This is a challenging issue since in most cases the whole process is under economic and organizational constraints resulting in services with high fares to guarantee economic viability and standards of service which do not fit

J.C. Carrasco-Giménez (✉)
Escuela de Ingeniería, University of Malaga, C/Dr. Ortiz Ramos S/N,
29071 Málaga, Spain
e-mail: jccarrasco@uma.es

E. Maeso-González
Escuela Técnica Superior de Ingeniería Industrial, Universidad de Málaga,
C/Dr. Ortiz Ramos S/N, 29071 Málaga, Spain
e-mail: emg@uma.es

with citizen's demand. This situation was the starting point of a research project which began as a petition to solve transportation needs of Benalmádena's citizens who were unhappy with the performance of their public service. Furthermore, the economic viability of the service tendered was impossible since bus routes had been designed without any previous study to determine transportation demand in the city. Consequently the challenge was too fold. Firstly, Benalmádena City Council did not expect to invest a large amount of money in the transportation planning as economic resources were limited in the town and secondly, they needed results as fast as possible as they were losing money every day with the then tendered service. In other words, they required a cost effective and timely redesign of their transit service.

This scenario brought up new questions, for example:

- Where are the optimum locations to interview people?
- Should interviews only be conducted with current public transport users?
- Which data collection strategy is the most efficient?

2 Case Study in Spain

2.1 Methodology

The methodology applied is shown in Fig. 1. There are three key steps in every survey process: targeting the population of study, choosing a correct data collection method, and the processing of that data. The methodology is ended adapting the idealistic transportation plan to economic constraints.

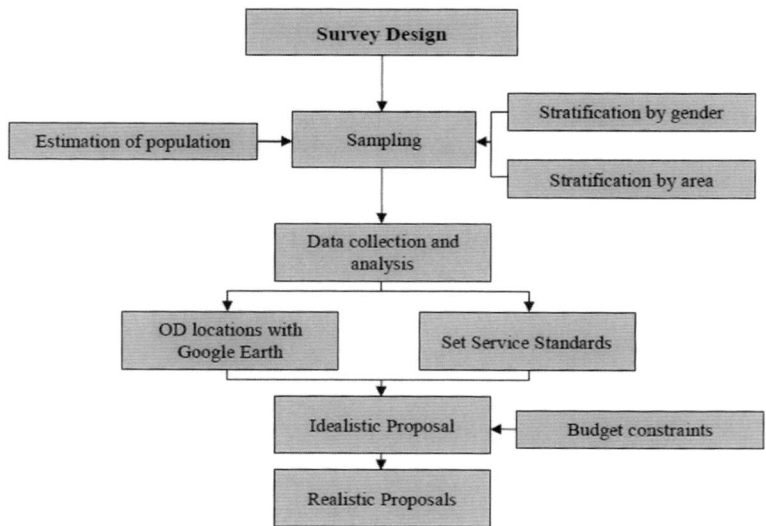

Fig. 1 Methodology

A survey must have an appropriate size. The index of responses is inversely proportional to the length of the survey. If the survey is too long and complicated, people may lose interest leaving most parts of it unanswered. Two short surveys in one were used to know both important destinations within the town and customer preferences. This helped to identify weaknesses of the previous transportation service and the way to improve it. At the end of every survey interviewees were asked for their opinion with an open question since sometimes there are relevant and particular aspects that are forgotten in the survey. The general process for a successful survey starts with the definition of the target population and continues with the construction of a sampling frame, determining the minimum sample size and choosing a sampling method. It ends with the decision of how to convert sample estimates to population parameters (Collis and Hussey 2009).

2.1.1 Target Population, Sampling Frame and Sample Size

The population was all citizens living in Benalmádena as there was a necessity in knowing the population's needs for transportation whether or not they were users of transit services and why. Benalmádena is a town located in the province of Malaga, part of the autonomous community of Andalusia in southern Spain. The town is located approximately 12 km to the west of the city of Malaga. Benalmádena has three main urban areas, named Benalmádena Pueblo, Arroyo de la Miel and Benalmádena Costa.

The most important thing when sampling is to identify different behaviour regarding transit service among the population. In the situation of Benalmádena, there were no noticeable differences between social groups. This is common in small and medium size towns where different areas can be considered economically homogeneous, whereas big differences regularly arise in large towns with the origin of depressed suburban areas. While economic differences are not very strong, the situation is completely the opposite when referring to gender and population. Small and medium size town are generally formed by the joining of different divisions or areas that previously existed and grew over time as it is the case of Benalmádena. It is very important to consider these parts as different because though the type of buildings and travel patterns are the same, corresponding population levels are different. There is also another key issue since the female population of the town represents a dominant varying proportion for each area, as their behaviour is different referring to transportation patterns; hence the best method of surveying is simple random sampling and then applying stratification, namely proportionate allocation to strata regarding the population areas and gender. The sample was calculated using the following equation (Abascal and Grande 2005):

$$n = \frac{N \cdot k^2 \cdot P \cdot Q}{e^2(N-1) + k^2 \cdot P \cdot Q} \quad (1)$$

where n is the number of surveys to be conducted, N is the population of Benalmádena, P is the portion of people who have the characteristic. Q = 1 − P, e is the error (assessed in a 5 %) and k = 2 for a precision of 95.5 %.

Once the sample has been estimated the following step is the determination of the stratification by gender and area:

$$\rho_i = n \cdot \alpha_i \cdot \beta_i \quad \forall i \in N \quad (2)$$

where ρ_i is the sample size for each area considered in the Town.

α_i is a factor representing the proportion of people living in the area in comparison with the total population and β_i is a factor representing the distribution of gender in the Town.

2.1.2 Data Collection Method

There are different ways of collecting data that can be grouped into four categories: Telephone, interpersonal, electronic and by mail (Malhotra 2010). Among them the method used was the telephonic survey. The main reason was the relatively short time allowed to do it. The mix needed between speed and accuracy when surveying was decisive to choose the phone to survey. It was a cheap way as the interviewer would not have to go house to house to collect data as in the interpersonal method. Electronic is a good way to do it as it is cheaper than mail, but there were no email addresses available to send them to. Distributing the survey by mail would take a long time in the delivering and collection process and the index of response it is not very high (Otero and Ladoux 1987). In telephone households surveys undercoverage occurs because no telephone sampling frame includes people in households without telephone lines (Groves et al. 2009); however, this methodology is commonly used and fits perfectly to the aim of this project. Furthermore, there was no risk of missing a particular type of interviewee because of to this void of phone lines. Surveys were conducted throughout one month during weekdays, mornings and evenings, and Saturday mornings to get a general view of regular trips during a month.

2.1.3 Data Processing: Travel Demand Modelling

Problems in mobility studies are frequently linked to data processing since the amount of data produced is huge (Otero and Ladoux 1987). This issue was especially critical with the travel demand modelling in which every street in the town

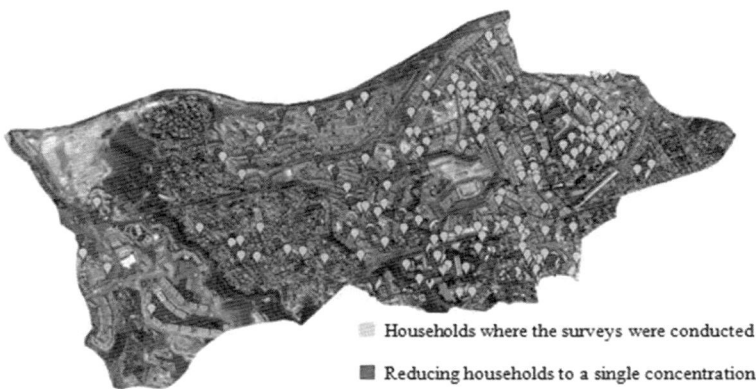

Fig. 2 Concentration of origins

was considered a location, the result was many different origins and destinations spread across the town. This problem was overcome with the readily available free GIS software Google Earth. To get an overview of the transportation demand, all these targets were precisely located into a map of the town using addresses provided during the survey. The result was a wide spread of targets as shown in Fig. 2. With so many different locations, the following step was the establishment of a weighting system scale for each location on the map taking into account the number of trips generated and the distance between them. Each group of near origins was replaced by a new one in which trips generated by the previous locations were summarized. New locations are shown in Fig. 2 which correspond to the urban area of Arroyo de la Miel.

This translates into a reduction in the number of rows and columns the matrix which provides a list of the most important locations at which the number of trips generated is high, but this is not enough and an extra step is needed. The last step in the process aims to get a set of critical points which the new transportation service has to go through. In this way, every area within the town was analysed regarding density of population, situation of speed ways, and the situation of the previous points of interest obtained before. As a result, a new group of points of interest was obtained. They are shown in Fig. 3 together with the old locations.

Finally, the result was a group of both origins and destinations for journeys which are the nuclei of different catchment zones through the town designed considering transportation demand, density of population and desired routes.

2.1.4 Data Processing: Multiple Choice Study and Opinion Research

Questions were designed as a mix based both on revealed and stated preferences (Ortúzar and Willumsen 2002). In this way, not only was a screenshot of the current

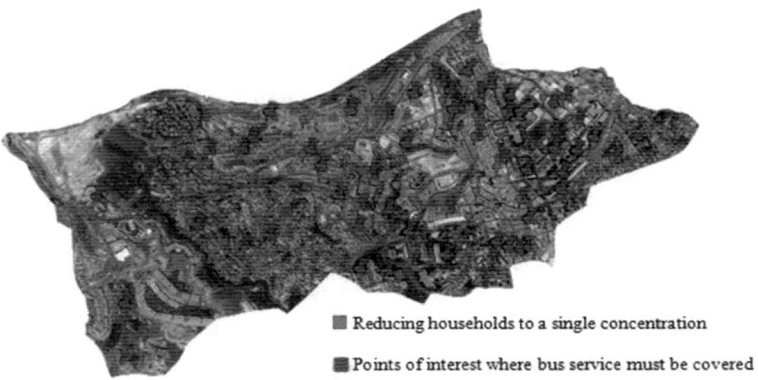

Fig. 3 Concentration of destinations

transportation demanded by the population obtained, but also the knowledge of common preferences and patterns of public transportation systems and quality standards for the success of improving the system in the future.

2.2 Results

2.2.1 Habits of Use of the Public Transportation System

There exists an incredibly high percentage of population who have never used transit and a very low number of frequent users. The 84.5 % of the interviewee declared not to use the public transportation system and only the 4.5 % of them affirmed a daily usage.

This is very common in small and medium size towns in Spain (ATUC 2009). Distances are short and historically cars have led the response to transportation demands. Consequently the index of motorisation is high and people are reluctant to use transit service, but it is also needed as a service since not everybody owns a car. This situation also creates a negative impact since the more people that drive cars, the higher the standard of service that is requested.

Concerning the standards within transportation systems, the most important is the frequency of service. Common car users find it frustrating waiting for the bus, although the frequency of service was too high. Hopefully, this only happens with a small percentage of the population. It is difficult to satisfy high expectancy demands for transportation with a limited budget. Knowing the frequency of service will help us and give an idea of the time people are willing to spend waiting for the bus.

Non-users of public transportation service are more demanding than regular users. The average frequency of service for users was 20.41 min and 19.21 for non-users. It is remarkable that the 54.1 % of users and the 47.7 % of non-users would be satisfied with a time of frequency of 20 min. Hence, that frequency of service was used in the design process of the transport plan.

2.2.2 Habits of Displacement in the Town

The reason of the travel is also very important. It gives us an idea of both the reasons people use public transportation and the normal activity in the town. Small and medium size towns are frequently used as dormitory communities if they are close to big cities, the case of Benalmádena, which results in a low demand of transport inside the town. Work is the third most popular reason for a journey with shopping and leisure being two main reasons for travelling in the town.

2.2.3 Shortcomings of the Current Public Transportation Service and Ways to Improve It

The best way to improve the service is to strengthen its weaknesses and get a depiction of the situation. The majority of car drivers will rarely shift their preferences towards other modes of transport for short trip distances as the information gathered reveals. There are however people who are forced to drive due to the bad service tendered and who would like to avoid using private vehicles. The main problem was the bad frequency of the service, followed by the number of people who use Benalmádena as a dormitory town. It is also remarkable that the fare of the service was not an access barrier to the service. Gaining potential users to the service is a hard task, however the study shows that if the frequency was better, bus stops were nearer, routes were optimally designed, the connectivity between modes of transport improved, fare reduced and other minor issues improved; there would be an increase of 39.3 % of trips done by bus.

This shows that potential users exist and are waiting for improvements in the public transportation service and a possibility of bringing a high number of new users to the system. Everything is dependent on the existing limited economic resources; therefore, it is difficult to make changes. Regarding this issue, a comparison of different reasons for journeys shows that there is not special trend in any category except for education. In general, half of transit can be done by bus if the situation is improved.

The last step in the process was the design of the transportation plan after collecting all the data. In this way, proximity to bus stops and frequency of service were determining factors. The frequency of design was set in 20 min and the distances between stops in 400 m (438 yd) maximum. The idealistic proposal designed was reduced to four different solutions managing the budget limitation and providing the number of required customers to achieve the viability of the system in order to rate the possibility of implementation for each proposal. Results showed that with the initial budget estimated by the Town Council, the type of service requested by the public was impossible to achieve. As a result, the only way was the increase of the frequency speed to around 30 min and maintaining the distance among bus stops. All of this represents the necessity of conducting policy changes. New redistributions of transport time/money budgets will be seen, all of that will probably be linked to how value of transport time and WTP evolve (López-Ruiz 2010).

3 Conclusion

This paper provides a pragmatic approach to transport planning in small and medium size towns, considering a range of typical patterns due to their economic constraints, and provides a simple but effective methodology to support the planning and implementation of effective transportation solutions. This study has highlighted the vital points that need to be considered when redesigning deficient public transportation services. The possibility of speeding up the process for effective public transportation has been shown using new sampling considerations, OD matrices with unfixed origins and destinations, and the usage of free GIS software. The results showed that there exists the possibility of increasing the level of use of public transport in this type of towns where private cars are leaders, and suggests the necessity of changing this trend in order to avoid future traffic congestions problems. This analysis points to the importance of improving quality standards as a way of attracting more customers to the public transport system. Difficulties found to satisfy the public with current estimated budgets for transportation in small and medium size towns suggest that perhaps policy makers should change tax regulation to overcome this lack of funding. Such regulations have already been introduced by some public transportation organizations in different countries.

References

Abascal E, Grande I (2005) Survey analysis. ESIC, Madrid
Association of Public Transportation Companies of Spain (ATUC) (2009) Efficient Management of Public Transportation Services. ATUC, Madrid
Collis J, Hussey R (2009) Business research: a practical guide for undergraduate and postgraduate students. Palgrave Macmillan, New York
Groves RM, Fowler FJ, Couper MP, Lepkowski JM, Singer E, Tourangeau R et al (2009) Survey methodology. Wiley, New York
López-Ruiz HG (2010) Recommended practice in adaptation strategies for long-term sustainable transport planing. In: Proceedings of the 13th international IEEE annual conference on intelligent transportation systems
Malhotra NK (2010) Marketing Research: an applied orientation. Prentice Hall, New Jersey
Ortúzar JD, Willumsen LG (2002) Modelling transport. Wiley, London
Otero JM, Ladoux M (1987) Mobility study in the city of Málaga. University of Málaga Press, Málaga

Methodology to Manage Make-to-Order and Make-to-Stock Decisions in an Electronic Component Plant

Julien Maheut, David Rey and José P. Garcia-Sabater

Abstract Supply chain management is a crucial part of any production system, therefore many strategies and concepts have been developed in order to help firms make the most out of their manufacturing capacity and to fulfil the best service level for customers. However, companies still face the problem of supply chain failures. Some of the main strategies to fulfil orders that exist nowadays are make-to-order (MTO), make-to-stock (MTS), assemble-to-order (ATO) or engineer-to-order (ETO) policies. In this paper, we review the literature about methods to help stakeholders to decide if a product has to be MTO or MTS. A case study of a hybrid MTO/MTS system on an electrical component manufacturing plant, that has to consider new decision factors and constraints, is introduced.

Keywords Make-to-stock · Make-to-order · Case study

1 Introduction

Several supply chain policies to fulfil orders have been defined over the years. The most applicable ones are make-to-stock (MTS), make-to-order (MTO), make-from-stock (MFS), assemble-to-order (ATO), manufacturing-from-stock (MFS) and engineer-to-order (ETO). Each strategy comes with its benefits and drawbacks and has preferential types of product for which they perform better. The choice of a strategy can be a complex decision as it depends on a lot of criteria such as demand volume, variability and frequency of the orders, lead times, production capacities, and the different costs that can be involved. Nowadays customers ask for higher

---

J. Maheut (✉) · D. Rey · J.P. Garcia-Sabater
Dpto. de Organización de Empresas, Edificio 7D, Universitat Politecnica de Valencia,
Camino de Vera S/N, 46021 Valencia, Spain
e-mail: juma2@upv.es

customisation in products and companies have to manage this complexity (Maheut et al. 2014). This affects the production level and also in the distribution and storage level because pure MTS or MTO systems are not viable anymore, and hybrid strategies seems to be the most suitable (Zaerpour et al. 2008).

In this paper we concentrate on the decision making of an electrical component manufacturing plant. We attempt to determine the factors that are the most important ones when sorting out the products between two order fulfilment strategies. Two new factors (average order lot size and mean time between orders) and one new constraint (MTS/MTO mix-ratio limit) have been identified as critical to enhance the operation decision process.

The paper is structured as follows. Existing literature about models and methodology is reviewed in Sect. 2. Section 3 introduces a case study of a manufacturing plant that produces electrical components in a hybrid MTO/MTS system where some new parameters and constraints entered have to be considered. And, finally Sect. 4 proposes a conclusion.

2 Literature Review

The literature on the issues of MTS versus MTO goes back to the 1960s (Rajagopalan 2002), however few have really taken into account the hybrid MTO/MTS aspect in the production system. Authors (e.g. Kingsman et al. 1996; Silver et al. 1998) have addressed the issues in pure MTO and pure MTS policies.

The pure MTS policy proposes the best service level and the shortest delivery time possible by keeping stock levels up and avoiding shortage. The items are produced following the economic order quantity which guarantees the lowest production costs. However this policy generates stock in the plant or in storage places so an inventory management system is required, thus generating stock holding costs. This strategy is suitable for products with high and stable demand volumes, with standard or low customisation possibilities. It is necessary to forecast demand levels in order to define the suitable order fulfilment strategy (safety stock level, expected customer service levels, etc.). Other limitations of this policy are items that have high holding costs and products that have a low shelf-time or perishable, which cannot be stored for too long.

The pure MTO policy implies that the production process starts when a purchase order is received and accepted. In this strategy no inventory is held so no stock holding costs are generated. The product variety that can be held is broader, so a mass customisation is made possible. However products are generally more expensive than standard products produced on a MTS basis because this can increase setup costs and the cost to manage complexity. One consequence of MTO strategy is the longer delivery time that can lead to lower customer service levels. Some authors (e.g. Soman et al. 2004) say MTO policy is suitable for products of low demand volume, highly perishable products, items with long production times, or expensive products. To support MTO policy, the supply chain requires flexibility in the capacity planning and scheduling.

Soman et al. (2004) point out managerial differences of both systems. MTS and MTO do not seem compatible because the objectives are conflictive: MTS implies high customers service levels with high holding costs whereas MTO implies eliminating stock and introducing longer delivery times.

In a hybrid system, new MTO orders can create a certain distortion at production and supply levels. The act of switching from a product to another can require setup times and generates extra costs. If the number of MTO orders received increases, that can reduce the overall productivity. And in some cases, less production capacity is dedicated to producing MTS items which can lead to shortage periods. To avoid them, inventory levels must be brought up, thus generating higher holding costs, or some MTO orders must be backordered or rejected. This also generates additional costs due to backorder penalties or sales loss. The levelling between MTO versus MTS is critical for the hybrid system to run smoothly since a wrongly classified item can generate unwanted stress on the production line or by taking up unnecessary space in the inventory.

Several papers in the literature propose a decision making approach for hybrid MTO/MTS systems and most of them tackle the scheduling problem. In this paper we concentrate on the decision making and attempt to determine what factors are the most important ones to take into account when sorting out the products between the two strategies.

Li and O'Brien (2001) propose a quantitative analysis between product types and supply chain strategies. Items are classified by their value-adding capacity and demand uncertainty. The performance of three different supply chain strategies, MTO, MTS and MFS are compared. Performance is defined as the gap between the expected values and the final values of profit, lead times and delivery promptness. This comparative simulation takes into account factors such as the finished product sale price, maximum and planned production lead time, the expected demand and the expected inventory levels fulfilling demand. However it does not take into account setup costs and no hybrid systems are considered.

Van Donk et al. (2005) provide a decision aid for the matter in the food processing industry. A methodology to choose whether to produce a product using MTO or MTS policies is proposed. The particularity of the food industry is that items have a comparatively low shelf-life, consumers behaviour is erratic and expected lead times are relatively low. Manufacturers have to opt for a hybrid MTO/MTS system and their main focus is on reducing costs while keeping high service levels. In the model, the overall production and holding cost is estimated for each item and the MTO/MTS repartition is carried out with the objective of reducing costs subject to the production capacity installed. The model is a good basis for decision making however it deals with a constant annual demand for each item and setup times or shelf-life is not taken into account. They also point out the decision, though complex, should generally be taken every six months or every year.

Kerkkänen (2007) deals with the decision of what MTO products can be MTS ones in a steel mill company. In this activity sector, most of the products are MTO and the setup costs have a great impact. The objective is to define what products are eligible for MTS in order to reduce the number of setups. The decision procedure starts by picking out, of all the products offered, the most standard grades and that are produced in larger quantities. Several factors are taken into account such as setup costs, steel grades that can be assimilated to product types or customisation degree of the product, production volume and customer fidelity. The choice is constrained by feasibility criteria such as material testing (if the material is made-to-stock) or potential losses of material. Once the product has been cut to the size of the order the material that is left is sometimes too small and must be rebound.

Zaerpour et al. (2008) propose a decision tool based on a hybrid AHP-SWOT approach. The procedure firstly identifies SWOT factors that define MTO production by carrying out surveys at different levels of a fir. These factors are later divided in four groups: strengths, opportunities, weaknesses and finally threats. The next steps define the weights of each of these groups using fuzzy pair-wise comparison. If strengths and opportunities are greater than weaknesses and threats then a product can be carried out as an MTO product otherwise MTS is suitable. The method is complete and can be easily applied and computed. However the data is subjective and the firms' capacity is not taken into account.

Rajagopalan (2002) proposes a heuristic model to solve the MTO/MTS decision problem. The model helps to define the best MTO/MTS repartition and for MTS products to define stock levels and mean queue time for each product. The objective is to minimise production costs considering each items: holding cost, the expected demand with its standard deviation, processing time and demand probability. The constraints are related to capacity utilisation and the desired lead time. This model does not deal with setup costs and it would be very interesting to implement different prioritising policies for MTO products over MTS product when a MTO order is received.

A brief summary of the different factors present in the reviewed articles is presented in Table 1.

The factors that are regularly considered in these previous articles are holding costs and product lead times. These factors allow the authors to define a strategy in a hybrid MTO/MTS system. In a hybrid environment stock levels might need to be higher to make so time available for MTO production. MTO items might need to be produced with a longer lead time if MTS items are being produced when a MTO order is received.

Table 1 Factors used for decision making

	Li (2001)	Van Donk (2005)	Kerkkänen (2007)	Zaerpour et al. (2008)	Rajagopalan (2002)
Product price	X			X	
Material purchase	X				
Expected demand volume	X	X	X		X
Demand uncertainty	X			X	X
Product lead time	X	X		X	X
Delay penalties	X			X	
Manufacturing costs		X			
Inventory levels and safety stock	X	X	X		X
Inventory holding cost	X	X	X	X	X
Setup time		X	X		X
Setup costs		X	X		

3 Case Study

3.1 Current Company Situation

The case company manufactures switches, electrical components, in a Valencian plant. A description of the production process and product can be found in Maheut et al. (2013). In one of its sixth assembly lines, 232 different end-products are assembled in a hybrid MTO/MTS strategy; 97 items are MTO and 135 are MTS.

The decision of whether to classify an item as MTO or MTS is done using the ABC/FMR classification (Oberlé 2010). ABC defines the products value (A is for products of high value and C for lower value products). FMR describes the frequency of the demand (F is for a very frequent demand, M for a medium demand frequency and R corresponds to products that have a low demand frequency). Products are then classified according to the corresponding ranking in the two groups, AR, BR, CR etc. Products that have a highly frequent demand and low value, such as CF and CM class products, are encouraged to be produced under the MTS policy. Products that present high value and low frequency of demand should be made to order. High value items with a highly frequent demand are stocked but the evolution of these products in the inventory is watched closely.

The decision is made once when the product is introduced to the market and no updates of the classification is done. The company has a large quantity of historical data concerning the sales orders. The customer order date and the quantity ordered for each item during the past number of years is known. Over the year MTS production has taken up an average of 95.6 % of the total capacity and the 4.4 % left is used to process MTO orders. The average lot size for a MTS product is 31 whereas MTO product have a mean lot size is 22. Since 2008 the company suffers a very unstable demand and during some working days the capacity required to produce MTO items oscillates between 1 and 9.8 % of the daily production capacity. Due to this uncertainty and instability, the supply chain is failing to replenish stock level of MTS items. The objective of the company is to determine another way to decide whether a product should be MTS or MTO to smooth out any unnecessary inventory.

3.2 Problem Description

The production system has to face "tensions" because of the demand volatility which means that some shortages can occur for MTS items. After an analysis of historical demand, some items' policies need to be updated due to demand evolutions. For example, some MTS products are at the end of their life cycle and should be produced with a MTO policy. The demand frequency of some MTO products is comparable or even higher than some MTS products. Some products are ordered in high volumes but only once a year and they should be MTO. In the production line, MTO items have priority over MTS products. The number of MTO references is increasing and consequently more shortages appear for MTS products and more tension appears in the production lines. Overall the number of products has increased 20 % from the last year and should increase another 50 % over the next two years. However the MTS/MTO status of the products is not up to date, and this will cause the supply chain to fail more often in the future.

A few years ago the company has moved from a pure MTS policy to a hybrid system in order to cut its inventory expenses. Simultaneously, the company has suffered an increase in the product complexity (number of end-product and reduction in packaging size). On the other hand, some references have a relatively high and stable demand whereas other products have a very low demand, or receive only one order per year and this is in constant evolution. A pure MTO system is not viable as the demand cannot be smoothed and is uncertain.

The MTO/MTS repartition follows the ABC/FMR logical as the MTS products have a higher demand in general. MTO products even though they represent 42 % of the items, only represent 4.4 % of the production volume.

However after an analysis, it has been found that the difference between the average order lot-size of each class is quite similar. Moreover, some MTS products have been produced in small quantities before being stored and some are only ordered once a year. These products consume capacity when more important MTS orders could be processed, they should be moved to MTO items.

Also it is observed that some MTO products are produced in high volume because the demand is higher than expected. If the demand volume and frequency of MTO items is comparable or higher than some MTS items then the policy must be updated. An update will smooth out the capacity consumption by high demand MTO products and avoid unnecessary setups for low demand MTS items.

Another issue to which confronted the company is the product MTO/MTS mix ratio to be produced. If more than 7 % of the capacity is dedicated to MTO production, the production line in incapable of dealing with so much MTO orders because of supply constraints. This issue case happened several times over the year and explains why the production lines suffer high tensions in some periods of the year. The MTS/MTO mix ratio is an important factor to take into account which to date has not been dealt with by the literature.

3.3 New Factors and Constraints

The previous models reviewed take into account parameters like standard deviation. However here, in order to measure the demand variations or uncertainty, new parameters need to be considered.

First the average order lot size (AOLS) is defined as the volume divided by the number of received orders. A low AOLS is due to low volume of demand and high numbers of orders. Products with a larger AOLS should be granted a MTS status whereas low AOLS products should be made to order.

Second the mean-time-between-orders (MTBO) is defined as the mean time between days with consecutive orders for a product. It is equal to zero when only one order is received in the year. A high MTBO means that there is a lot of time between two consecutive orders therefore, if the product is made to stock, it will spend more time in the inventory, generating holding costs. A low MTBO means that the product is frequently demanded, therefore the stock turnover is higher and a MTS strategy is suitable.

The plant needs to control a new constraint, the maximum product-mix ratio that is acceptable by the system. This constraint ensures that enough capacity is dedicated to manufacturing MTS products, ensuring in the process that stocks do not run out.

4 Conclusions

In this paper we present a literature review that deals with the decision of whether to manufacture and serve an item to order or to stock in a hybrid production system. The first observation is that there are few articles with fewer case studies. Next, a case study of a Valencian manufacturing plant for electrical components is introduced. In this case study, new considerations must be considered because product policy must be dynamic and the existing models are not directly applicable to this case. Two new factors, average order lot size and mean time between orders, and a new constraint, MTS/MTO mix-ratio limit, need to be introduced in the decision process. This research can be used as a base to design a new flexible optimisation tool to help stakeholders to define and update periodically the strategy of each product in the system based on historical data.

Acknowledgments The research leading to these results has received funding from the Spanish Ministry of Science and Innovation within the Program "Proyectos de Investigación Fundamental No Orientada" through the project "CORSARI MAGIC DPI2010-18243" and from the Project "Path Dependence y toma de decisiones para la selección de herramientas y prácticas de Lean Manufacturing" (PAID-06-12-SP20120717) of the Universitat Politècnica de València.

References

Kerkkänen A (2007) Determining semi-finished products to be stocked when changing the MTS-MTO policy: case of a steel mill. Int J Prod Econ 108:111–118

Kingsman B, Hendry L, Mercer A, de Souza A (1996) Responding to customer enquiries in make-to-order companies problems and solutions. Int J Prod Econ 46–47:219–231

Li D, O'Brien C (2001) A quantitative analysis of relationships between product types and supply chain strategie. Int J Prod Econ 73:29–39

Maheut J, Garcia-Sabater JP, Garcia-Sabater JJ (2013) Integrated production and simulation scheduling tool to solve the mix model assembly line problem considering Heijunka and operational constraints: a case study. In: Proceedings of the 7th international conference on industrial engineering and industrial management, pp 381–388

Maheut J, Besga JM, Uribetxeberria J, Garcia-Sabater JP (2014) A decision support system for modelling and implementing the supply network configuration and operations scheduling problem in the machine tool industry. Prod Plann Control 25:679–697

Oberlé J (2010) Air sea ratio reduction initiative. In: Proceedings of the World energy congress, pp 1–13

Rajagopalan S (2002) Make to stock or make to stock: model and application. Manage Sci 48:241–256

Silver EA, Pyke DF, Peterson R (1998) Inventory management and production planning and scheduling, 3rd edn. Wiley, New York

Soman CA, van Donk DP, Gaalman G (2004) Combined make-to-order and make-to-stock in a food production sytem. Int J Prod Econ 90:223–235

Van Donk DP, Soman CA, Gaalman G (2005) A decision aid for make-to-order and make-to-stock classification in food processing industries. In: Proceedings of the EurOMA

Zaerpour N, Rabbani M, Gharehgozli AH, Tavakkoli-Moghaddam R et al (2008) Make-to-order or make-to-stock decision by a novel hybrid approach. Adv Eng Inform 22:186–201

Effects of the Implementation of Antequera Dry Port in Export and Import Flows

Guadalupe González-Sánchez, Mª. Isabel Olmo-Sánchez and Elvira Maeso-González

Abstract Currently many companies are forced to go outside due to the economic situation and in many cases, as the sole alternative to ensure their survival. In this context the logistic performs a leading role. To promote and to take advantage of an efficient logistic management is key to exports. In this sense, dry ports constitute the required nexus for road-rail-sea intermodal transport. The establishment of a dry port in Antequera would boost the export and import flows to and from Andalusia with the rest of the world.

Keywords Freight transport · Intermodality · Dry port

1 Introduction

The market globalization and the current industry's offshoring grant a fundamental role to the goods flow. Means of transport are nowadays essential in the trade of products as a backbone of the economy. A poor transportation system would entail a lack of customer confidence and a decrease of sales, which is an important business conditioning for the company. Satisfying the transport necessities requests different solutions of different degree of complexity.

G. González-Sánchez · Mª.I. Olmo-Sánchez
Cátedra de Gestión del Transporte, Universidad de Málaga, Estación de Autobuses,
Pso. de los Tilos s/n, 29006 Málaga, Spain
e-mail: ggonzalez@uma.es

Mª.I. Olmo-Sánchez
e-mail: m.olmo@uma.es

E. Maeso-González (✉)
Cátedra de Gestión del Transporte, Escuela Técnica Superior de Ingeniería Industrial,
Universidad de Málaga, C/Doctor Ortiz Ramos S/N, 29071 Málaga, Spain
e-mail: emaeso@uma.es

© Springer International Publishing Switzerland 2015
P. Cortés et al. (eds.), *Enhancing Synergies in a Collaborative Environment*,
Lecture Notes in Management and Industrial Engineering,
DOI 10.1007/978-3-319-14078-0_17

The analysis of the transport flow demand, the logistics hubs location, and their effects on the region, is essential to improve logistics and to rebalance the modes in order to get a sustainable transport system (Maeso et al. 2012). Carrying out a study and a planning leads to infrastructures and modes of transport properly developed and, consequently, to strengthened regions with a stronger offer and provision capacity.

The awareness that the future of freight transport lies in combining the different modes of transport, each one within its scope, involves that today we highly bet for the railway and maritime modes, as a mode to be developed in such a way to complement the rest of the modes.

Among the policy actions that promote it are the new White Book about Transport Policy (COM 2011) where it is indicated, as an objetive to transfer the current road mode of freight transport to the one by railway or river navigation 30 % (2003) and more than 50 % (2050) by relying on the development of a proper infrastructure formed by effective and ecological corridors. Another initiative is the Trans-European Network of Transport, a basic European network of transport to connect the several EU countries so that, among other objectives, the different modes of transport and the environmental aspect of the network are integrated.

In Spain, in The Strategic Plan of Infrastructures and Transport (PEIT 2005–2020), railway is especially promoted so that an integrated transport system aimed to obtain greater economic and environmental efficiency is implanted. The objectives marked for maritime transport include, among others, the development of: The ports as key elements of intermodality, Short Sea Shipping and land access to ports, especially by railway, through ad hoc systems of management and cofinancing (MFOM 2005).

At a regional level, in the Plan of Infrastructures for sustainability of transport in Andalusia, PISTA (COPT 2008), the peripheral position of Andalucía with respect to Europe and the lack of transport infrastructures have been identified as the determining causes for its lesser development and its ports hinterland. Some strategies of this Plan are: To promote intermodality, to prioritize the most sustainable modes and to develop the Network of Logistic Areas of Andalucía.

Dry ports are the necessary link for the maritime-railway-road intermodality. These logistic infrastructures are characterized by being directly linked to a port mainly by railway and by enlarging the port's area of influence inland, by permitting that the port freights arrived in an integrated way to the dry port, where all the logistic operations needed for distribution are carried out; in turn, it also act as receiving centre of the freight bound for the sea port.

2 Objective

The objective of this work is to study the appropriateness of the implementation of Puerto Seco de Antequera (PSA) and its possible effects on the export and import flows.

3 The "Puerto Seco de Antequera"

The PSA is one of the most important actions of the logistic hub, Central Area of Andalucía, considered in PISTA due to its geographic-strategic location as the nerve centre of Andalucía, in a strategic communication junction at regional, national and international level. Located between the Santa Ana AVE (High Velocity) station and the Bobadilla railway junction, a communication junction with traditional gauges (freights) and high velocity ones (passengers). It is directly linked to the ports of Algeciras, Málaga, Seville, Granada and Córdoba. The link to Algeciras is part of the Spanish and European railway hub for freight transport what permits connecting the freight traffic from the ports in the South of Spain with North Europe. It is an industrial area of 338 Ha for all activities related to logistics, transport and distribution and production of freights in the widest sense and, especially, those activities with added-value in railway sector.

3.1 Effects on the Freight Maritime Transport

The significant increase of the Andalucía ports' freight traffic, greater than the average of the Spain port system, has permitted consolidating the traffic recovery started in 2010. All the Andalucía ports have increased their traffic, being the outstanding ones Málaga (130 %), Huelva (20.59 %), Almería (19.24 %) and Bahía de Algeciras (17.89 %) (APPA 2012). This increase in the port activity makes evident the PSA successful potential, since it is directly linked to the ports of Algeciras, Málaga and Seville. Among the significant Andalucian ports it stands out the port Bahía de Algeciras, the port with the greatest transport demand (More than 65 MT). In relation to the traffics of general freight in containers, the ports of Algeciras and Malaga are the ones with a greatest in transit percentage (95 % over the total of freights in containers). It exists the possibility of working to increase the percentage of freights in containers with origin in/destination to the hinterland of the ports in Andalucía (both in Algeciras as in Málaga it is only a 5 %).

Logistics, efficiency, cost, intermodality and infrastructures are key elements for the future of the ports. "The ports that do not develop towards logistic ports will have many difficulties to be part of the more and more competitive maritime market; the same will happen to those which are not efficient, economic and able to manage the great volume of freights that will be moved through by increasingly bigger ships". (Mr Jose Llorca, President of Port of the State 2013).

With respect to the land traffics related to the ports, in 2009 just a 5.2 % was transported via railway (8.8 Mt). In Andalucia, in 2009, some of the ports with significant freight volumes did not recorded railway movements. In Algeciras it was of 0.62 %. It is necessary to increase the use of railway transport related to ports. In Spain, the development of intermodal routes depends, to a large extent, on the ports' freight management capacity. The main chargeports located in the Antequera's area of influence (Algeciras and Málaga) have significant traffics. The creation of a dry port in Antequera will facilitate the future development of freight traffic in the mentioned ports and their Hinterland extension towards efficient logistic spaces. "Only the ports having infrastructures and links will survive, and in the Spanish case, it is necessary to improve the link to the railway". (Expansión 2013). The PSA connects perfectly with these necessities since it is strategically complemented by the port Bahía de Algeciras located "zero-deviation maritime route Round the world". The changes in the maritime flows is giving more and more prominence to Asia and to the emerging countries. These flows will change even more when the extension of the Panama Channel start operating. The Ministerio de Fomento (Ministry of Public Works) is already working so that the Mediterranean ports can get traffics with destination North Europe by the Suez Channel route, as they would save four or five days if they entered in Spain via railway. The central location of the Spanish ports can benefit not just the port of Algeciras but also those located in the Atlantic, to the detriment of others Mediterranean European ports. The Andalucía's Mediterranean and Atlantic coasts and their ports are strategically located in the big inter-oceanic corridors of maritime worldwide transport: East-West axis (Asia/Europe), Americas Atlantic coast/South of Europe/Mediterranean and North-South axis (Europe-West Africa mainly). One of the strategic focus for the port Bahía de Algeciras development is its conformation as "Europe South Gate". To achieve that, it is promoted the container traffic, not only those in-transit but also the export-import ones. Another aspect not to be forgotten is the trade between Spain, nearly exclusively, and the African continent through Andalucía. Along the same line, it is anticipated another PSA potential in direct collaboration with the ports of Málaga, Almería, Motril and Algeciras: operating in import and export and capillary distribution inside Andalucia and, by extension, in Spain.

It is then a strategic sector where initiatives such as the Dry Port help to improve the services, penetrability and capacity of the Andalucian ports and, in concrete, of the port Bahía de Algeciras.

3.2 Effects on the Freight Railway Transport

Spain is the fourth European country with a lower railway freight transportation rate (Eurostat 2013). This traffic represents 4.5 % of the land transportation (year 2011), still far from the European objective of rebalancing and stabilizing modal distribution. The Strategic Plan to promote the railway freight transport in Spain marks as an objective reaching a 8–10 % rate in 2020 (MFOM 2010). When comparing to other ways of freight transport, we observe that rail is significantly below the road and maritime modes (Fig. 1). Even if recession effects have reduced every mode of freight transport (OSE 2011), the transport by rail has particularly decreased up to 2.6 % (MFOM 2012).

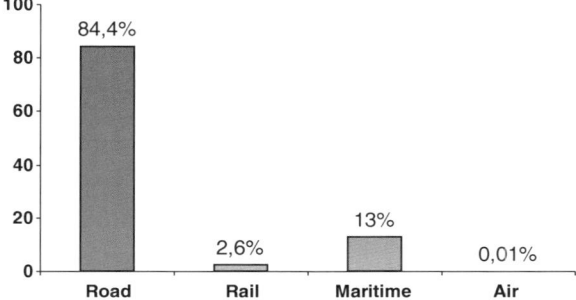

Fig. 1 Distribution of national freight traffic according to means of transport in Spain (2011)

When considering the most significant freight flow by railway (>100,000 t), intermodal transport and conventional wagon, the area of concentration is in the Northeast of the Peninsula, with branches towards others areas of the territory, such as the Central-South (Madrid-Andalucía) corridor or the Mediterranean (Mediterranean littoral) corridor.

In Andalucía, among the traffic installations in conventional wagon, it stands out: Huelva and Bobadilla. This encourages the adequacy of developing the PSA. Among the terminals with greatest activity of intermodal transport (2008) it stands out in Andalucia: San Roque and Seville. Currendy, it is not possible the intermodal transport in Bobadilla.

The liberalization of the railway freight sector is a definitive tool to promote this kind of traffic since it introduces the competition principles and market rules in a sector which is traditionally very inflexible and unable to compete with the road transport rate. CONTREN, one of the mercantile company belonging to RENFE, is the one assuming the activity of intermodal business although it does not export because it just has connections with concrete companies in countries such as Portugal and France. The Multiclient Network does not work out of Spain. Therefore, if it is wanted to transport to Europe, it has to be done through one of its clients, which act as a forwarding agent, being Combiberia the main of them (CCP 2013).

Combiberia is a combined transport wholesaler via railway. The freight movement in Andalucia is carried out through the infrastructures for intermodal exchange of San Roque-La Línea, Seville-La Negrilla and Córdoba-El Higuerón. In Fig. 2, it can be observed the existing links among the intermodal terminals in the Andalucía region, and the rest of Spain and Europe (ADIF 2011). In the route followed by the trains from the southernmost point of the peninsula, San Roque-La Línea, until their exit from Spain, the Bobadilla station is strategically located in the Andalucian network. Currently, all the Andalucía freight routes are channeled through Madrid.

The PSA offers a unique logistic opportunity thanks to its location in the junction of the Mediterranean and Atlantic railway corridors. PSA and port Bahía de Algeciras represent the South vertex of the Mediterranean and Atlantic railway axes within the scope of the Transeuropean Transport network of the EU (TEN-T). Currently, there is little import-export railway traffic in port Bahía de Algeciras. Until September 11,863 containers were transported by railway through the line

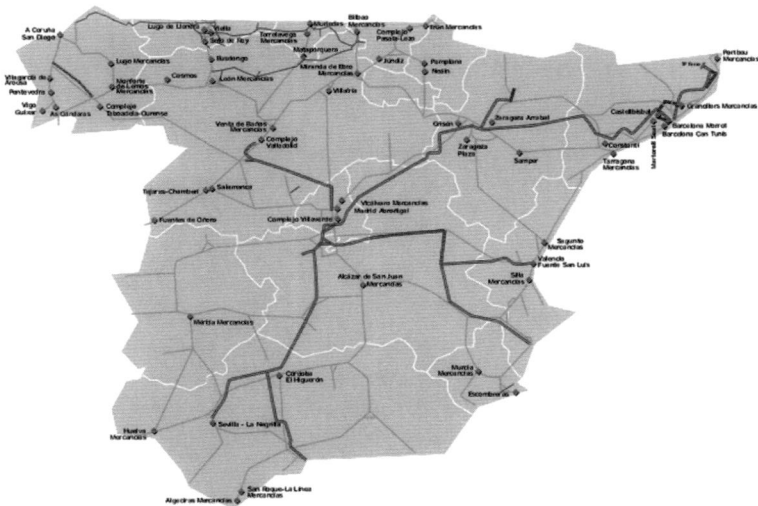

Fig. 2 Lines layout on national territory

Algeciras-Bobadilla (Antequera). Around 10 and 12 trains per week. In the whole year, a total of 153 trains have had origin and destination in the port area (APBA 2012). To improve these numbers it is essential to invest in a competitive railway which permits the successful linking of the port Bahía de Algeciras to its hinterland in the center of the peninsula and to the French border (Europasur 2013). The PSA clearly goes for the railway mode by incorporating in the future development all those activities with added-value in the railway sector.

However, the railway growth cannot be posed in an isolated way. The railway must be the center of a logistic model combining the different transport modes, as it be advisable, so that it is adapted to the client necessities and achieve an efficient, effective and sustainable system. Railway, with limitations as regards its logistic service offer, needs to collaborate with other modes, especially with the road one. In Spain, in general, the transport centers linked to road and railway terminals have set up two independent networks with lacking points of contacts; the opposite happens in other countries around us. Another factor is that when designing the railway networks it has not been sufficiently taken into account the importance of terminals which, according to experts, are even more important than the very own railway lines for the quality of railway transport and, clearly, for the integration with road. Among the initiatives to improve the railway-road modality we have the creation of new intermodal terminals located in strategic points next to the freight's origin/destination, with good access and connection and with the proper equipment. In that sense, the suitability of the PSA is already corroborated. Another initiative is focused on that the large railway axes should link the main road centers with the areas of logistic activity by constructing a real intermodal networks as the one already existing in the center of Europe. In this sense, Antequera is already located in these large transeuropean transport networks by the institutions of Andalucía, Spain and Europe. From the infrastructure point of view, to its strategic railway links the PSA adds its advantageous road location.

The intermodality development is still poor due to problems (time restrictions, flexibility, penetration, etc.). In this sense, a dry port implies a significant logistic structure which permits carrying out all the procedures external to freight transport with the greatest guarantees for security, productivity and quality. Thus, the PSA implementation represents a base for the growth of the logistic system in Andalucía as it is an essential piece for the intermodality advance and for attracting international companies operating in Spain.

In turn, the area of influence of Andalucía as a whole logistic platform from the point of view of regional market (Spain, Portugal, South of France, Morocco and Algeria) represents around 100 million of possible consumers. Combining the land, railway and maritime connections is essential to serve those possibilities. The most outstanding companies located in the PSA area are those related to olive oil sector, an economic sector essential in Andalucía. In concrete, Deoleo, the first world packaging company of olive oil, and Hojiblanca, the first producer, they already have a leading position in the world market which will be strengthened by the PSA creation (UMA and US 2013). A key company for the PSA development is Acotral. A leading company in the sector of national and international freight transport, with the main headquarters in Antequera. Currently, Acotral moves around 700 trains per year (21,000 TEUs) in the routes Seville-Antequera-Valencia, Seville-Antequera-Tarragona, Tarragona-Valladolid, Valladolid-Seville and Madrid (Acotral 2012). The PSA development would be essential so that this company increases the railway transport flow because, as it has been described, it is very limited geographically with few and distant load terminals. In concrete, this company's forecast is to use the PSA with 3 trains/week in 2015, and to increase 1 train per year until reaching 8 trains/week in 2020.

The PSA activity will make possible the load consolidation/deconsolidation with different origins/destinations, the effective management of the capacity of round trip railway transport, the explotation of the trains departing from Antequera to the North with imported products through the Ports and export products from Andalucía to bring these trains back loaded with importations to be distributed in the South of Spain and with products from the rest of Spain and Andalucía to be exported through the maritime terminals.

4 Conclusions

The analysis of the effects of the introduction of PSA leads us to a number of associated benefits. From the business point of view, would be a space for the installation of companies which would count with all the services of logistics operators and infrastructure necessary for its international expansion. Moreover, its direct connection to the Andalusian ports would increase the capacity and area of influence of these further increasing its maritime freight traffic. Another advantage of its collaboration with the Andalusian ports, would improve the import and export operation and regional and national capillary distribution in its trade with the African continent. The PSA would improve and promote the presence of the rail freight and, therefore, a better use of their strengths in terms of energy efficiency, external costs and emissions, and may further contribute to the development of an efficient economy and low carbon. Thus, it is appreciated that the existence of the Dry Port in Antequera favor flows of imports and exports, so would not only be positive locally by its effect on business and economic development, but it would strengthen the market over increasingly demanding changes in demand for goods, promoting intermodality, attracting international companies with global logistics vocation positioning and enabling the growth of the Andalusian and Spanish logistics system.

Acknowledgments Puerto Seco de Antequera for their collaboration in the realization of this article.

References

Acotral (2012) Presentación Corporativa 2012. Compañía Logística Acotral
ADIF (2011) Actualización Declaración Sobre la Red 2011. Administrador de Infraestructuras Ferroviarias, Madrid
APBA (2012) Memoria Anual 2012. Puerto Bahía de Algeciras, Autoridad Portuaria de la Bahía de Algeciras
APPA (2012) Agencia Pública de Puertos de Andalucía Noticias, 13 Feb 12
CCP (2013). Sale of Renfe-Operator shares in the Intermodal Transport Society Combiberia S.A. and Depot TMZ Services S.L. Advisory Council of Privatizations, Madrid
COM (2011) White paper. Roadmap for a single European area of transport. A competitive and sustainable politics for transport, E.C. Brussels
COPT (2008) Infrastructure Plan for the sustainability of transport in Andalusia (PISTA 2007-2013). Consejería de Obras Públicas y Transportes, Junta de Andalucía, Sevilla
Europasur (2013) El puerto de Algeciras alcanza en junio los tráficos de import/export de todo 2010
Eurostat (2013) Modal split of freight transport. Railway. European Comisión, Brussels
Expansión (2013) Los puertos que no mejoren logística difícilmente entrarán en el mercado
Maeso E, Caballero J, Sánchez FA (2012) Impacto del Puerto Seco de Antequera en el Sistema Logístico Andaluz. Revista de Estudios Regionales 3:149–174
MFOM (2005) Strategic Plan for infrastructure and transport 2005-2020. Fomento, Madrid
MFOM (2010) Strategic Plan to promote freight rail transport in Spain. Fomento, Madrid
MFOM (2012) Transports and infrastructures. Year report 2011. M. Fomento, Madrid
UMA y, US (2013) Plan de Acción para promover el crecimiento económico y mejorar la innovación en el ámbito del transporte del CEI Andalucía Tech. Universidad de Málaga y Universidad de Sevilla

An Iterative Stochastic Approach Estimating the Completion Times of Automated Material Handling Jobs

Gunwoo Cho and Jaewoo Chung

Abstract Recently manufacturing facilities increasingly use automated material handling systems (AMHS) owing to the advancement in control technology. They provide various material handling functions such as transportation, storage, sorting, and identification with quicker and precise devices. This paper studies an estimation method encountered in manufacturing facilities operated by automated material handling systems. A heuristic method based on a stochastic approach is used to estimate the arrival times of transportation jobs to their final destinations in a manufacturing facility. To analyze the performance of the new method, the authors collected a set of actual transportation data from the industry and the analysis shows that the new method outperforms an existing method that uses simple statistics based on historical data.

Keywords Automated material handling system · Arrival time estimation · Stochastic method

1 Introduction

Material handling in manufacturing facilities is automated by state of the art technology that helps not only to reduce labour costs but also increase job standardization. In turn, it increases the visibility of operations in facilities and supports various engineering activities indirectly to improve product quality. The advantages of production automation are described in the literature (Chung and Jeng 2004; Campbell and Laitinen 1997). Among the many advantages of automation, fast transportation with the right job on the right location and at the right time is

G. Cho · J. Chung (✉)
School of Business Administration, Kyungpook National University,
Daegu 702-701, South Korea
e-mail: chung@knu.ac.kr

the most fundamental reason why more and more manufacturing facilities keep investing in expensive automation systems.

The objective of this research is to develop an analytical method to estimate the arrival times of transportation jobs moved by automated material handling systems (AMHSs) in manufacturing facilities. The proposed method uses an iterative heuristic based on a stochastic approach that provides a realistic estimation method for the transportation jobs. Information about the estimated arrival time of a transportation job is important for production managers to determine the priority of the job. An individual transportation job has a due time that prevents starvation of a process toolset due to delays in transportation. If the estimated arrival time of a job is too late compared to its due time, a production manager who monitors the AMHS can change the transportation priority of the job so that the job can arrive on time (Suh and Chung 2013). This is illustrated in Fig. 1. Both the starvation risk based on the completion time estimation and process urgency based on process information are high, the material control system will serve the job with top priority, and vice versa.

Additionally, the estimated transportation time can be used to increase the preciseness of production scheduling or AMHS scheduling (Poppenborg et al. 2012; Li et al. 2005). If the ready times of jobs are an important constraint in a scheduling problem (Pfund et al. 2008), the arrival time information would be a useful input to obtain a reliable solution for the problem. Hence, the estimated completion time could be used as input data of the production scheduling system in actual practice.

This paper first proposes a mathematical model that estimates the arrival times of transportation jobs and provides an extensive analysis to measure the performance of the proposed method. The superiority of the new method is that it is able to account for the waiting time on the route more precisely than the existing method with a relatively simple computing method. It considers not only currently waiting jobs on the route but also waiting jobs in the future. The new method was implemented in an AMHS through an academic-industry collaboration project. The data used to analyze the performance of the new method were collected after the actual implementation.

Fig. 1 Transportation priority depending on completion time

	3rd priority	Top priority
	Low priority	2nd priority

Process urgency (y-axis) / Starvation risk (x-axis)

2 Problem and Estimation Model

2.1 Problem Definition

To explain the problem, a graph is used shown in Fig. 2 which represents a transportation problem of a manufacturing facility. For simplicity, only the main material handling systems are displayed in the figure. In the graph, nodes S_1 to S_6 represent material handling systems (MHS) that move *cassettes* or *transportation jobs* from one location to another. The arcs with directions indicate that a cassette can be moved from one MHS to other directions of MHSs. For example, assume that a job should be moved from S_2 to S_6, then two routes, S_2–S_3–S_6 or S_2–S_5–S_6 seems applicable. If the job uses route S_2–S_3–S_6, the material handling device (e.g., automated guided vehicle or automated storage and retrieval system) used in S_2 moves the job to S_3 and then the job is moved to S_6 by the material handling device S_3. Two transportations are required to complete the transportation, and if there are N number of nodes in a route, the required transportations become $N - 1$ since a transportation is not necessary in the final node.

A node indicates a processing toolset, automated storage and retrieval system, lift system or overhead shuttle system in the graph. If a node represents a toolset, it has a material handling device serving transportation. If it is a material handling system such as a MHS system or overhead shuttle system, it has shelves to store cassettes inside. Hence, all nodes can be a source or destination and they have their own material handling device. There are hundreds of toolsets and several tens of material handling systems in actual production and a transportation problem becomes much more complicated than the one in Fig. 2.

In general, a route for transporting a job includes two to a few tens of nodes in actual production. One transportation job takes a few minutes to several tens of minutes depending on the processing and waiting times at the nodes in the route. The variance in transportation times within the same route is relatively large because the capacity of AMHS is limited due to high investment cost, and job arrivals are often concentrated during a certain period in this reference. When a job arrives at a node, it waits until the material handling device of the node is available

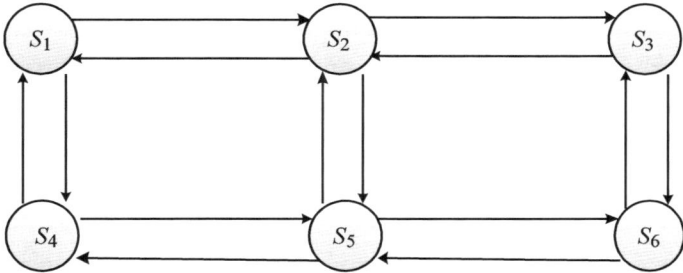

Fig. 2 A graph representing transportation problem in a manufacturing facility

and this time is called the *waiting time* of the job at the node. If the device is available to serve the job, it takes time to pickup, travel and drop-off and this time is called the *processing time* of the job at the node in this paper. In most cases, a waiting time is much longer than a processing time. The sum of the processing time and waiting time becomes the *sojourn time* of the node.

2.2 Approach

There are two issues for the estimation problem. One is how to estimate the processing time of a node and the other is estimating the waiting time of a node. To devise methods for estimating the processing and waiting times, respectively, the author of this study investigated the distributions of actual transportation times with different nodes. About three thousand actual transportation times were sampled from the reference site. The result showed that the other nodes have a similar pattern as the one. There are three different types of distributions, which are the processing time, waiting time and total transportation time (i.e., sojourn time). An individual sojourn time was decomposed into a processing time and a waiting time.

The distribution for processing times has a much smaller variance than the waiting time. Note that the mean and standard deviation of the processing times are 1.56 and 0.524 min, respectively. Assuming it follows a Normal distribution, 2σ is approximately one minute (2×0.524) and it means that more than 95 % of the processing times can be estimated with a 1-minute tolerance if the mean value is used as an estimator of the processing time for a job. The precision level was generally accepted by production managers in the reference site; hence, the authors decided to use the mean processing time of the node to estimate an actual processing time at a node.

Using the mean value for estimating the waiting time of a job could create a higher deviation since the waiting time distribution is more widely scattered than the processing time. Hence, this study introduces a more detailed estimation model that could increase the estimation accuracy as follows. As mentioned before, a route includes a sequence of nodes for each job. The model iteratively calculates the sojourn times of each node one at a time, and when the job arrives at a node, it has the cumulative sojourn times that it took to reach the node. The sojourn time at a node is calculated by considering the number of waiting jobs at the time of estimation, expected arrival jobs, and expected processed jobs during the cumulative sojourn time to reach the node in the route of the job. This new method is named stochastic iterative time estimation (SITE) and an explanation on the more detailed model is omitted due to the limited space of this paper.

3 Performance Analysis

This paper compared the average and standard deviation of the estimated transportation times with those of actual ones along with the different numbers of nodes. It was observed that jobs with 3–4 nodes were most frequent. The estimated times were very close to the actual ones regardless of the number of node. One interesting observation is that the actual transportation times were slightly but consistently larger than the estimated times. This study initially tried to adjust for this difference; however, production managers wanted to ignore the difference because the difference was considered to be minor for individual jobs. The standard deviations were slightly higher than the actual ones seen in the third and fifth columns of the table. This is natural because SITE considers only a limited number of factors of the AMHS while the actual transportation times are influenced by more factors such as short and sudden stops or job priorities as mentioned.

The performance measure used in this paper is the hit rate defined below.

$$\gamma = \frac{\sum_{i=1}^{N} H_i}{N} \qquad (6)$$

γ hit rate of estimated jobs
N total number of transportation jobs
A_i actual transportation time of job i
H_i indicator variable of estimation accuracy for job i

$$H_i = \begin{cases} 1, & \text{if } |E_i - A_i| \leq 1.0 \min \text{ or } \frac{|E_i - A_i|}{A_i} \leq 0.1 \\ 0, & \text{otherwise} \end{cases}.$$

The hit rate, γ in Eq. (6), measures the accuracy of the estimation by comparing the estimated transportation time and actual transportation time. As defined above, H_i is one; i.e., regarded as accurate, if the difference between the estimated and actual times of a job is less than one minute or the difference is less than 10 % of the actual transportation time. The reason why a difference less than one minute is considered as accurate is that if the difference is less than one minute, the job is fast enough to arrive on time even though its error rate, $\frac{|E_i - A_i|}{A_i}$ is large. Note that $\frac{|E_i - A_i|}{A_i}$ becomes large when the actual time A_i is small.

The authors regret that this paper cannot disclose more detailed information including the name of the firm since confidentiality is a top concern in the industry due to strong competition. The data were collected for one month before and after SITE was implemented. One month before SITE was implemented, the existing simple average transportation estimation (SATE) was used in the fab. A total of

30,000 transportation jobs were randomly observed out of approximately 230 thousand jobs. While sampling the jobs, we eliminated some of the transportation jobs taking a much longer time than the estimated ones due to unexpected disturbances such as device errors and machine breakdowns.

The authors also compared the overall accuracies between the two methods based on scatter plots for which data were collected from individual jobs with two to nine nodes in their routes. Each scatter plot shows how the actual times are distributed on the y-axis for the given estimated times on the x-axis. As observed, the actual times were more strongly related to the estimated times with SITE, which means that the estimated times with SITE are more accurate than that of existing method.

The authors compared the hit rate of SITE with that the exiting one over different numbers of nodes as shown in Fig. 3. The hit rate is devised to measure the accuracy of the estimation, which counts how many jobs are validly estimated out of the total transportation jobs. An individual job is judged whether it is validly estimated based on the difference between the estimated time and the actual time. The performance of the new method is compared with an existing simple estimation method used in practice based on the hit rate. The hit rate of SITE varied from 83.3 to 90.8 % while the hit rate of the existing one was as low as 44.2 % when the number of nodes was two or three and it increased as the number of nodes increased. In other words, SITE is much more accurate when the number of nodes is smaller. When the number of nodes is greater than 10, the performances of the two methods become similar. The reason for this observation is to explain the distributions of the actual transportation times and estimated times by SITE for the jobs with different numbers of nodes. When the number of nodes is smaller, both distributions tend to be right skewed. As the number of nodes increase, the

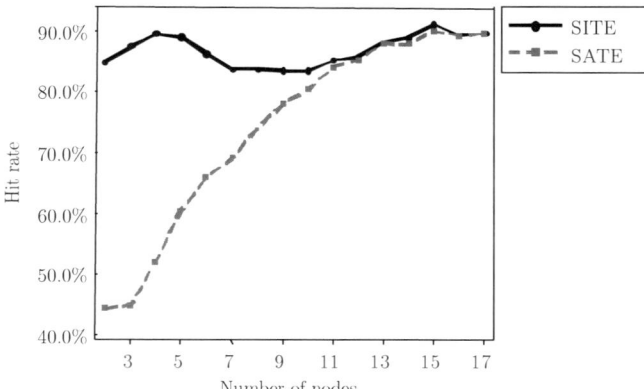

Fig. 3 Hit rate comparison between the new method and existing method

distribution becomes more symmetrical and it eventually approaches to a normal distribution. Hence, we assume that as the number of nodes increases, the estimated time by the existing one becomes more accurate due to the central limit theorem.

4 Conclusion

This paper has developed an arrival time estimation method for transportation jobs based on a stochastic approach for automated material handling system (AMHS) used in manufacturing facilities. The estimated times are provided for production managers through the material control system (MCS) to determine the appropriate priorities of processing jobs so that they are delivered to the next toolsets on time to reduce the starvation of bottleneck toolsets and increase the throughput of the production. The results show that there are improvements in the hit rate which measures the accuracy of the estimation compared to the existing method used at the reference site. Beyond the results reported above, the method has been used for various improvement tools. For example, the procedure helped system engineers to detect errors in the system. This could be readily available by comparing estimated transportation times and actual transportation times. The MCS recorded actual transportation times together with the estimated transportation times when all transportation jobs were completed. At the beginning of the implementation, a non-negligent portion of the transportation jobs had a large gap between the estimated time and its actual delivery time. Many of them turned out to be not from general estimation errors but from bugs in the MCS. The system engineers found several critical errors by investigating the reasons for the gaps and fixed the problems.

References

Campbell PL, Laitinen G (1997) Overhead intrabay automation and microstocking—a virtual production case study. In: Proceedings of IEEE/SEMI advanced semiconductor manufacturing conference, pp 368–372

Chung SL, Jeng M (2004) Productionulous MESs and C/Cs: an overview of semiconductor production automation. IEEE Robot Autom Mag, March 2004, pp 8–18

Li B, Wu J, Carriker W, Giddings R (2005) Factory throughput improvements through intelligent integrated delivery in semiconductor fabrication facilities. IEEE Trans Semicond Manuf 18 (1):222–231

Pfund M, Fowler JW, Gadkari A, Chen Y (2008) Scheduling jobs on parallel machines with setup times and ready times. Comput Ind Eng 54(4):764–782

Poppenborg J, Knust S, Hertzberg J (2012) Online scheduling of flexible job-shops with blocking and transportation. Eur J Ind Eng 6(4):497–518

Suh Y, Chung J (2013) Logistics priority management under multiple performance metrics. J Appl Sci 14(3):245–251

Part III
Strategy and Entrepreneurship

Transportation Infrastructure and Economic Growth Spillovers

Herick Fernando Moralles and Daisy Aparecida do Nascimento Rebelatto

Abstract The present study aims to analyze the spatial spillover effects of transportation infrastructures on the economic growth in Brazilian states through the application of a spatial econometric technique (Spatial Durbin Model). The results demonstrate that road infrastructure has the greatest impact on economic growth, and that the spatial unit neighboring highways are most relevant to their growth than their own infrastructure. Still, some infrastructures showed insignificant or negative spillovers.

Keywords Transportation infrastructures · Economic growth · Spillovers · Spatial econometrics

1 Introduction

The concept of spillover basically stems from the interest to develop theoretical models that include interactions between economic agents, not seeing them as isolated decision makers (Anselin 1988).

The effect of spillover neighborhood is already known in the literature since Jaffe (1986) and Cohen and Levinthal (1989) with respect to the external knowledge. Later, Borjas (1995) and Glaeser et al. (1996) developed the first works that had made efforts to measure spillover effects in the context of spatial models.

H.F. Moralles
Department of Production Engineering, Federal University of São Carlos,
Rodovia Washington Luís, Km 235-SP-310, São Carlos, São Paulo, Brazil
e-mail: herickmoralles@dep.ufscar.br

D.A. do Nascimento Rebelatto (✉)
Department of Production Engineering, University of São Paulo,
Av. Trabalhador São-Carlense, 400 CEP, São Carlos, São Paulo 13566-590, Brazil
e-mail: daisy@sc.usp.br

Although some studies consider the transportation infrastructure as a determinant for economic success are relatively few studies that take into account spatial effects. Among these are Cohen and Monaco (2009) who, using data from counties, evaluated the impact of port infrastructure for retail sales in California through a spatial econometric model.

Such study, besides taking into account issues of spatial and transport infrastructure, also measured the spillover effect (diffusion or external) between counties, noting that the 1 % increase in investments on transport infrastructure of a county causes an addition of 0.22 % in exports of retailers in surrounding counties. This is the essence of the spillover effect, which is equivalent to the effect of one location policies for its neighbors.

Besides, transport infrastructures can be seen as facilitators of economic growth because efficient transport infrastructure expand the productive capacity of a nation by increasing the mobilization of existing resources, as well as improving their productivity (Pradhan and Bagchi 2012).

Also, is necessary to consider the existence of spillovers associated infrastructure to economic growth. Specifically, transport infrastructure spillovers can be negative as pollution, accidents and congestion, or positive, such as facilitation of exports, spatial movement of workforce, improving income distribution.

Studies have increasingly shown the influence of externalities caused by transport infrastructure. More specifically (Hu and Liu 2010) found that an 1 % increase in transport infrastructure causes an elevation of around 0.28 % in Chinese GDP, and of these, 0.22 % is the direct effect, and 0.06 % refers to the external. In the same way, Baum (1998) states that 25 % of German GDP comes from positive externalities related to investment in transport infrastructure. Likewise, Xu and Yang (2007) found high positive correlation between economic development and Chinese transport infrastructure. Specifically, the study revealed cointegrated series and bidirectional relationship between the variables for the years 1992–2005.

Also making use of cointegration tests Zhang (2011) finds a positive relationship between development of civil aviation and economic growth, also checking a bidirectional relationship between variables and elasticities of 0.91 % for the expansion of air in relation to the variation of 1 % in Chinese GDP and elasticity of 0.97 % for the opposite.

In this context, the transport infrastructure acts as a network that interconnects regions, allowing diffusion effects, whether positive or negative, since the migration of factors makes it possible to economic growth in a region result in the decrease of others, by agglomeration effects, for example. This occurs because, to create advantages for the investor region, there is the possible migration of productive inputs to adjacent regions to the investor one (Zhang 2008).

Thus, the knowledge of spillovers effects or externalities occupies a central role in various fields of the economy and it is possible consider, for example, the existence of spillovers of economic growth, spillovers, pollution, and knowledge (Beer and Riedl 2011).

Also, by the perspective of the estimated parameters quality in econometric models, authors like Anselin (1988) show that characteristics of spatial order, such as spatial autocorrelation can lead to estimation problems. Then, beyond the issue of accuracy of estimates, the knowledge of spillover effects (indirect effects or externalities) in the context of spatial econometric models are of great use for public planning (LeSage and Dominguez 2012).

View of these facts, the present study has as main objective to analyze the relationship between economic growth, and transportation infrastructure in the Brazilian states and their spillover effects through the Spatial Durbin Model (SDM), a spatial econometric technique. Such tool will not only enable the calculation of spillover effects, but also provide more accurate parameter estimates.

The choice of transport infrastructure is due to economic and social welfare that provides, as it appears among the main of the main obstacles to economic growth and social development in emerging economies.

The problem to be investigated consists in the verification of the relationship between those variables, which lead to formulate the following research questions: The transport infrastructure of a state significantly affects its own economic growth and the growth of other nearby states? How much? In view of that background, it is clear the intention of the proposed work in measuring not only the relationships between variables, but its dissemination effect for geographically nearby regions (spillover effect).

2 Spatial Durbin Model (SDM)

The SDM was initially found in Anselin (1988) via suggestion Durbin (1960), for the case of serial correlation in time series. Here, there is an additional variable representing the average effect of neighboring locations in one, or a set of independent variables (LeSage and Pace 2009). This specification can be exemplified in matrix form as (2.1).

$$Y = \rho WY + X\beta + WX\gamma + \varepsilon \qquad (2.1)$$

wherein, β, ρ and γ are parameters.

In particular, the γ parameter at WX computes the average impact of the variable X relating to the neighbors of an unit i for variable Y at unit i. Thus, for example, would be equivalent to saying that investment in technological innovation (variable X) of the neighbors of i, affects economic growth (Y) at i.

Starting in 2005 with Brasington and Hite (2005), and Ertur and Koch (2007) model passes to receive more emphasis to the measurement of spillover. However, only in 2010 the publications that used the SDM to measure externalities have intensified, being possible to cite studies like Autant-Bernard and LeSage (2011), LeSage and Dominguez (2012) and Beer and Riedl (2011).

The general form shown in (2.1) could be reduced to other models depending on the statistical significance of the parameters. Thus, if on (2.1) $\gamma = 0$, than the model reduces to a SAR model, if $\gamma = -\rho\beta$ there is a SEM model, and finally, if $\rho = 0$

there is an traditional econometric model (Beer and Riedl 2011; Burridge 1981). However, if the parameters are significant, the model (2.1) computes both direct and indirect impacts (neighboring effect). Here, then, the form of spatial econometric estimation of spillover effects.

The only consideration about this model is that sometimes it may suffer from multicollinearity, so its occasionally advisable to withdraw of $X\beta$ in order to avoid collinearity (Viton 2010; LeSage and Dominguez 2012; Beer and Riedl 2011).

Finally, the natural extension of the SDM to panel data allows increased degrees of freedom and variability of the sample, as well accounting for constant omitted variables in the model, which improves the parameter estimates.

Thus, one way proposed by Beer and Riedl (2011) to control the individual heterogeneity within panels on SDM can be performed via fixed effects, thereby allowing the individual heterogeneity to be correlated with the regressor. Also, it's possible to treat the panel issues via sample demeaning (Viton 2010).

3 Methods

This paper proposes the construction and subsequent estimation of an econometric model with two equations, whose variables are donated in matrix notation, which will aim to answer the stated research questions.

Here, the model to be estimated will be a complete spatial model and therefore encompasses error and spatial lag (SAR and SEM), along with the variables that compute the spillover via spatial Durbin model (SDM).

Given this, the model is based on the logic of the production function, controlling for spatial effects, since studies such as Aiyar and Dalgaard (2009) have shown that the Cobb-Douglas remains an interesting approach. Therefore, the model is presented in (3.1).

$$\ln C_{it} = \rho W_1 \ln C_{it} + \alpha_1 \ln X + \alpha_2 \ln T_{it} + \alpha_3 W_1 \ln X_{it} + \alpha_4 W_1 \ln T_{it} + Fe_i + \varepsilon_1 \quad (3.1)$$

where,

$$\varepsilon_1 = \lambda_1 W_2 \varepsilon_1 + u_1$$

C   Gross Domestic Product (GDP) as economic growth (endogenous variable that requires the use of instrumental variables techniques). Source: IPEA DATABASE

T   Matrix of transportation infrastructures, taking into account road network, rail, ports, and airports. Source: Exame magazine Annual Survey of Infrastructure

W_1 Spatial weighting matrix representing the inverse of the distance between the capitals of each state in order to base the calculation of spillover effects and spatial autocorrelation (SAR)

W₂   Binary neighborhood matrix that trait the spatial error (SEM)
Fe   Represents the panel individual heterogeneity (fixed effect)
X   Matrix of control variables to avoid specification bias in the estimated parameters

Despite not being present in the objectives of this research study, the inclusion of the variable X is of extreme importance to the reliability of the estimated parameters in the model, because the omission of such factors could cause specification bias, which leads to inconsistent parameters model.

The data were collected for each federative unit from years 2004 to 2009 for the variables in question on the Exame magazine Annual Survey of Infrastructure, IPEA DATABASE, and the Brazilian Ministry of Science and Technology (MCT).

The proposed model has characteristics of simultaneous equations with endogenous variables, panel data, and allows the existence of factors such as the characteristic spatial error and spatial lag.

Thus, for estimation of the model carried the following steps.

1. Treatment of the individual heterogeneity of the panel via sample demean.
2. Removal of heteroscedasticity via use of natural logarithms, which also allows us to interpret the estimated coefficients in terms of elasticities.
3. Construction of spatial weighting matrices for panel data. At this stage we use the software STATA MP12, which has a routine capable of transforming a cross-section spatial weight matrix into an array for panel data (Drukker et al. 2013). For example, considering the matrix of binary neighborhood of three general locations A, B, and C to data cross-section in Fig. 1.

Nevertheless, for panel data, taking into account a three-year time series for each entity, the resulting matrix is the block diagonal one on Fig. 2.

	A	B	C
A	0	1	0
B	1	0	1
C	0	1	0

Fig. 1 Neighborhood binary matrix for cross-section data

Fig. 2 Neighborhood binary matrix for panel data

	A1	A2	A3	B1	B2	B3	C1	C2	C3
A1	0	0	0	1	0	0	0	0	0
A2	0	0	0	0	1	0	0	0	0
A3	0	0	0	0	0	1	0	0	0
B1	1	0	0	0	0	0	1	0	0
B2	0	1	0	0	0	0	0	1	0
B3	0	0	1	0	0	0	0	0	1
C1	0	0	0	1	0	0	0	0	0
C2	0	0	0	0	1	0	0	0	0
C3	0	0	0	0	0	1	0	0	0

So that for each entity, the panel data matrix grows with the length of the time series in rows and columns; noting also that the diagonal blocks are entirely composed of zero elements.

4. Estimation of the model with the addition of variables that capture the spillover effects via maximum likelihood using MATLAB 2009. Spatial econometric models are usually estimated using the maximum likelihood logic, as described in Anselin (1988) in an iterative algorithm similar to the procedure Cochrane-Orcutt for correlation in time (Cochrane and Orcutt 1949; Gujarati 2006).

In fact, there are programs developed for the purposes set, which should act as drivers. These are packages for spatial econometrics on STATA® and MATLAB® are the basis for the script used in this study.

4 Results

The estimated parameters of the model are shown in Table 1.

In Eq. (3.1) with respect to the own transport infrastructure, there was a positive and significant effect of highways 0.188 and a significant and negative effect of railroads. The port infrastructure results indicates insignificance for economic growth, however, the airport infrastructure had positive and significant effect of the order of 0.020 on average for the Brazilian states.

Now, with respect to spillover effects, there were statistically significant variables like highways 0.751, railways (−0.124), and airports 0.137. These results demonstrate that, for example, a 1 % increase in length of the highway infrastructure investments from one state causes an increment of 0.188 % in its GDP, and an impact of 0.751 % in GDP of their neighbors in average to Brazil.

Table 1 Model (3.1) estimated parameters via maximum likelihood

Explanatory variables	Growth (C)	
	Coefficient	p-value
Road extension	0.188	0.020
Railways extension	−0.019	0.016
Cargo movement in ports	−0.001	0.638
Cargo movement in airports	0.02	0.017
W*Road extension	0.751	0.070
W*Railways extension	−0.124	0.007
W*Cargo movement in ports	−0.008	0.300
W*Cargo movement in airports	0.137	0.008
Rho	0.082	0.133
Lambda	−0.002	0.954

This greater neighbor effect to GDP is probably due to the fact that neighboring infrastructure configures a restriction in the flow of production of a spatial unit. Thus, the surrounding road infrastructure can configure a *sine qua non* condition for economic growth in a given Brazilian state.

Lastly, for the existence of spatial effects like spatial lag and spatial error, the spatial effects were not significant at the usual level of 10 %, however, for the term of spatial dependence, the p-value was 0.133, and therefore, unadvisable its exclusion, in the sense of causing specification bias. Accordingly, the interpretation of this term follows the logic that an increase of 1 % in the GDP of a state unit causes an increase of the order of 0.082 % on their neighbors GDP.

5 Conclusions

This study is one of the few to use panel data in a spatial model, also relatively new area in the literature and lack of references. Along with the spillovers and using a full model (with lag and spatial error), it's possible to verify the intent to address a more complete approach proposed by this investigation, as well as his effort to consider these characteristics in the estimates.

Regarding transport infrastructure, the road network was, as expected, the infrastructure with greater elasticity for economic growth, both for its own infrastructure, and for the spillover effects. These results reflects the historical decision and the importance of this infrastructure for the Brazilian socio-economic organization, although other forms of transport have advantages widely known.

Thus, the negative results for the railways infrastructure, hence, may have been justified by the deterioration of such infrastructure, as well as its underuse over the highways. Therefore, the railway infrastructure should not be disregarded as an alternative for transport in Brazil.

For Eq. (3.1) of economic growth, both effects error and spatial lag were insignificant. However, the spatial lag returned a p-value of 0.133, which although insignificant to traditional standards of 10 %, should not be overlooked, showing hence that the increase in GDP of a region impacts the growth of its neighbors. Thus, the increase in economic activity in a region causes positive effects on neighboring economies.

References

Aiyar S, Dalgaard CJ (2009) Accounting for productivity: Is it OK to assume that the world is Cobb-Douglas? J Macroecon 31(2):290–303
Anselin L (1988) Spatial econometrics methods and models. Springer, Dordrecht
Autant-Bernard C, LeSage JP (2011) Quantifying knowledge spillovers using spatial econometric models. J Reg Sci 51:471–496
Baum JAC (1998) Social benefits of road transport. Mimeo, Cologne

Beer C, Riedl A (2011) Modelling spatial externalities in panel data: the spatial durbin model revisited*. Papers in Regional Science 91(2):299–318

Borjas GJ (1995) Ethnicity, neighborhoods, and human-capital externalities. Am Econ Rev 85:365–390

Brasington DM, Hite D (2005) Demand for environmental quality: a spatial hedonic analysis. Reg Sci Urban Econ 35:57–82

Burridge P (1981) Testing for a common factor in a spatial autoregression model. Environ Plann A 13:795–800

Cochrane D, Orcutt GH (1949) Application of least squares regression to relationships containing auto-correlated error terms. J Am Stat Assoc 44:32–61

Cohen WM, Levinthal DA (1989) Innovation and learning: the two faces of R&D. Econ J 99:569–596

Cohen J, Monaco K (2009) Inter-county spillovers in California's ports and roads infrastructure: the impact on retail trade. Lett Spat Resour Sci 2(2):77–84

Drukker DM, Peng H et al. (2013) Creating and managing spatial-weighting matrices with the spmat command. Stata J 13(2):242–286

Durbin J (1960) Estimation of parameters in time-series regression models. J Roy Stat Soc Ser B (Methodol) 22:139–153

Ertur C, Koch W (2007) Growth, technological interdependence and spatial externalities: theory and evidence. J Appl Econometrics 22:1033–1062

Glaeser EL, Sacerdote B, Scheinkman JA (1996) Crime and social interactions. Q J Econ 111:507–548

Gujarati DN (2006) Econometria Basica, Campus, Rio de Janeiro

Hu A, Liu S (2010) Transportation, economic growth and spillover effects: the conclusion based on the spatial econometric model. Front Econ China 5:169–186

Jaffe AB (1986) Technological opportunity and spillovers of R&D: evidence from firms' patents, profits, and market value. Am Econ Rev 76:984–1001

LeSage J, Dominguez M (2012) The importance of modeling spatial spillovers in public choice analysis. Public Choice 150:525–545

LeSage JP, Pace RK (2009) Introduction to spatial econometrics. CRC Press, Boca Raton

Pradhan RP, Bagchi TP (2012) Effect of transportation infrastructure on economic growth in India: the VECM approach. Research in Transportation Economics 38:139–148

Viton PA (2010) Notes on spatial econometric models. City Reg Plann 870(03)

Xu HC, Yang JL (2007) Relationship between highway transportation and economic development in China. J Chang'an Univ (Soc Sci Ed) 9:8–13

Zhang X (2008) Transport infrastructure, spatial spillover and economic growth: evidence from China. Front Econ China 3:585–597

Zhang WB (2011) Interregional economic growth with transportation and residential distribution. Ann Reg Sci 46:219–245

Influence of the Environmental Factors in the Creation, Development and Consolidation of University Spin-offs in the Basque Country

E. Zarrabeitia Bilbao, P. Ruiz de Arbulo López
and P. Díaz de Basurto Uraga

Abstract The university entrepreneurship, together with its highest representative, the university spin-off, are a key instrument to technological innovation and economic development of a territory. Therefore, due to the importance of these companies for the future sustainable development of the regions, the study of the factors that influence the creation, development and consolidation of university spin-offs is really important. Specifically, the aim of the present study is to analyze the importance of various environmental factors in the creation and development of Basque university spin-offs, and the presence of these factors in the specific environment of the Basque Country, in general, and according to the companies development stage, in particular.

Keywords University spin-offs · Entrepreneurship · Basque Country

1 Context of Research

In today's knowledge society, universities and research centres play an increasingly important role in the economic growth (Corti and Riviezzo 2008). This is due, in large part, because in the era of knowledge economy, society requires the university more than the creation, transmission and dissemination of knowledge. Society also demands the university that its capacity of generating scientific and technical knowledge not remain in the closed circle that represent public institutions of higher research, but that it becomes a source of revitalizing the entrepreneurship and the transfer of knowledge to a productive economy (Dirección General de Política de la Pequeña y Mediana Empresa 2006).

E. Zarrabeitia Bilbao (✉) · P. Ruiz de Arbulo López · P. Díaz de Basurto Uraga
University or the Basque Country, Alameda Urquijo s/n, 48013 Bilbao, Spain
e-mail: enara.zarrabeitia@ehu.es

Precisely, at a time when social values that sustain us are breaking because of a multidimensional crisis in the institutional field, and in a complex situation in the economic field (De Pablo 2011), it is not enough that universities carry out research and publish the results. Universities must take the mission to promote economic development of the geographical environment in which they are located, being necessary to do so, the commercial application of knowledge generated (Condom and Valls 2006).

In this regard, the modalities traditionally used by the university to transfer technology to the private sector, and ultimately to society, have been patents and research contracts. However, in recent decades, a pathway that has been gaining importance because of its contribution to knowledge transfer, and its economic impact, has been the creation of companies (Rodeiro 2008; Trenado and Huergo 2007).

These companies, known as university spin-offs, have particular interest because they drive local economic growth and development, generate economic returns for the university, make possible the commercialization of certain technologies developed at the university, increase the interaction between the university and its environment, reorient education and research, and bring changes in the culture of the university (Rodeiro et al. 2011; Shane 2004).

These reasons, among others, have promoted that the university entrepreneurship, along with its highest representative, the spin-off, is recognized as a key instrument for the technological innovation and economic development of a territory, and the increase of policies and fostering activities for this type of university companies, from the public administration and the universities themselves (Rodeiro 2008).

However, although it is something accepted that the creation of university spin-offs in a region is a source of wealth and development, the fact is that if different regions are analyzed, it shows that the development of this type of companies is disparate (González and Álvarez 2006). This is because in addition to the desire to create university spin-offs, on the one hand, there are a number of ingredients that generate a optimal broth that fosters their creation, installation and development in a region (Goñi and Madariaga 2003); and on the other hand, there are certain critical factors, inherent of universities and spin-offs, that encourage the birth, growth and proliferation of these companies.

Thus, the creation and successful consolidation of university spin-offs, is a phenomenon influenced, on one hand, by the particular conditions of each university and its environment, and moreover, by the characteristics and particularities of the own spin-offs and entrepreneurs (Etzkowitz 2003) (see Table 1).

2 Justification and Research Goals

Although the creation of new businesses in general, and university spin-offs in particular, is the result of the decision of an entrepreneur to undertake an entrepreneurial activity, these companies are created and developed in environments

Table 1 Determinants of the emergence and development of university spin-offs

MACRO FACTORS: *Environmental characteristics of the university*
–Legal framework in force
–Supporting mechanisms from public institutions
–Industrial, technological and entrepreneurial environment
–Venture capital
MESO FACTORS: *Institutional determinants and organizational resources of the university*
–Cultural norms of the university
–Nature of the research
–Level and nature of financial resources for research
–Prestige of the university and quality of the research carried out
–Mechanisms of sensitization, education and attracting entrepreneurs
–Strategies and structures for the development and management of university spin-offs
MICRO FACTORS: *Determinants of the skills and attitudes of entrepreneurs and of the development of university spin-offs*
–Socio-demographic characteristics of entrepreneurs
–Psychological characteristics of entrepreneur
–Characteristics of university spin-offs
–Development stage of university spin-offs

Source Aceytuno and Cáceres (2009), Aceytuno and de Paz (2008), Djokovic and Souitaris (2008), O´Shea et al. (2008)

where act other actors that facilitate or encourage these processes more or less directly. Therefore, and due to the importance of the spin-offs for the future sustainable development of the Basque Country, the study of the factors that influence the creation, development and consolidation of the Basque university spin-offs, is an important topic.

Specifically, with the present work it is intended to analyze the importance of various environmental factors in the creation and development of Basque university spin-offs, and the presence of these factors in the Basque Country, in general, and according to the companies development stage, in particular.

3 Scope of the Study and Research Methodology

First, on the one hand, it should be noted that the Basque University System is composed of three universities: the University of the Basque Country (UPV/EHU), the University of Deusto and the University of Mondragon. However, these universities have a history and a different influence within the educational framework of higher education of the Basque Country, being significantly higher the UPV/EHU in the number of students, programs offered, faculty, administration personnel and knowledge generated (Departamento de Educación, Universidades e Investigación

Table 2 Technical details of the research

Sampling universe	91 Spin-offs from the University of the Basque Country
Sampling procedure	Computer assisted survey through a Web formulary
Response rate	44 %
Sample size	40 spin-offs
Sampling error	11.66
Confidence level	95 %
Field work	October 2012–December 2012

2012; Departamento de Educación, Universidades e Investigación 2009). Moreover, it is the only university with a presence in the three provinces of the Basque Country.

Also, even though the three universities have resources to promote and support the creation of university spin-offs, the number of spin-offs created in the UPV/EHU is significantly higher than in the other two universities.

For all these reasons, this study is limited in its scope and extension to the 91 university spin-offs generated from the UPV/EHU, up until October 2012. However, the results of this research can be largely extrapolated to the other two universities of the Basque University System (Table 2).

Moreover, thanks to those responsible of the university entrepreneurship programs at the UPV/EHU, it has been able to identify the entrepreneurs who were sent a questionnaire. That is, it has identified the researchers that have decided to create a spin-off from the UPV/EHU and are linked to the company (provided the spin-off is active or has not been sold). In cases where the spin-off has more than one entrepreneur, it has been selected a researcher who has discovered, evaluated and decided to exploit the business opportunity, i.e., the researcher who has been more involved in the whole process, from the generation of the idea, to the creation and development of the spin-off.

4 Results

4.1 Importance of external factors for the development of Basque university spin-offs and the presence of them in the environment of the Basque Country

Figure 1, shows the results of the Basque university spin-offs in relation to the importance of the external or environmental factors highlighted in the literature for the generation and development of these enterprises, as well as the presence of these factors in the environment of the Basque Country.

It is observed that Basque university spin-offs consider the factors related to the legal framework, the supporting mechanisms from public institutions, the industrial,

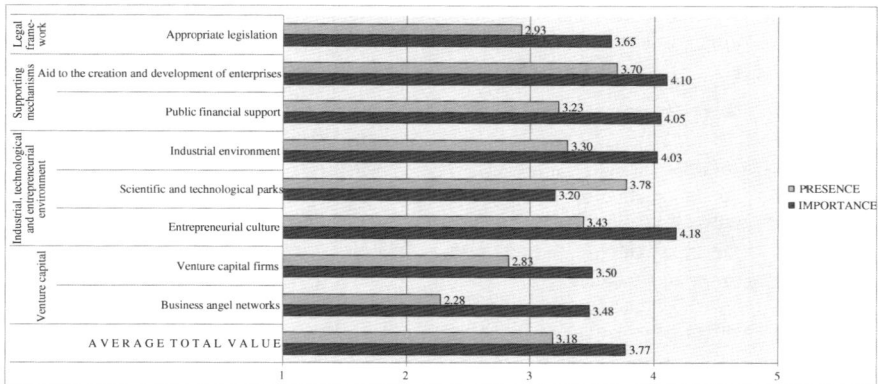

Fig. 1 Importance of environmental factors in the creation and development of university spin-offs and their presence in the Basque Country

technological and entrepreneurial environment, and the venture capital, are important for the creation and development of these enterprises. All the factors analysed get a score higher than 3 points in a Likert scale from 1 to 5, from not very important to very important (see Fig. 1).

Specifically, the environmental factors considered the most important environmental factors for the creation and the growth of these firms are: the entrepreneurial culture, the strategies and programs to support the creation and development of enterprises, the public financial support, and the existence of a suitable industrial environment (with a value higher than 4 points).

However, Fig. 1 shows a clear difference between the importance of the factors analyzed and the valuation of its presence in the Basque Country. Specifically, stands out with a difference of more than one point, the importance and presence of business angels networks.

4.2 Importance of external factors for the development of Basque university spin-offs, and the presence of them in the environment of the Basque Country according to the development stage of the company

In the study of the importance and presence of environmental factors depending on the development stage of the companies, it is observed, on one hand, that the spin-offs that are in the creation phase are the ones that give more importance to the environmental factors analyzed; and nevertheless, on the other hand, the companies that are on the other extreme, i.e., those that are consolidated, are the ones that value less the presence of these factors in the environment of the Basque Country (Tables 3 and 4).

Table 3 Importance of environmental factors in the creation and development of university spin-offs according to the development phase

Environmental factors (importance)		Development stage of the university spin-off			
		Creation	Growth and consolidation	Consolidated	Without activity
Legal framework	Appropriate legislation for the creation of university spin-offs	4.00	3.70	3.20	3.67
Mechanics	Strategies and programs to support the creation and development of enterprises	4.50	4.00	4.60	4.00
	Public financial support	4.50	4.07	4.20	3.33
Environment	Industrial environment	4.50	4.03	4.20	3.33
	Scientific and technological parks	3.50	3.27	2.60	3.33
	Entrepreneurial culture	5.00	4.03	4.80	4.00
Venture capital	Venture capital firms	3.50	3.50	4.20	2.33
	Business angel networks	4.00	3.50	3.80	2.33
Average total value		**4.19**	**3.76**	**3.95**	**3.29**

Table 4 Evaluation of the presence of environmental factors in the Basque Country according to the development phase of the spin-offs

Environmental factors (presence—Basque Country)		Development stage of the university spin-off			
		Creation	Growth and consolidation	Consolidated	Without activity
Legal framework	Appropriate legislation for the creation of university spin-offs	3.00	2.93	2.40	3.67
Mechanics	Strategies and programs to support the creation and development of enterprises	3.50	3.73	3.00	4.67
	Public financial support	3.00	3.13	3.40	4.00
Environment	Industrial environment	4.50	3.33	2.80	3.00
	Scientific and technological parks	3.50	3.90	3.40	3.33
	Entrepreneurial culture	2.50	3.43	3.20	4.33
Venture capital	Venture capital firms	2.50	2.97	2.40	2.33
	Business angel networks	2.00	2.33	1.80	2.67
Average total value		3.06	3.22	2.80	3.50

5 Conclusions

The phenomenon of the Basque university spin-offs is considered a reflection of the organizational behaviour and the organizational resources of the university, because the creation of these companies is not a fact that spontaneously happened in different faculties. However, the process of spin-offs, since the idea developed or materialized to its launch and its subsequent development, is a complex process where become very important various factors that are external to the university, like: the entrepreneurial environment, the programs to support the creation and development of enterprises, the financial support, and the industrial environment.

However, if a true direct contribution is desired by the University of the Basque Country to the economic development of the territory through the promotion of knowledge-intensive companies, it will be necessary to develop appropriate policies and actions to reduce the gap between the important environmental factors for these enterprises and the presence of these environmental factors in the Basque Country.

A key factor being the financial resources, especially in the case of companies that are small, recently created and with innovative or technological nature, conditions that generally meet university spin-offs (Rodeiro 2008), according to the opinion of entrepreneurs of the spin-offs, stand outs in the environment of the Basque Country the scarce presence of business angel networks and venture capital companies. These companies face particular difficulties in obtaining financing through traditional instruments, due to the inherent risks in new technologies that they developed, their limited supply of capital, and the lack of tangible assets that can serve as additional guarantee (Aguado et al. 2002). Therefore, other forms of financing such as formal or informal venture capital are necessary.

In this regard, it is noteworthy that the Law 16/2012 of 28 June, to Support Entrepreneurs and Small Business of the Basque Country, regulates the figure of private informal investors "business angels" through a directory which eliminate possible uncertainties in this form of financing, and supports networks of business angels that will act as points of encounter and of adjustment between the supply and the demand for capital among those who invest and those who undertake. Thus, it is expected that the valorisation of the presence of the networks of business angels in the Basque Country will increase, and also the access to them as a funding source. Nevertheless, as of today no action has been carried out thereon by the Basque Government. There is a lack of policies or actions that will help or foster these networks that will facilitate the development of spin-offs.

Also, though it is in the creation phase when is given more importance to the environmental factors that are considered important for the successful development of spin-offs, the companies that are consolidated are the ones that least value their presence. So, it should be kept in mind that these companies, despite being companies that can provide significant added value, are fragile companies due to their small size, which must continue be helped once they are established. The various actions should be addressed not only to the creation of these companies, but also to the development and successful growth of them.

References

Aceytuno MT, Cáceres FR (2009) Elementos para elaboración de un marco de análisis para el fenómeno de las spin-offs universitarias. Revista de Economía Mundial 23:23–52

Aceytuno MT, de Paz MA (2008) La creación de spin-off universitarias. El caso de la universidad de Huelva. Economía Ind 368:97–111

Aguado R, Congregado E, Millán JM (2002) Entrepreneurship, financiación e innovación. La situación en la Unión Europea. Economía Ind 347:125–134

Condom P, Valls J (2006) La creación de empresas spin-off en las universidades emprendedoras. Alta Dirección 241–242:192–202

Corti E, Riviezzo A (2008) Hacia la universidad emprendedora. Un análisis del compromiso de las universidades italianas con el desarrollo económico y social. Economía Ind 368:113–124

De Pablo I (2011) Universidad emprendedora, sociedad emprendedora. Asociación para el progreso de la dirección 268:20–21

Departamento de Educación, Universidades e Investigación (2009) Información general del Sistema Universitario Vasco. Servicio central de publicaciones del Gobierno Vasco, Vitoria

Departamento de Educación, Universidades e Investigación (2012) Plan Universitario 2011–2014. Servicio central de publicaciones del Gobierno Vasco, Vitoria

Dirección general de política de la pequeña y mediana empresa (2006) Iniciativas emprendedoras en la universidad española. Ministerio de Industria, Turismo y Comercio, Madrid

Djokovic D, Souitaris V (2008) Spinouts from academic institutions: a literature review with suggestions for further research. J Technol Transfer 3:225–247

Etzkowitz H (2003) Research groups as quasi-firms. The invention of the entrepreneurial university. Res Policy 1:109–121

González O, Álvarez JA (2006) Análisis de los factores que explican la creación de spin off en las universidades españolas. XV Jornadas de la Asociación de Economía de la Educación, Granada, pp 139–152

Goñi B, Madariaga I (2003) Las empresas innovadoras de base tecnológica como fuente de desarrollo económico sostenible. V Congreso de Economía de Navarra, 19–20 de noviembre, Pamplona

Ley 16/2012, de 28 de junio, de Apoyo a las Personas Emprendedoras y a la Pequeña Empresa del País Vasco; BOE no 172 de 19 de julio de 2012

O´Shea RP, Chugh H, Allen TJ (2008) Determinants and consequences of university spinoff activity. A conceptual framework. J Technol Transfer 6:653–666

Rodeiro D (2008) La creación de empresas en el entorno universitario español y la determinación de su estructura financiera. Tesis Doctoral, Universidad de Santiago de Compostela

Rodeiro D, Fernández S, Vivel MM (2011) Universidad emprendedora. La creación de spin-offs. In: Fernández S, Rodeiro D (coords) El emprendimiento femenino en el sistema universitario español y gallego. Un análisis económico-financiero. Universidad de Santiago de Compostela, 73–122

Shane S (2004) Academic entrepreneurship: university spin-offs and wealth creation. Edward Elgar, Cheltenham

Trenado M, Huergo E (2007) Nuevas empresas de base tecnológica. Una revisión de la literatura reciente. Centro para el Desarrollo Tecnológico Industrial y Universidad Complutense. Documento de trabajo 03

Conceptual Model for Associated Costs of the Internationalisation of Operations

Ángeles Armengol, Josefa Mula, Manuel Díaz-Madroñero and Joel Pelkonen

Abstract We propose a conceptual model for a representative cost structure associated to the internationalisation of operations. The main contribution of this paper for researchers and industry practitioners is the identification of cost issues in order to reduce the number of unexpected costs related with companies that have internationalised their operations. This conceptual model could serve as a starting point for cost challenges and issues that firms will face when they get involved in the internationalisation of operations.

Keywords Internationalisation of operations · Production networks · Cost structure

1 Introduction

The internationalisation of operations has become a common trend for companies and this has further confirmed the necessity for strategies that continuously enable the renewal of capabilities to adapt to this global competitive environment (Mediavilla et al. 2012). McIvor (2005) shows that the cost of transferring manufacturing to other locations often is underestimated, as well as the increased logistic costs for transportation and stock. Other studies have considered these topics in different approaches such as Hammami et al. (2008), Stringfellow et al. (2008) and Sethi and Judge (2009).

This research has been funded by the Ministry of Science and Education of Spain project, entitled 'Operations Design and Management in Global Supply Chains (GLOBOP)' (Ref. DPI2012-38061-C02-01).

Á. Armengol · J. Mula (✉) · M. Díaz-Madroñero · J. Pelkonen
Centro de Investigación en Gestión e Ingeniería de Producción (CIGIP),
Escuela Politécnica Superior de Alcoy, Plaza Ferrándiz y Carbonell, 2,
03801 Alcoy, Alicante, Spain
e-mail: fmula@cigip.upv.es

Due to lack of experience, knowledge or reasons given above firms going abroad often face unexpected costs and surprising phenomena, barriers or resistance (Forsgren 2002; Johanson and Vahlne 1977) when internationalizing operations. The management of an integrated global operations network could be an upcoming challenge for the operations management (Ernst and Kim 2002). It was stated by Errasti (2011) that several facts have made it necessary to configure and manage more complex production and logistics networks.

Here, a particular emphasis is made regarding to the associated costs, the purpose being to summarize and gather in a conceptual model the cost structure of internationalized operations in the supply chain. The intention of this conceptual model is to identify the main costs based, mainly, on a literature review related to internationalisation of operations.

2 Literature Review

Colotla et al. (2003) suggest the need for developing strategy processes to help companies reconcile both levels, addressing the strategic implications of the interaction and interplay of factory and network capabilities, and the strategic nature of capability building over time. The assessment role, for example, can be determined by using the framework by Ferdows (1997). According to Christopher (2005), cost accounting is a tool for managers to allow them plan and control costs identifying cost drivers and supporting decision making in the internationalisation process. In Errasti (2011) is stated that businesses suffer lack of visibility of costs as they are incurred through the logistics pipeline. The advantage cycle proposes that the size and composition of the package changes over time (Johanson and Vahlne 1990). It is known that direct costs in the internationalisation concept are easier to detect and define than indirect costs. For example, the costs relating to labour, transportation, material, energy and facilities are very basic, and many reviewed papers name and describe them. Basically, customs tariffs and money exchange rates belong to the same group, despite them varying substantially. It is important to highlight that monetary risks could have a sudden and unpredictable impact in the internationalisation of operations (Elango 2010; Errasti 2011). It is important to take into account the product life cycle analysis (Asiedu 1998). Moreover, these costs can increase as a consequence of the internationalisation of operations. Now, we describe with further detail the main cost aspects identified and incorporated to our model.

Home country costs. When internationalisation of operations and offshoring projects are carried out, original employees might lose their jobs in their home country, as stated by Sommer and Troxler (2007). Bock (2008) especially considers lower skill level tasks since they are easier for the firm to transfer. On the other hand, interesting insight regarding multi-location versus relocation and the effects in the employment are provided by Luzarraga and Irizar (2012). Furthermore, according to Hammami et al. (2008) and Gomes Salema et al. (2010), if a company leaves a domestic part of its supply chain outside the network because of internationalisation, this is negative for the country and the firm must also face the costs incurred in relation to this issue. Sethi and Judge (2009) realise that costs, or at least strategic limitations to a subsidiary abroad, are incurred by its connection with the parent multinational enterprise. Hutzschenreuter et al. (2011) propose that international operations may have such an impact on the firm's future due to the dynamic adjustment costs in the new environment. This phenomenon is also known as the Penrose effect.

Interaction distance costs. Stringfellow et al. (2008) develops a conceptual framework were interaction distance is included, which considers cultural distance, language distance and geographical distance. Moreover, Luo and Shenkar (2011) examine the cultural friction concept and note that the entry mode to the foreign environment has an impact on the cultural friction level. As Stringfellow et al. (2008) also state, invisible costs are influenced by interaction intensity and interaction distance.

Host country costs. Cuervo-Cazurra et al. (2007) note a disadvantage where, for instance, a government can discriminate the foreign firm based on its nationality or by transferring technical and managerial systems routines. Lack of complementary resources also belongs to the host country category. Another root cause for more indirect costs is the lack of local roots and relationships in the host country environment. Sethi and Judge (2009) note that this forms part of liabilities of foreignness and proves costly for firms. Also, the authors state lack of legitimacy and economic nationalism against foreign firms as indirect internationalisation costs. This is also mentioned by Kedia and Mukherjee (2009), who talk about economic prejudice. Moreover, there are also political risks of the region or nation (Errasti 2011) as well as pollution and environmental costs, in addition to the costs of monitoring trade policies and governmental issues. These aspects are examined more specifically by Krikke et al. (2003) and Sommer and Troxler (2007), respectively.

Firm infrastructure and facilities location costs. There are some costs drivers as for example when deciding to relocate a production in a low cost country. Aybar and Ficici (2009) note loss of advantage in their research and they suggest that the causes for this phenomenon can be very simple; e.g., lack of experience, and organizational inertia or laziness. Errasti (2013) has out lighted that the process and management system modification and adaptation to local characteristics such as labour costs, equipment, maintainability, product demand variety and volumes, suppliers local network, is a need not always taken into account.

Human resource management costs. Given the need for deeper relationships, collaborative initiatives (Klassen and Vereecke 2012) might require a much longer time horizon to develop, implement and yield performance benefits than monitoring capabilities. It is important to take into account literature streams related to social issues in the supply chain (Klassen and Vereecke 2012).

Technology costs. Cuervo-Cazurra et al. (2007) examine the loss of advantage in international expansion. Concepts are divided into firm-specific losses, meaning that the value of a resource is advantageous in any location just a limited time, and also into non firm-specific losses, where the resources transferred to a new country are not as advantageous given the new environment. Fuchs and Kirchain (2010) note that because of the increased cost-advantage of a prevailing technology used abroad, the firm's technology development generally lowers during operations.

Procurement costs. Purchasing policies depend on the potential risk/benefit of the purchase decision and the power relation between suppliers and purchasers (Errasti 2011). The factors to be considered are: cost of non delivery; supply interruption probability due to logistical route, supplier social or economical problems or quality failure; benefits due to logistical, quality and information and communication technology integration; and benefits due to better product specification or new developed products. Another connectivity cost was highlighted by Errasti (2011) who applies to this category the costs of generating new competitors from new suppliers.

Inbound and outbound logistics costs. The transportation cost has a significant impact on optimizing the global supply chain when transportation occurs across borders (Lee and Wilhem 2010). Errasti (2011) states the total cost of ownership is a methodology and philosophy which looks beyond the price of a purchase to include many other purchase related costs such as procurement and quality.

Operations costs. Cuervo-Cazurra (2011) discusses that international expansion is easier and faster for smaller firms in general, because they have exhibit established routines and more flexibility in their operations. The authors (Mendibil and MacBryde 2005; Rudberg and West 2008) state that the economical and social sustainability of operations depends mostly of the adopted productive models and plant specialization, they also mention five types of productive models: volume, volume and variety, continuous cost reduction, innovation and flexibility, variety and flexibility. Some processes and cost drivers are identified in Errasti (2011).

Marketing sales costs. As we indicated earlier there is a growing dissatisfaction with conventional cost accounting, another of them is that there is a general ignorance of the true costs of servicing different customer types/channels/market segments.

Service costs. Nordin (2005), searching for the optimum product service distribution channel, examines the actions of five industrial firms. The author suggests that the more the service offer is strategic and customized, the more important is "to retain service processes internally or to align with external partners in close relationships" However, setting up a totally owned network to meet customers' requirements can be very expensive, especially in case of very large customer bases.

Social responsibility costs. Vallaster et al. (2012) claim that multiple conceptualizations of corporate social responsibility (CSR) exist. CSR is about companies going beyond legal obligations and their own interests to address and manage the impact their activities have on society and the environment (Reverte 2012).

Customer costs. There is also a growing dissatisfaction with conventional cost accounting, particularly as it relates to logistics management. The one that is related to the current category is that companies understand product costs but not customer costs.

3 Modelling Methodology

In order to build the proposed cost model a constructive research approach has been applied. Regarding to Malhotra and Grover (1998), the research methodology used to create the proposed conceptual model is the result of a combination of conceptual, descriptive, empirical, exploratory cross-sectional, exploratory longitudinal research based on the literature review. It relies on experience and observation from the cases studies used in the review. It sets a general structure of future research lines in order design tools that help calculating and assessing parameters, assisting in the decision making, calculation of performance and other themes identified in the proposed cost model. In this case, 2 phases (research and model construction) are accomplished. The information used to build and shape the proposed cost model is based in data extracted from the reviewed papers. Here, it is considered that one member of the supply chain is got the main power. In this assumption, the supply chain purpose is to improve the performance of the total supply chain.

4 Conceptual Model

Two levels are created in the current model. Level 0 entails the general fields that allocate most of internationalisation costs (Fig. 1). There is also a detailed Level 1, where costs are classified and described (Fig. 2). At level zero, three main bodies are defined. Starting from the top, in the trapezoid, some aspects and features that define an enterprise can be found. In the middle part, the internationalisation

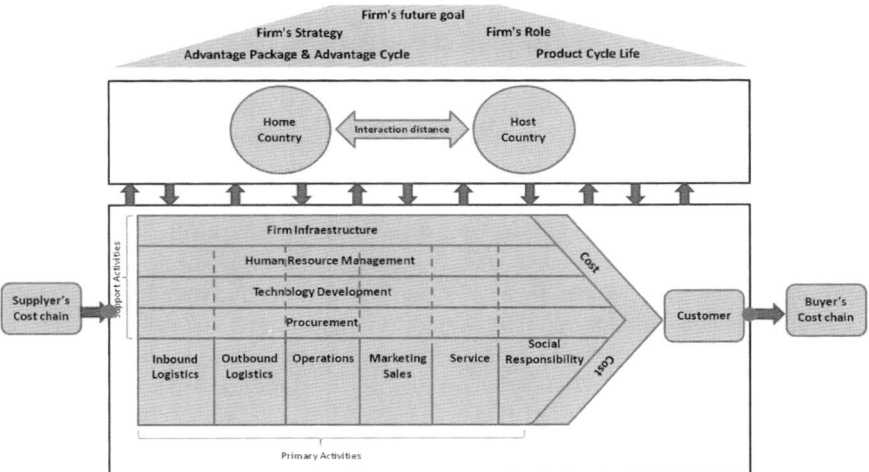

Fig. 1 Cost structure model for internationalisation of operations. Level 0

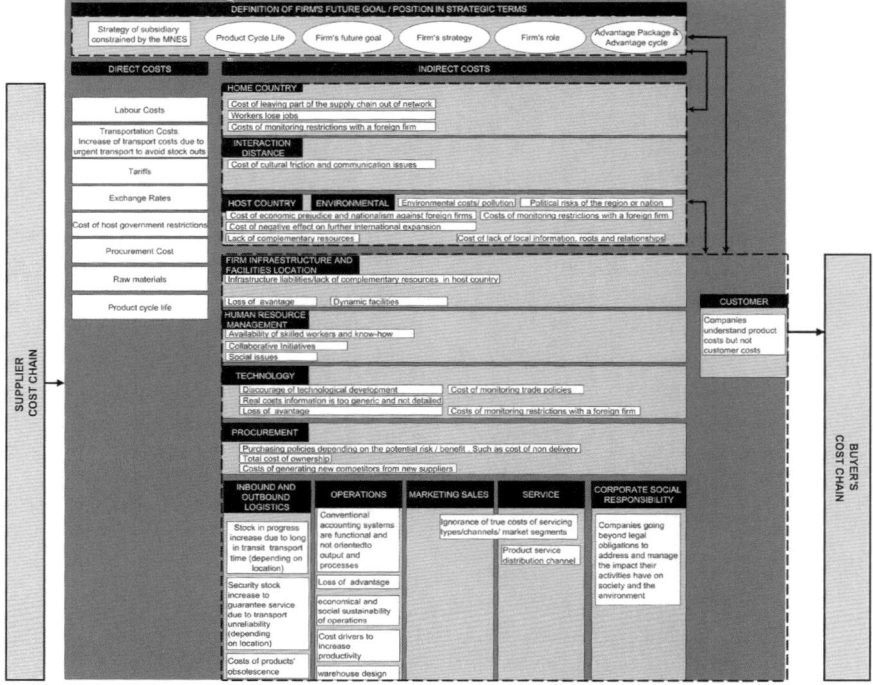

Fig. 2 Cost structure model for internationalisation of operations. Level 1

approach is introduced, and at the third body other generic activities that are carried out in a traditional company are described. In this case, were the international environment is taking place, the value system is used as a basis to start building the cost model (Porter 1991). Nevertheless, in the current model the concept of value is replaced by costs. Therefore, in the first body, general decision making aspects and facts relating to the costs in the internationalisation of operations are described. These decisions have an effect in all levels in the company and all levels in the proposed cost model. In the second and third body, the costs are linked in more direct way to activities. Costs are described in the detailed model. At the first part, the aspects that define the enterprise are: firm's future goal, firm's strategy, firm's role, advantage package and advantage cycle and the last but not less important is the product cycle life. At the second body, the home country, host country and interaction distance aspects can be found. These aspects are described in the detailed level. The third body is based on the primary activities and support activities by Porter (1985). Meanwhile in the proposed model, it is considered that the same activities that create value, they create costs too. At this stage, one activity was added to the primary activities, which is social responsibility.

5 Conclusions

Over the past few decades, internationalisation literature has contributed to defining and characterizing the internationalisation process, as well as indirectly informing about its costs and consequences, either in service or manufacturing environments. However, how to calculate certain costs had remained largely unexplored in the internationalisation of operations context. The features of the proposed model are extracted from a previous literature review. This is a general model that can be applied to any firm that is involved in the internationalisation of operations or wants to start it. Moreover, due to difficulties and their root causes in internationalisation, and since this research area is very broad, some individual aspects are examined merely on the surface level. Thus more in-depth research into these areas could be the basis for further research. Therefore, further research could be oriented to internationalisation and quantitatively-analysed decision tools based on the identified costs.

References

Asiedu Y (1998) Product life cycle cost analysis: state of the art review. Int J Prod Res 36(4):883–908
Aybar B, Ficici A (2009) Cross-border acquisitions and firm value: an analysis of emerging-market multinationals. J Int Bus Stud 40(8):1317–1338
Bock S (2008) Supporting offshoring and nearshoring decisions for mass customization manufacturing processes. Eur J Oper Res 184(2):490–508
Christopher M (2005) Logistics and supply chain management: creating value-adding networks. Pearson Education, New Jersey
Colotla I, Shi YJ, Gregory MJ (2003) Operation and performance of international manufacturing networks. Int J Oper Prod Manag 23(10):1184–1206
Cuervo-Cazurra A, Maloney MM, Manrakhan S (2007) Causes of the difficulties in internationalisation. J Int Bus Stud 38(5):709–725
Cuervo-Cazurra A (2011) Selecting the country in which to start internationalization: the non-sequential internationalization model. J World Bus 46(4):426–437
Elango B (2010) Influence of industry type on the relationship between international operations and risk. J Bus Res 63(3):303–309
Ernst D, Kim L (2002) Global production networks, knowledge diffusion, and local capability formation. Res Policy 31(8–9):1417–1429
Errasti A (2011) International manufacturing networks: global operations design and management. In: Eusko Jaurlaritzaren, pp 29–56
Errasti A (2013) Global production networks: operations design and management, 2nd edn. CRC Press, Boca Raton
Ferdows K (1997) Making the most of foreign factories. Harvard Bus Rev 75(2):73–88
Forsgren M (2002) The concept of learning in the Uppsala internationalisation process model: a critical review. Int Process Firm 11(3):257–277
Fuchs E, Kirchain R (2010) Design for location? The impact of manufacturing offshore on technology competitiveness in the optoelectronics industry. Manage Sci 56(12):2323–2349

Gomes Salema MI, Barbosa-Povoa AP, Novais AQ (2010) Simultaneous design and planning of supply chains with reverse flows: a generic modelling framework. Eur J Oper Res 203(2):336–349

Hammami R, Frein Y, Hadj-Alouane AB (2008) Supply chain design in the delocalization context: relevant features and new modeling tendencies. Int J Prod Econ 113(2):641–656

Hutzschenreuter T, Voll JC, Verbeke A (2011) The impact of added cultural distance and cultural diversity on international expansion patterns: a penrosean perspective. J Manage Stud 48 (2):305–329

Johanson J, Vahlne J (1977) Internationalisation process of firm—model of knowledge development and increasing foreign market commitments. J Int Bus Stud 8(1):23–32

Johanson J, Vahlne J (1990) The mechanism of internationalisation. Int Mark Rev 7(4):11–24

Kedia BL, Mukherjee D (2009) Understanding offshoring: a research framework based on disintegration, location and externalization advantages. J World Bus 44(3):250–261

Klassen RD, Vereecke A (2012) Social issues in supply chains: capabilities link responsibility, risk (opportunity), and performance. Int J Prod Econ 140(1):103–115

Krikke H, Bloemhof-Ruwaard J, Van Wassenhove LN (2003) Concurrent product and closed-loop supply chain design with an application to refrigerators. Int J Prod Res 41(16):3689–3719

Lee C, Wilhelm W (2010) On integrating theories of international economics in the strategic planning of global supply chains and facility location. Int J Prod Econ 124(1):225–240

Luo Y, Shenkar O (2011) Toward a perspective of cultural friction in international business. J Int Manag 17(1):1–14

Luzarraga JM, Irizar I (2012) La estrategia de multilocalización internacional de la Corporación Mondragon. Ekonomiaz: Revista vasca de economía 79:114–145

Malhotra MK, Grover V (1998) An assessment of survey research in POM: from constructs to theory. J Oper Manag 16(4):407–425

McIvor R (2005) The outsourcing process: strategies for evaluation and management. Cambridge University Press, Cambridge

Mediavilla M et al (2012) Value chain based framework for assessing the Ferdows' strategic plant role: an empirical study. In: Frick J, Laugen BT (eds) Advances in production management systems. value networks: innovation, technologies, and management. IFIP advances in information and communication technology. Springer, Berlin, pp 369–378

Mendibil K, MacBryde J (2005) Designing effective team-based performance measurement systems: an integrated approach. Prod Plann Control 16(2):208–225

Nordin F (2005) Searching for the optimum product service distribution channel: examining the actions of five industrial firms. Int J Phys Distrib Logistics Manag 35(8):576–594

Porter M (1985) Competitive Advantage, Free Press, New York

Porter ME (1991) Towards a dynamic theory of strategy. Strateg Manag J 12(S2):95–117

Reverte C (2012) The impact of better corporate social responsibility disclosure on the cost of equity capital. Corp Soc Responsib Environ Manag 19(5):253–272

Rudberg M, West BM (2008) Global operations strategy: coordinating manufacturing networks. Omega-Int J Manag Sci 36(1):91–106

Sethi D, Judge W (2009) Reappraising liabilities of foreignness within an integrated perspective of the costs and benefits of doing business abroad. Int Bus Rev 18(4):404–416

Sommer C, Troxler G (2007) Outsourcing and offshoring: the consultancies' estimates. In: Meyer B, Joseph M (eds) Software engineering approaches for offshore and outsourced development. Springer, Berlin, pp 109–113

Stringfellow A, Teagarden MB, Nie W (2008) Invisible costs in offshoring services work. J Oper Manag 26(2):164–179

Vallaster C, Lindgreen A, Maon F (2012) Strategically leveraging corporate social responsibility: a corporate branding perspective. Calif Manag Rev 54(3):34–60

Relations Between Costs and Characteristics of a Process: A Simulation Study

J. Fortuny-Santos, P. Ruiz de Arbulo-López and M. Zarraga-Rodríguez

Abstract Operations managers have to take decisions where costs have to be taken into account. However, the amount of those costs depends on the costing system in use. In consequence, the aim of this paper is to analyse the influence of the costing system on the production process and vice versa. To assess the importance of the different factors, a computer-based simulation experiment has been carried out (a four-stage production system has been modelled) and results have been analysed with a full factorial experimental design. Two different experiments are explained in this paper: The first one analyses the variation of the average manufacturing cost as a function of the costing system and the cost distribution pattern. The second one, analyses the influence of the lot size and the product family on the manufacturing cost.

Keywords Cost · Simulation · Scheduling · Lean manufacturing · Design of experiments

1 Introduction

Operations management often relies on cost-based decisions. A costing model can be defined as a set of procedures and rules to assign costs to products and other cost objects such as processes or customers. The cost of a product includes the costs of

---

J. Fortuny-Santos
Technical University of Catalonia, Manresa, Spain
e-mail: jordi.fortuny@upc.edu

P. Ruiz de Arbulo-López (✉)
University of the Basque Country (UPV/EHU), Bilbao, Spain
e-mail: patxi.ruizdearbulo@ehu.es

M. Zarraga-Rodríguez
University of Navarra, Donostia/San Sebastián, Spain
e-mail: mzarraga@unav.es

materials and labour (direct costs) and a share of manufacturing overhead. Each costing method has different allocation/apportionment criteria and thus the resulting costs (i.e. of a product) depend on the costing method. Such result may drive different decisions on the manufacturing process. It seems logical that the differences resulting from different cost approaches depend on the characteristics of the process (how it is implemented and managed).

On the one hand, traditional absorption costing methods focus on direct labour (the most important cots a 100 years ago) and those results in considering overhead as a fraction of labour costs (and inconsequence overhead is considered a volume related cost). When indirect costs were small, such system was reliable but currently many times indirect costs are bigger that direct ones (Son 1993) and therefore it is necessary a costing systems that takes into account the cause and effect relation (Miller and Vollmann 1985). On the other hand, while traditional costing promotes manufacturing in large series (mass production), new manufacturing approaches such as lean manufacturing stress the importance of short runs, continuous improvement instead of adherence to standards and non-financial performance indicators (because costing systems do not reflect the costs caused by issues like delays or bottlenecks, that should be addressed in manufacturing).

In consequence, a company should choose the right costing method for its manufacturing structure. However, Karmarkar et al. (1990) state that the reasons why a company prefers one or another costing system are not known. They empirically prove that the costing system is related to some characteristics of the process (such as its complexity, the number of products or the layouts) and the frequency of reporting.

As a natural consequence of the previous lines, this paper tries to empirically analyse the relations between some characteristics of a manufacturing process and different costing systems. In order to get the necessary flexibility to test how different characteristics of the process affect the total cost, a discrete simulation model has been implemented. The combination between simulation, costs and manufacturing systems for research purposes has been previously used by Takakuwa (1997), Spedding and Sun (1999) or Mehra et al. (2005). The results of each experiment have been analysed using a design of experiments approach (a factorial design).

2 Methodology

In order to assess the importance of each factor, extensive simulation experiments have been conducted using the Extend (Krahl 2008) package. The model represents a manufacturing process made up of several stages, which operates according a pull technique (Fig. 1). Several costing techniques are used to compute the total manufacturing cost. Scheduling is based on three different priority rules (two of them based on cost rules while the third one is the typical first-come-first-served). The manufacturing process takes place in a multiproduct job shop. A job shop is defined

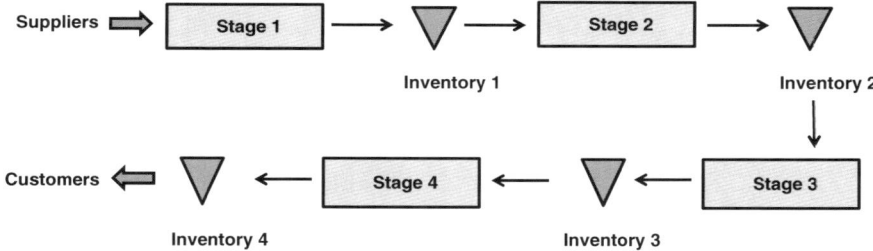

Fig. 1 Production process that is simulated

as a functional organization whose work centres are organized around particular types of equipment or operations. Tasks have a linear sequence. Products flow through departments in batches corresponding to individual orders. All products go through the same stages, but the added value of each operation depends on the product family. Machines and associates are assigned to different work centres with a specific production capacity. In each centre all machines are identical. An operation may be assigned to one or more centres. All employees assigned to a specific centre have equal productivity. There is another type of operators (auxiliary operators) that can be assigned to different centres to perform auxiliary tasks such as setup, changeover, maintenance or transportation. Task times are stochastic. Besides, they depend on the product family. Depending on the manufacturing stage, complementary setup and changeover operations exist. Each random variable's associated probability distribution or probability density function is the most commonly used in literature.

The simulation model takes into account the customary characteristics of the job shop model (Bitran et al. 1983), although some modifications are necessary in order to analyse the issues related to cost management:

- Each machine is available for scheduling along the scheduling horizon (H), which has a value of 3,500 h.
- All the tasks in a job are set in order, according to a lineal sequence.
- Make-to-order policy. Lot size is a variable in this study. Three product families are considered (F1, F2 and F3).
- No operation can be interrupted.
- No two tasks of a same job can be done at same time.
- No machine can be doing two operations at same time.
- Transportation time between workstations is not negligible.
- Changeover time is independent of the sequence but it depends on the product family that is going to be processed and on the type of machine (stage number). Changeover time is not incorporated in task time.
- Two scarce resources are considered: machines and operators.
- No alternative routes are allowed.
- Idle time is possible.

Table 1 Attributes for each entity (job or order) in the job shop simulation

Attribute	Description
Rk	Date order k is released
Ck	Date job k is completed
Dk	Date order k is delivered to the customer
Fki	Products in order k belong to family i

Table 2 Attributes related to the product family in the job shop simulation

Attribute	Description
Nui	Number of units (products in family i) in order
PVi	Sales price per unit (products in family i)
$LPji$	Manufacturing lot size at stage j for articles in family i
$LTj\text{-}j+1, i$	Transfer lot size between stage j and stage $j+1$ for articles in family i
$Tpji$	Task time for an article in family i in stage j
$Tcji$	Time for setup and changeover in stage j for articles in family i

In the simulation model, the jobs (orders) are considered entities. Customers can order products from three different families. Allowed order quantities depend on the family (200 units for family 1, 400 units for family 2 and 600 units for family 3). There are no relevant differences between the different articles from a single family. The attributes in Table 1 are the characteristics that define each job.

When a family i is assigned to an order (Fki), several more attributes related to the family type are assigned to that order (Table 2).

Finally, some assessment metrics have been considered: number of completed jobs per family; machine utilization and worker utilization; average wait time; inventory between workstations.

Besides, the system requires some more parameters to start the simulation:

- Customer orders arrive at the factory according to an exponential distribution with an expected value of 10 h.
- Every order is given a 136 h delivery time.
- Task times depend on the product family and the stage number. They follow a normal distribution. Setup and changeover times are different for each stage and follow a normal distribution.
- Manufacturing lot size (Q) per family is specified for each run: (1) Q = order size (same family); (2) Q = order size/2; (3) Q = order size/4.
- The system requires priority rules for scheduling whether based on costs or not.
- Five different manufacturing cost distribution patterns are considered (Table 3). 32 million monetary units are divided among variable direct costs (DV), fixed direct costs (DF), variable indirect costs (IV) and fixed indirect costs. Table 4 shows how the resulting cost types are assigned to each stage of the process.

Table 3 Proportion of variable direct (DV), fixed direct (DF), variable indirect (IV) and fixed indirect (IF) costs in each cost distribution pattern (being 100 % = 32 million currency units)

Cost type > pattern	DV (%)	DF (%)	IF (%)	IV (%)	Total (%)
1	25	25	25	25	100 % = 32 million c.u.
2	10	40	40	10	100 % = 32 million c.u.
3	40	10	10	40	100 % = 32 million c.u.
4	40	40	10	10	100 % = 32 million c.u.
5	10	10	40	40	100 % = 32 million c.u.

Table 4 How each type of cost—variable direct (DV), fixed direct (DF), variable indirect (IV) and fixed indirect (IF) costs—is divided among the different stages of the process

Cost type > stage	DV (%)	DF (%)	IF (%)	IV (%)
1	63	12.5	21.5	21.5
2	25	27.5	28.5	28.5
3	12	25	21.5	21.5
4	0	25	28.5	28.5
Total	100	100	100	100

Methods used to compute the cost (C) of a product are those described by Baguer and De Zárraga (2002): (1) Full costing (FC); (2) Direct costing (DC); (3) Homogeneous functional groups (HFG) or cost centres; (4) Activity-based costing (ABC); (5) Modern HFG.

3 Results

Results of the different simulation runs have are the inputs of an experimental design test. The type of analysis required in each experiment is the full factorial design in order to understand the effect of two or more independent variables upon a single dependent variable. The number of factors and their levels depend on each specific experiment. Treatments are combinations of the levels of cost model, cost pattern, lot size and priority rule. There are many examples of the conjoint application of simulation and design of experiments in literature. Nazzal et al. (2006) use simulation and experimental design to compute costs in a decision-making process.

To validate the model, the main assumptions of factorial design have been tested: residuals are independent and follow the normal distribution. SPSS package has supplied all the statistical support in this research. The runs test (or Wald–Wolfowitz test) has been used to test the hypothesis that the residuals are at random and mutually independent. Histograms and Kolmogorov-Smirnov test have been used to test normality. Scatter diagrams have been used to test the hypothesis of independence and Levene's test has been used to assess the homogeneity of variances.

Our experiments are distributed in several groups. In this paper, we focus on group I which is devoted to analysing the total manufacturing cost for each stage of the process and for each product family. Two different experiments in group I follow:

In the first experiment, we analyse how the average manufacturing cost $CF_{j,i}$ in each stage j of the process and for each family i depends on the costing model (C) in use and the cost distribution pattern (DC). As an example, Fig. 2 shows the results from stages 1 and 2 (for family 1). Lines show how marginal averages change with the levels of a factor. Parallel lines mean no interaction between factors, while non-parallel lines show interactions.

In experiment 2, we test the influence of manufacturing lot size (Q) and family (F) on the manufacturing cost of stage 2. Figure 3 shows the results.

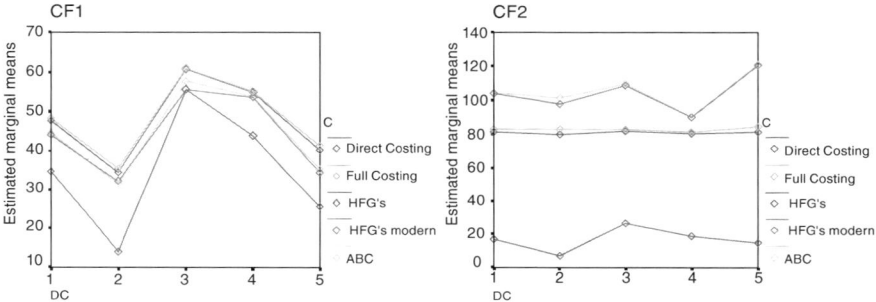

Fig. 2 Estimated marginal means graphs: average manufacturing cost for stages 1 and 2 (for family 1) as a function of the costing model (C) and the cost distribution pattern (DC) in Table 3

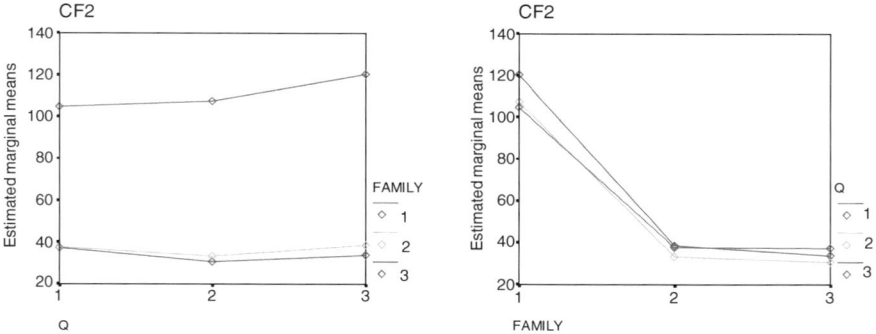

Fig. 3 Estimated marginal means graphs: average manufacturing cost for stage 2 as a function of the product family and the manufacturing lot size (Q)—full, half or quarter order

4 Conclusions

Both experiments in this article show the existence of relations between cots and technical characteristics of the process. From experiment 1, we conclude that:

- The average manufacturing cost of any stage, and for any family, depends on the cost distribution pattern and the costing method. Besides, there is interaction between both factors.
- When moving from a costing system based on HFG to a full costing system, in most cases, there is no difference (Bonferroni method). In a similar way, there is no difference when moving from modern HFG to ABC. The difference is significant when moving from other costing methods to ABC.
- Direct costing yields the lowest costs whatever the stage and the family.
- Factors C and DC have the same effect on any family in stage 1, because task times are the same for all the articles.
- Variable costs have a great influence on Direct Costing. When cost pattern number 2 is in place (variable costs = 20 %), total cost is the lowest while when using cost pattern number 3 (variable costs = 40 %), total costs reach their maximum value.
- Indirect costs have a great influence on full costing and HFG costing. For full costing, fixed indirect costs have the greatest influence, while for HFG costing, variable indirect costs are decisive.
- Indirect costs also influence ABC and modern HFG.

From the second experiment, we conclude that:

- Manufacturing cost has two parts. One is caused by the operation itself while the other is caused by setup and changeover time and therefore it depends on the lot size.
- Family 1 will always have the highest manufacturing cost because it has the highest task times and the smallest order size. As the order size increases for families 2 and 3, the setup and changeover cost per unit decreases.
- When splitting the manufacturing lot (factor Q) does not entail a decrease in the setup and changeover time (for example, for family 1), the average manufacturing cost will increase if calculated with an ABC system.

References

Baguer A, De Zárraga M (2002) ¡Dirige! Manual de conceptos prácticos y necesarios para la gestión empresarial. Ediciones Díaz de Santos, Madrid

Bitran G, Maqbool D, Sison L (1983) A simulation model for job shop scheduling. Working paper (Sloan School of Management) WP 1402-83, Massachusetts Institute of Technology, Cambridge

Karmarkar U, Lederer P, Zimmerman J (1990) Choosing manufacturing production control and cost accounting systems. In: Kaplan RS (ed) Measures for manufacturing excellence. Harvard Business School Press, Boston, pp 353–396

Krahl D (2008) Extendsim 7. In: Mason S, Hill R, Mönch L et al (eds) Proceedings of the 2008 winter simulation conference, pp 215–221

Mehra S, Inman A, Tuite G (2005) A simulation-based comparison of TOC and traditional accounting performance measures in a process industry. J Manufact Technol Manage 16(3):328–342

Miller J, Vollmann T (1985) The hidden factory. Harvard Bus Rev 63(5):142–150

Nazzal D, Mollaghasemi M, Anderson D (2006) A simulation-based evaluation of the cost of cycle time reduction in Agere systems wafer fabrication facility—a case study. Int J Prod Econ 100(2):300–313

Son Y (1993) Simulation-based manufacturing accounting for modern management. J Manuf Syst 12(5):417–427

Spedding TA, Sun GQ (1999) Application of discrete event simulation to the activity based costing of manufacturing systems. Int J Prod Econ 58(3):289–301

Takakuwa S (1997) The use of simulation in activity-based costing for flexible manufacturing systems. In: Andradottir S, Healy K, Whiters D et al (eds) Proceedings of the 1997 winter simulation conference, pp 793–800

Critical Success Factors on Implementation of Customer Experience Management (CEM) Through Extended Marketing Mix

A. Arineli and H. Quintella

Abstract If Customer Experience Management (CEM) is a strategy to focus on operations and processes of a business around the experiences of the customers with the company, it is essential to seek grants to structure it. Through research with managers of retail chains Brazilian clothing, it was possible to identify the Critical Success Factor (CSF) and the relation between Extended Marketing Mix (EMM) and CEM. To promote superior customer experiences, CEM can be used as a guiding matrix management. Specific characteristics of the segment suggest that is crucial that each industry scan your own variables as a basis for CEM.

Keywords Brazilian fashion retail · Customer experience management · Extended marketing mix · Retail management · Critical success factor

1 Contextualization

The quest for differentiation is true for the current competitiveness, which may be driven by the competence of the management to concentrate their efforts on the experience provided to its consumers. The Customer Experience Management (CEM) becomes an essential tool for the management of companies seeking sustainability for your business and can generate an unparalleled competitive advantage.

Grewal et al. (2009) argue that due to the current economic scenery and the competitive retail environment, the customer experience should be the focal point to the companies that want to compete effectively. Undeniably that customer experience in retail is systemically crucial for business. Customer participation in the

A. Arineli (✉) · H. Quintella
Laboratório de Tecnologia, Gestão de Negócios e Meio Ambiente (LATEC) da Universidade Federal Fluminense, R. Passo da Pátria, 156/329-A, 24001-970 Niterói-RJ, Brazil
e-mail: adriana@experiencia.com

H. Quintella
e-mail: hquintel@unisys.com.br

procurement process puts the focal point of the company ceasing to be a passive recipient of influences to an active decision maker. Who decides what is perceived value and quality, is the customer.

As Schmitt (2003) states that there are phenomena of sale that can only be explained by the value of the experience offered to the consumer, Cohen (2006), says that advertising was once largely responsible for building a brand, and currently loses gradually space for the customer experience. You need to break through barriers and understand the systemic form, the dimension of the experience offered to the customer. Companies that deliver a memorable experience, are more competitive and perennial brands. If the Marketing continue to neglect the experiences of the consumer, it will limit its potential for innovation and ultimately waste opportunities to generate significant differential current for customers.

According to Palmer (2010), managers who seek sustainable competitive advantages need to think on how they will develop a flow of experiences over time. Berry (2004) also states that companies should implement a system that audits experiences. For Prahalad and Ramaswamy (2004), winning strategies are those that satisfy its customers through the design of the experiences, however what are the key points that define success or failure for companies? Witch Critical Success Factors (CSF) should support planning and strategic actions of the clothing retail business? For Smith (2006) if customers are looking for happy experiences, it is mandatory to identify how to create and repeat happy experiences, defining when, where and how often it happens.

Novak et al. (2000) state that the absence of serious systematic research brings tremendous impact on the comprehension of the customer experience. The lack of research requires a solid conceptual study, with conceptions and results. With this knowledge, companies can orchestrate an integrated set of directions that meet or exceed the intangible needs and expectations of the people. So you can show patterns and build quality requirements of the customer experience.

2 Problem Overview and Discussion

The main objective of this study is to investigate and collaborate with the structure of CEM, verifying the possibility of identifying the CSF through the EMM. The objective was to define the vital elements that promote superior experiences to leverage competitiveness and sustainability for retail businesses.

Three conceptual pillars structure the study of the problem: the management experience and its importance to businesses, the variables of the EMM and CSF.

One can conceptualize the experience as internal and subjective response that customers may have from any contact with the company. Therefore, the CEM will be determined by the ability of the company to be active in the management of these experiences. Customers want memorable experiences and according to Dickson et al. (2009) they are willing to pay for experiences in which quality is perceived.

It is easy to say that a business is oriented to the customer when there is no data to prove otherwise. Kamaladevi (2010) states that it is necessary to move, measure and manage business performance and activities that deliver the required expertise and consumer leads to having brand preferences.

For Haeckel et al. (2003) companies must learn how to manage the emotional components with the same vigor that manage products and services. While Yen and Yuan (2010) argue that the design of the service experience is a way to promote high positive emotions for customers, Botha and Rensburg (2010) belief is that it is necessary to define a model that integrates traditional approaches to process improvement with CEM design.

According to Rockart (1979), CSF are areas of management organization whose performances are essential to achieve the objectives. These aspects involve the customer experience and are essential for this study.

In order to understand the extent of the experience and its impact, the marketing mix has been adopted. The 4 P's of McCarthy (1960) proposes the fragmentation of marketing aspects as a method to organize the management. It is one of the most important references in retail management. They are: Product (product mix, variety, quantity); Place (location, business location, visibility, access); Promotion (communication, publicity, advertising etc.); Price (margin, pricing policies, payment methods).

Through an holistic view of the customer experience, it is possible to consider a new conceptual model. Thus, for the purpose of this study, the Extended Marketing Mix definition, of Booms and Bitner (1981) is the more appropriate. The authors extended the variables to seven by adding the following P's: People (directly or indirectly involved in providing the service), Process (flow of activities and procedures involved in customer service) and Physical Perception (involves the environment that care is provided).

3 Methodology

This study was based on the hypothesis H1: The CSF of the CEM can be scaled through the EMM; which was dully tested.

Data collection taking as universe, managers from three Brazilian companies, market leaders in the women's clothing fashion segment, with its own brand creation, high representation throughout the national territory, own stores and franchises, class A public and significant dedication to the elements involved in the CEM. The companies were not identified separately.

The field research was conducted through a survey, using an adaptation of the Servperf instrument by Cronin and Taylor (1992), composing 23 questions. The survey was designed to be self-filled with fixed questions in 5 choices of predetermined responses, using a Likert (Lakatos 2003) five-point scale from strongly agree to strongly disagree. Surveys were sent to 30 managers, 10 of them completed it. The rate of return is not representative of all managers.

The items were identified according to the 7 P's, as shown on the survey below. The categorization of P's was defined as P1 = Product; P2 = Place; P3 = Promotion; P4 = Price; P5 = Processes; P6 = People and P7 = Physical Perception (Table 1).

4 Found Results and Analysis

There was no relevant discrepancy between the methods that showed significant in all the elements that compose the MME, except items 7, 11, 22 and 23. According to the data analysis, the variables behaved differently, in regards to 'Price', that were rejected as a significant variable in all analysis and the other items considered significant (Table 2).

Regarding frequency analysis: items 3, 6, 9 and 16 showed high level of agreement, item 17 presented unanimously in all items considered positively. Item 3 refers to 'Physical Perception', 6 matches 'People', while 9 refers to 'Place' and

Table 1 Survey items for research

No.	P	Questão
1	P7	The olfactive perception (smell brand) increases the time of permanence of customer in the store
2	P7	The way that the customer sees the product in the store is decisive to buy it
3	P7	The look of the store has to always be in synergy with the product
4	P7	The played playlist reinforces the concept of the brand and increases the time of permanence of customer in the store
5	P7	Tastings at the point of sale (coffee, candies, champagne etc.) increases the time of permanence of customer in the store
6	P6	The customer appreciates a fast, effective, personalized and differentiated service
7	P6	A store with more attendants offers a higher personalized service
8	P6	A nice attendant appearance is a differential for winning the customer
9	P2	Good store location adds perceived brand value
10	P2	Access for the disabled is favorable and taxable for winning customers
11	P2	Integrated transportation system facilitates access to customers
12	P3	Advertising in magazines, newspapers and internet increases customer confidence in the brand
13	P3	Launching collection increases brand visibility and hence the sales of products
14	P3	Use of new digital technologies (sms and social networks) makes relationship between customer and brand, closer
15	P1	The use of sustainable materials increases the perceived value of the brand
16	P1	The packaging impacts on perceived quality
17	P1	The "treats" like extra buttons fixed on product or detailed product instructions, increases the product's perceived value
18	P5	A flexible exchange process increases the perceived quality of the brand
19	P5	The store that grants discounts to frequent customers increases the chance of loyalty
20	P7	The appropriate climatization on store impacts on time of permanence
21	P5	The more accessible information about the store to the customer (Google, Facebook, Twitter, SMS), the fastest is customer return
22	P4	The price is the main instrument for the company
23	P4	Sales discounts must be offered throughout the year

Table 2 General classification of itens

P's	Item	Importance			Agreement		Paraconsistent logic		True
		Moderate importance	Good importance	High importance	Full agreement	Parcial agreement	Belief degrees	Disbelief degrees	
P1	17			5.0	100	0	1.000	0.000	t
P6	6			4.9	90	10	0.975	0.025	t
P2	9			4.9	90	10	0.975	0.025	t
P7	3			4.8	80		0.950	0.050	t
P1	16			4.8	90	0	0.950	0.050	t
P3	13			4.7	70	30	0.925	0.075	t
P7	1		4.6		60		0.900	0.100	t
P6	8		4.6		70	20	0.900	0.100	t
P7	5		4.5		70	10	0.875	0.125	t
P7	20		4.5		50	50	0.875	0.125	t
P7	4		4.4		40	60	0.850	0.150	t
P3	12		4.3		40	50	0.825	0.175	t
P7	2		4.1		60		0.775	0.225	t
P1	15		4.1		50	20	0.775	0.225	t
P3	14	3.8			40	30	0.700	0.300	At → t
P5	18	3.8			40	20	0.700	0.300	At → t
P5	19	3.7			30	40	0.675	0.325	At → t
P2	10	3.4					0.600	0.400	At → t
P5	21	3.3					0.575	0.425	At → t

t true, *At* almost true
The bold shows the items highlighted in the three statistical methods used

Table 3 Classification of P's

P's	Items	Average score of the items	Minimum	Maximum	Median	Inter quartile range[a]
Product	15	4.1	4	5	4.833	0.6667
	16	4.8				
	17	5				
Place	9	4.9	2.67	4.67	3.667	1.3333
	10	3.4				
	11	2.5				
Promotion	12	4.3	3	5	4.125	1.5
	13	4.7				
	14	3.8				
	21	3.3				
Price	22	2	1	3.5	1.875	1
	23	3.8				
Process	18	1.6	1.5	5	3.75	0.5
	19	3.7				
People	6	4.9	3.33	5	4	1
	7	2.9				
	8	4.6				
Physical perception	1	4.6	3.83	5	4.583	0.833
	2	4.1				
	3	4.8				
	4	4.4				
	5	4.5				
	20	4.5				

[a] Based on Tukey Fences

items 16 and 17 correspond to variable 'Product'. Items 7, 11, 22 and 23 do not contribute positively to the scale of measurement. Item 7 corresponds to variable 'People', item 11 to 'Place' and items 22 and 23 correspond to variable 'Price' (Table 3).

Regarding score: according to the statistical description below, it is clear that 'Price' stands below the average score, with median 1,875. The other items are above average score, which ranked as the median, are presented in the following order: 'Product' 4,83; 'Physical Perception' 4,58; 'People' 4,00; 'Promotion' 4,12; 'Place' 3,66; 'Processes' 3,75.

Regarding Paraconsistent Logic: items with more certainty (true and almost true) were: 'Product', 'Place', 'Promotion', 'Processes', 'People' and 'Physical Perception', represented in the following figure. They can be considered CFS (Fig. 1).

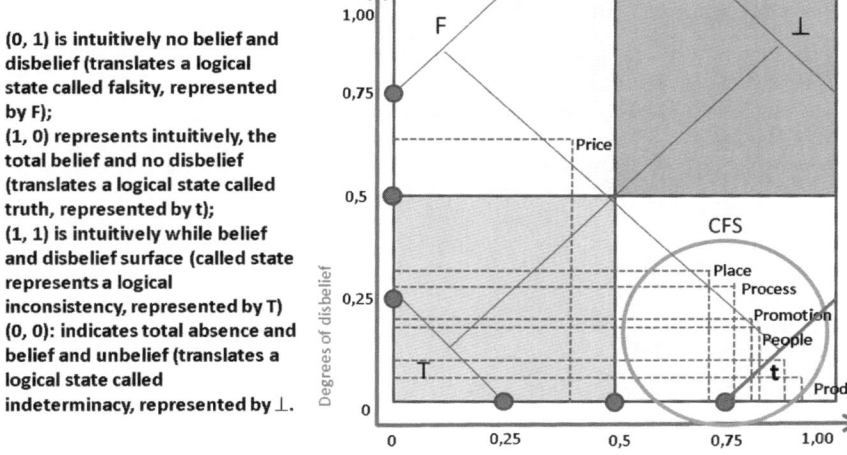

Fig. 1 Paraconsistent logic quadrant of P's

5 Conclusions

The aim of this study was to identify the CSF's involved, through the P's of the EMM, to the application of CEM. It can be concluded that the 7 P'S shape the dynamics of the experience.

It is not possible to generalize conclusions through this study, due to its low sampling and consequent not representative conclusions. There is evidence, however, as per the strength of the data obtained, that the particularities of each retail segment will determined the amount of items to be managed. In this study, for example, of women's fashion retail customers, class A, the low importance of the variable 'price' may have been affected by the target audience profile of the companies analyzed. We conclude that the dynamics of the CEM can be observed through the 7 P's but should be managed considering particularities of the target audience and the segment. It will impact the relevance of the different aspects.

Positive experiences, even those that do not result in immediate sales, contribute significantly to the consolidation of the brand, which is a key asset in the competitive landscape in which retail businesses are. The store demonstrated to be essential. The ambiance and the sensations, in synergy with the brand, can create longer relationships, more pleasurable experiences, in addition to make brand values tangible.

Companies seeking sustainability in their business and customer loyalty should understand that management must manage customer experiences by creating and strategically structuring a design experience that can be replicated as a way to add value and show significant differences from the point of view of the customer.

To invest in CEM may seem unusual, since we live in a world where talking about intangible and of subtle aspects of the experience, can be an obscure challenge to management. Measure what has no obvious and accessible numbers and statistics can be scary or revealing, depends on the legitimate desire to meet competitive advantages that go along with the contemporary customer's life style and its complexity.

References

Berry L (2004) Discovering the soul of service. The Free Press, New York
Booms B, Bitner M (1981) Marketing strategies and organisation structures for service firms, marketing of services. American Marketing Association, Chicago
Botha G, Renseburg A (2010) Proposed business process improvement model with integrated customer experience management. South Africa J Ind Eng 2(1):45–57
Cohen A (2006) Follow the other hand. Martin Press, New York
Cronin J, Taylor S (1992) Measuring service quality: a reexamination and extension. J Mark 56 (3):55–68
Dickson P, Lassar W, Hunter GK, Chakravorti S (2009) The pursuit of excellence in process thinking and customer relationship management. J Pers Selling Sales Manag 29(2):111–124
Grewal D, Levy M, Kumar V (2009) Customer experience management in retailing: an organization framework. J Retail 85(1):1–14
Haeckel S, Carbone L, Berry L (2003) How to lead the customer experience. Mark Manag 12 (1):18–24 1
Kamaladevi B (2010) Customer experience management in Retailing. Bus Intell J 3(1):37–54
Lakatos E, Marconi M (2003) Fundamentos de metodologia científica. São Paulo, Atlas
McCarthy EJ (1960) Basic marketing: a managerial approach. Irwin
Novak TP, Hoffman LD, Yung FY (2000) Measuring the customer experience in online environments: A structural modeling approach. Mark Sci 19(1):22–42
Palmer A (2010) Customer experience management: a critical review of an emerging idea. J Serv Mark 24(2–3):196–208
Prahalad CK, Ramaswamy V (2004) The future of competition: co-creating value with customers. Harvard Business School Press, Boston
Rockart JF (1979) Chief executives define their own data needs. Harvard Bus Rev 57(2):81–93
Schmitt B (2003) Customer experience management: a revolutionary approach to connecting with your customers. Wiley, New York
Smith S (2006) Defining CEM. White paper Shanghai *GCCRM*
Yen H, Yuan H (2010) Modeling service experience design processes with customer expectation management: a system dynamics perspective. Kybernetes 39(7):1128–1144

Financing Urban Growth in Aging Societies: Modelling the Equity Release Schemes in the Welfare Mix for Older Persons

David Bogataj, Diego Ros-McDonnell and Marija Bogataj

Abstract The paper examines the role of Equity Release Schemes (ERS) in the welfare mix for seniors. The concepts for asset-based welfare, where the housing owned by the occupant is part of his portfolio comprised of defined contribution (private) pension and residential property, are examined. We present how the variance of expected senior's income is reduced by using residential property as the pillar of the pension system, as proposed in the EC Green Paper on Pensions (2010). The study is focusing on modelling the decumulation of the housing equity and the defined contribution private pension, incorporating insurance mechanisms for management of longevity. Development of longevity insurance in the framework of reverse mortgage products is important in European Union due to lack of government insurance of reverse mortgage products as developed in United States under the HECM insurance program. Here we propose a new model in which periodic payout that the beneficiary receives is the difference between the amount drawn and the annuity premium for longevity insurance. The paper shows how the drawing amount in the loan model ERS (reverse mortgage) is decreasing with the increasing interest rate, while the pension arising from defined contribution systems is increasing with the increasing interest rate. The optimal combination of sources in the portfolio which minimize volatility induced by volatile interest rate is derived and discussed. The paper intend to show that regarding his/her old-age welfare protection, young person when employed should consider buying a home instead of rent, to gain the optimal structure of pension portfolio in retirement.

D. Bogataj
The European Faculty of Law, Nova Gorica, Delpinova 18b, Slovenia
e-mail: dbogataj@actuary.si

D. Ros-McDonnell
Universidad Politécnica de Cartagena, Murcia, Spain
e-mail: diego.ros@upct.es

M. Bogataj (✉)
MEDIFAS, Mednarodni prehod 6 Vrtojba, 5290 Sempeter pri Gorici, Slovenia
e-mail: marija.bogataj@guest.arnes.si

© Springer International Publishing Switzerland 2015
P. Cortés et al. (eds.), *Enhancing Synergies in a Collaborative Environment*,
Lecture Notes in Management and Industrial Engineering,
DOI 10.1007/978-3-319-14078-0_24

Keywords Actuarial mathematics · Equity release scheme · Housing · Longevity · Reverse mortgage, portfolio optimization

1 Introduction

In funded defined contribution pension systems, the amount of yearly pension depends on the accumulated amount in the individual retirement account and long-term interest rate at the moment of retirement. As presented by Shiller (http://www.irrationalexuberance.com, 2005; updated data, see Fig. 1), the long-term interest rates are highly volatile in USA, but in European Countries we have often fallen in hyperinflation, therefore secular trend is difficult to analyse. Therefore, the yearly amount of one's pension is uncertain. Markowitz (1952) has shown that such volatility of returns can be mitigated by adding negatively correlated assets in the portfolio. The paper examines the possibility of developing and implementing flexible Housing Equity Release Schemes (ERS) as a means of providing a more stable welfare provision for the elderly, adding housing wealth in a portfolio of pension instruments. This could provide a better welfare provision for older persons by stabilizing the total disbursement to the beneficiary, where the volatility depends on the volatility of the interest rate (funded pensions). The falling interest rates since 2009 have had adverse effects on funded pensions, and the austerity policies since 2010 have reinforced the inequalities, particularly among the 'income poor' but 'asset rich' older population. Pensions are often not sufficient to cover health expenses and other needs of older persons. Hence, the question of an optimal welfare mix for older people is significant. Therefore, the author seeks to examine the following research questions:

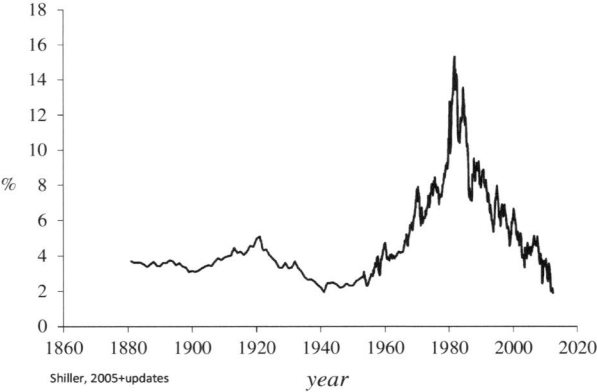

Fig. 1 Interest rate (in %: 1871 – 2012)

(a) How to ensure a more appropriate welfare mix for older people;
(b) What role could Equity Release Schemes play; and
(c) How can the financial industry develop attractive financial products to fit within the mix of other (private and public) welfare provisions, where the risks of poverty due to the falling interest rate can be mitigated to the benefit of older property owners?

Using actuarial mathematics with life contingencies, the paper will present how the reverse mortgage systems (ERS loan model) with the embedded insurance for longevity might improve the results of the senior housing provision and the satisfaction of inhabitants. Based on the presented findings and on the Portfolio Theory (Markowitz 1952; Tu and Zhou 2011), we can also show that the interest rate variation, which can reduce the income of older persons under the poverty line, has a significantly smaller impact on welfare of the elderly if pension pillars are combined with the ERS loan model. Such financial product would be a novelty in the insurance and banking industry.

The paper first describes the existing models of ERS, than it introduces the model for mitigating the credit default risk using longevity insurance. As an important contribution at the end, here is given a model which enable to study the optimal combination of assets in pension schemes, where the portfolio owner could decide the structure (private pension schemes and ERS) to prevent to fall under the poverty line in old ages. The paper intend to show that regarding his/her old-age welfare protection, young person when employed should consider buying a home instead of renting one, in order to accumulate real estate assets and not only to buy a private pension in proper structure of portfolio. A numerical example is also presented.

2 Modelling ERS

According to the clear Reifner's description, ERS transform fixed assets in owner occupied dwellings into liquid assets for private pensions. They thus enable a homeowner to access the wealth accumulated in the form of the home, while being able to continue to live in it. An illiquid asset becomes a source of liquidity, mainly for consumption needs. ERS can take two different forms:

(a) Loan Model ERS, also known as reverse mortgage, provide a loan that will be repaid from the sale after the death of owner, and
(b) Sale Model ERS, which involve an immediate sale of the property but provide for the right to remain in occupation and to use the cash price for income in retirement.

ERS must therefore:

(a) be a financial service;
(b) be a source of liquidity for the future;
(c) contain a strong entitlement to remain in occupation of the property; and
(d) rely solely on the sale of the property for repayment/payment of the funds released to be used as a retirement pension.

Payments take the form of a lump sum or periodic (monthly, yearly) income, and are either secured by means of a mortgage on the property or generated by an immediate sale. Under the Loan Model ERS, repayment is made from the proceeds of the sale of the property either after the death of the homeowner or when the property has become vacated for a longer time (Reifner et al. 2009).

The sale model is very straightforward and is similar to the general annuity model (funded pensions, i.e. 2nd and 3rd pillars). In the sale model, the whole value of the real estate is transferred to lifetime annuity at the moment of closing of the ERS contract. The value of the property is used for purchase of the lifetime annuity.

Let us use the standard actuarial notation in further equations. The amount of yearly pay-outs of the lifetime annuity therefore needs to cover the interest on the principal amount taken out and the yearly annuity paid to the beneficiary of the ERS, as is presented in Eq. (1):

$$\ddot{a}_{x+ps} = \sum_{j=0}^{110-x+ps} {}_j p_{x+ps} \cdot v^j = \sum_{j=0}^{110-x+ps} {}_j p_{x+ps} \cdot (1/(1+i))^j \qquad (1)$$

In Eq. (1) $\ddot{a}_{x+ps}$ is the actuarial notation for the present value of the prenumerando lifetime annuity of the amount 1 EUR for the person that is x years old (ps is the age correction conforming to the methodology of annuity mortality tables prescribed by the law, ${}_j p_{x+ps}$ is the probability that the person that is x years old will survive the next j years, and v is the discounting factor ($v = 1/(1+i)$), where i is the annual interest rate. The amount of lifetime annuity is calculated as the annuity factor $fr(x, i)$ multiplied with the net value of real estate, which is calculated as the value of the real estate minus the cost associated with the transaction (valuation costs, taxes, costs of sale). Annuity factor is:

$$fr(x, i) = 1/\{(1+\gamma_2) \cdot \ddot{a}_{x+ps}\} = 1/\left\{(1+\gamma_2) \sum_{j=0}^{110-x+ps} {}_j p_{x+ps} \cdot (1/(1+i))^j\right\} \qquad (2)$$

where rate γ_2 represents the costs associated with the pay-out of the annuity that the insurance company charges for each pay-out in the period of annuity. The yearly amount of annuity R is calculated according to the value of real estate VN and annuity factor $fr(x, i)$:

$$R = fr(x,i) \cdot (VN - C) = (VN - C) / \left\{ (1 + \gamma_2) \sum_{j=0}^{110-x+ps} {}_j p_{x+ps} \cdot (1/(1+i))^j \right\} \quad (3)$$

The loan model or "reverse mortgage" is a type of home loan that allows a borrower to open up a line of credit using his home as collateral. With the loan model the beneficiary draws liquid amounts in lump-sum or/and periodically from the value of the real estate in the form of loan secured by a mortgage on the real estate. With the part of this liquid amount that is drawn from the real estate the beneficiary purchases lifetime annuity in the form of a monthly (or yearly) premium. In this way the beneficiary insures his longevity so that if he lives longer than his life expectancy he will receive a lifetime annuity until his death. In the paper, we propose the ERS model with the insurance for longevity, where the periodic pay-out that the beneficiary receives is the difference between the amount drawn and the annuity premium for longevity insurance. In this way, if the beneficiary survives the drawing period of ERS (n years), he receives a lifetime annuity that covers the disbursement to the beneficiary and the interest on the outstanding loan. Generally, loan models allow the beneficiary to draw the value of the real estate in different ways:

(a) In lump sum at the closing of the ERS contract;
(b) In the form of line of credit so that he can draw it when necessary;
(c) In uniform periodic amounts in the period of life expectancy.

The maximum amount of loan (*MLA*) that can be drawn from the real estate is the value of the real estate (*VRE*) minus all the costs (*C*), i.e. those associated with the closing of the ERS contract C_1 and with the sale of the property after the death of the beneficiary C_2.

$$MLA = VRE - C = VRE - C_1 - C_2 \quad (4)$$

A life annuity consists of a series of payments which are made while the beneficiary (of initial age x) lives. The present value of the life annuity due with yearly payments in the amount of 1 EUR is denoted by $\ddot{a}_{x+ps:\overline{n}|}$, where the following equation can be written:

$$\ddot{a}_{x+ps:\overline{n}|} = \sum_{j=0}^{n} {}_j p_{x+ps} \cdot v^j \quad (5)$$

The present value of the life annuity deferred for n years with yearly payments in the amount of 1 EUR is denoted by ${}_{n|}\ddot{a}_{x+ps}$, where the following equation can be written:

$$_{n|}\ddot{a}_{x+ps} = {}_nP_x \cdot v^n \cdot \ddot{a}_{x+ps} = {}_nP_x \cdot v^n \cdot \sum_{j=0}^{110-(x+ps+n)} {}_jP_{x+ps+n} \cdot v^j \quad (6)$$

The premium rate for longevity insurance $prs(x, i, n)$ is:

$$prs(x, i, n) = (1 + \gamma_2) \cdot {}_{n|}\ddot{a}_{x+ps}/(1 + \gamma_1) \cdot \ddot{a}_{x+ps:\overline{n}|} \quad (7)$$

where γ_1 represents the rate of administration expenses that are charged against the policy in the period of premium payments and γ_2 represents the rate of administration expenses that are charged against the policy in the period of annuity payments. The yearly amount of premium (*PR*) is calculated as:

$$PR = prs(x, i, n) \cdot R \quad (8)$$

where R is the annuity amount. In this case, the yearly amount (*YDA*) that the beneficiary can draw from the real estate is:

$$YDA = i \cdot MLA/[(1+i)^n - 1] = YPA + PR = YPA + prs(x, i) \cdot (YPA + MLA \cdot i) \quad (9)$$

The main risks concerning ERS that can cause credit default are the uncertain longevity of the owner occupier, the risk of increase in interest rates and depreciation in the value of the property. Deferred annuity as an insurance for longevity is already used in insurance industry, but not in combination with reverse mortgage like proposed here. Sustainability and market consistency regarding longevity is one of the main concerns of Groupe Consultatif Actuariel Europeen (see the articles at: www.gcactuaries.org), to whom the author belongs.

Without an effective insurance for longevity, real estate cannot be used as the 4th pension pillar, because the equity release without longevity insurance presents a great risk for the provider of reverse mortgage (bank) and also for the beneficiary. For the reverse mortgage provider, there is the risk that the value of the loan together with accrued interest will be greater than the value of real estate in case of the death of the beneficiary. For the beneficiary, there is the risk that he will live longer than the agreed period of drawing liquid amounts that is defined in the reverse mortgage loan contract. To avoid exposure to these risks, a safe reverse mortgage contract also needs to include kind of insurance for longevity. This insurance can be provided in three ways:

(a) so that the risk is socialized by government (as is the case in the USA where The Home Equity Conversion Mortgages program is a clear example of such a scheme). Lenders under such program are protected against losses arising when the loan balance exceeds the equity value at time of settlement. But because of implicit government guarantees underlying this insurance, it may

become a serious drain on the fiscal government, as market expends after crisis, as described also at Wang et al. (2008) and Chen et al. (2010);
(b) in private sector with transfer of the risk to a joint stock insurance company. According to the results in Blake et al. (2013), the huge economic significance of longevity risk has begun to be recognized and quantified in their article, presenting the birth and development of the Life Market, the new market related to the transfer of longevity and mortality risks. The authors note that the emergence of a traded market in longevity-linked capital market instruments could act as a catalyst to help facilitate the development of annuity markets both in the developed and the developing world and protect the long-term viability of retirement income provision globally.
(c) the third way a mutual insurance company. The risk of longevity can be mitigated by the use of annuity insurance, where legal, obligatory mortality tables are based on mortality projection 2050, therefore insurance companies benefit on overestimation of longevity today. Therefore it is better to share the benefit of this overestimation among remaining seniors, but it is difficult to avoid the impact of the volatile interest rate.

3 Portfolio of Assets Based on ERS and Private Pension Scheme

From Shiller's time series of interest rate (2013) the most frequent moving average of interest rate was equal to 4, while the average and variance were over 5.

From (3) it follows that R of private pensions increases with increasing i. From (9) it follows

$$i \cdot MLA/[(1+i)^n - 1] = YPA + prs(x, i_a) \cdot (YPA + MLA \cdot i)$$
$$YPA = MLA \cdot i\{1/[(1+i)^n - 1] - prs(x, i_a)\}/(1 + prs(x, i_a)) \quad (10)$$

This yearly disbursement decreases with increasing positive interest rate i if $d\{i\{1/[(1+i)^n - 1] - prs(x, i_a)\}\}/di < 0$, therefore we can write:

$$1/[(1+i)^n - 1] - prs(x, i_a) - i \cdot n(1+i)^{n-1}/\left[(1+i)^{n-1}\right]^2 < 0 \quad (11)$$

From (11) it follows that the sufficient condition is $(1+i)^{n-1}[1 - (n-1)i] < 1$. Therefore, in the loan model of ERS, where it is always $i > 0$ and $n > 1$, YPA always decreases with increasing positive i. The curves (3) and (10) are close to the linear curve while at (3) correlation coefficient $\rho_{R,i}$ is always close to 1, and at (10) correlation coefficient $\rho_{YPA,i}$ is always close to -1. We can expect that there exists an optimal portfolio which can substantially reduce the variance of portfolio.

Theorem 1 *The minimum variance of yearly annuity $R_o = \alpha R + (1 - \alpha)YPA$ from private pension portfolio which consists of mathematical reservations for private pension and value of home is expected to be achieved if the share of asset in private pension scheme is equal to*

$$\alpha = \left[\sigma^2(YPA) = \rho_{R,YPA}\sigma(R)\sigma(YPA)\right]/\left[\sigma^2(R) + \sigma^2(YPA) - 2\rho_{R,YPA}\sigma(R)\sigma(YPA)\right]$$
$$\approx \left[\sigma^2(YPA) + 0.93\,\sigma(R)\sigma(YPA)\right]/\left[\sigma^2(R) + \sigma^2(YPA) + 1.86\sigma(R)\sigma(YPA)\right] \quad (12)$$

Proof Because R and YPA are correlated also αR and $(1 - \alpha)YPA$ are correlated, having the same correlation coefficient. In general, if the variables are correlated, then the variance of their sum is the sum of their covariance, therefore we can write:

$$\sigma^2(\alpha R + (1-\alpha)YPA) = \alpha^2\sigma^2(R) + (1-\alpha)^2\sigma^2(YPA) \\ + 2\alpha(1-\alpha)\rho_{R,YPA}\sigma(R)\sigma(YPA) \quad (13)$$

The minimum of variance $\min_{\alpha}[\sigma^2(\alpha R + (1-\alpha)YPA)]$ can be achieved when

$$d[\sigma^2(\alpha R + (1-\alpha)YPA)]/d\alpha = 0 \quad (14)$$

$$2\alpha\sigma^2(R) - 2(1-\alpha)\sigma^2(YPA) + 2(1-2\alpha)\rho_{R,YPA}\sigma(R)\sigma(YPA) = 0$$
$$\alpha[2\sigma^2(R) + 2\sigma^2(YPA) - 4\rho_{R,YPA}\sigma(R)\sigma(YPA)] = 2\sigma^2(YPA) - 2\rho_{R,YPA}\sigma(R)\sigma(YPA) \quad (15)$$

while the second derivative is always positive. We can see that

$$\sigma^2(R) + (\sigma^2(YPA) - 2\rho_{R,YPA}\sigma(R)\sigma(YPA) > 0 \quad (16)$$

is valid for any α if $i > 0$. From (15) it follows that for minimal variance is:

$$\alpha = \left[\sigma^2(YPA) - \rho_{R,YPA}\sigma(R)\sigma(YPA)\right]/\left[\sigma^2(R) + \sigma^2(YPA) - 2\rho_{R,YPA}\sigma(R)\sigma(YPA)\right]$$

By variation of $i > 0$ in (3) and (10) according to simulations based of the distribution presented in Table 1, $\rho_{YPA,R} = -0.93$ has been also calculated, which is nearly the same as in case of uniform distribution of i when $2\% < i < 13\%$. Changes of mortality tables as in the last 20 years in Europe or small changes of operational costs in the framework of variations among EU Member States do not change the correlation coefficient significantly. Therefore we can write:

$$\alpha = \left[\sigma^2(YPA) + 0.93\sigma(R)\sigma(YPA)\right]/\left[\sigma^2(R) + \sigma^2(YPA) + 1.86\sigma(R)\sigma(YPA)\right] \quad (17)$$
□

Table 1 Distribution of 22 years (264 months) moving averages for the yearly interest rate since 1871 till 2012 (22 years is an approximate life expectancy for man 65 years old, after 2050 according to German mortality table DAV1994R)

i [%]	No. of months	Relative frequency
3	229	0.16
4	559	0.39
5	212	0.148
6	85	0.059
7	74	0.052
8`	65	0.045
9	210	0.146

A numerical example of such a scheme was calculated where retirement age is 65, which is equal to the standard retirement age for man in many national legislations based on EU Directives. German mortality tables *DAV1994R* are used which are prescribed also in Slovenia. Based on variations of 22 years (264 months) moving averages for the interest rate since 1871 till 2012 (Shiller 2005–2014) and monthly income A we found that the minimum uncertainty of income after retirement is achieved when he decide to invest 1/3 to pension fund and 2/3 to housing if reverse mortgage is determined on the basis of the interest rate valid in the moment of retirement if there is no inflation and other taxes.

4 Conclusion

Equity Release Schemes transform fixed assets in owner occupied dwellings into liquid assets for private pensions. We have shown that the interest rate variation, which can reduce the income of older has a significantly smaller impact on welfare of the elderly if funded pensions, which have a positive covariance with the interest rate, are combined with the ERS loan model, where the correlation coefficient is negative. Because of volatility of the interest rate, volatility of pensions should be studied carefully and the proper combination of pensions and dynamics of ERS drawings has to be chosen to decrease the volatility of combined cash flows deriving from different pension pillars. Due to the negative correlation of cash flows from funded pensions with disbursements from ERS, the volatility induced by the volatile interest rate is reduced. Regarding their old-age welfare protection, young families should consider buying a home instead of renting one, in order to accumulate their assets. Therefore, governments should provide incentives to achieve this goal and should enable their citizens to buy properties based on mortgage financing.

References

Blake D, Cairns A, Coughlan G, Dowd K, MacMinn R (2013) The New Life Market. J Risk and Insur 80(3):501–558

Chen H, Cox SH, Wang SS (2010) Is the Home Equity Conversion Mortgage in the United States sustainable? Evidence from pricing mortgage insurance premiums and non-recourse provisions using the conditional Esscher transform. Insur Math Econ 46(2):371–384

Markowitz H (1952) Portfolio selection. J Financ, Am Financ Assoc 7(1):77–91

Reifner U, Clerc-Renaud S, Pérez-Carrillo EF, Tiffe A, Knobloch M (2009) Study on equity release schemes in the EU. Institut für Fi-nanzdienstleistungen e.V, Hamburg

Shiller RJ (2005) Irrational Exuberance, Princeton, New Jersey, updated data: http://www.irrationalexuberance.com, Cited 15 May 2014

Tu J, Zhou G (2011) Markowitz meets Talmud: a combination of sophisticated and naive diversification strategies. J Financ Econ 99(1):204–215

Wang L, Valdez EA, Piggott J (2008) Securitization of longevity risk in reverse mortgages. N Am Actuarial J 12(4):345–371

Sustainable Urban Growth in Ageing Regions: Delivering a Value to the Community

David Bogataj, Deigo Ros McDonnell, Alenka Temeljotov Salaj and Marija Bogataj

Abstract Europe is wealthy but unbalanced, and the flows of people, money, information, goods and services are suffering from this situation. There is a crucial question in today's European economy: "How to balance the European wealth and facilitate the flows to achieve wellbeing for all generations (job opportunities for the young and proper care for seniors)". The answer is in Smart Silver Economy, by developing a firm understanding of space of flows and space of places as considered by Castells. The idea is to develop the knowledge to embed it in SUGAR ICT toolkit, based on SUGAR technology platform (SUGAR TP), which would facilitate understanding to increase intensity of processes. It is the ambition of SUGAR to establish roadmaps for action at EU, national and international level, bringing together scientific knowledge, art (architecture), technological know-how (GIS, ICT), industry (health care, real estate and facility management, ICT development companies), regulators (banks, insurance, health, municipalities and regional spatial planning authorities), and financial institutions to develop a strategic agenda for leading technologies for smart urban growth in face of an ageing society. The core of this idea is presented in article here.

Keywords Housing market · Urban land rent · Migration flows · Mortgage financing · Housing stock

---

D. Bogataj (✉) · A.T. Salaj
European Faculty of Law, Delpinova 18b, Nova Gorica, Slovenia
e-mail: dbogataj@actuary.si; david.bogataj@evro-pf.si

A.T. Salaj
e-mail: alenka.temeljotov-salaj@evro-pf.si

D.R. McDonnell
Universidad Politécnica de Cartagena, Cartagena, Spain
e-mail: diego.ros@upct.es

M. Bogataj
MEDIFAS, Šempeter pri Gorici, Slovenia
e-mail: marija.bogataj@guest.arnes.si

1 Introduction

Europe is facing two main problems in cities: (a) ageing population and (b) more than 11 million of empty houses in towns (Guardian, February 23, 2014), many of them in the foreclosure and the bankruptcy procedures. Smart investment, also in research and innovation, is vital to maintain high standards of living while responding to pressing societal challenges and taking advantage of demands from EU and global housing and financial markets. The main aim is to show how to develop a decision support system for solving these two problems simultaneously. Some argue that globalization and modernization are destroying the "community" and this development left behind elderly without proper care. Involving them in the urban planning system may be a solution to this problem. When nearly one third of housing stock needs to be transformed to homes and facilities for services for the elderly, substantial financial resources are needed. The asset tied in real property, properly released, could provide additional sources for better coverage of these needs. Many Member States of European Union has to cope with demographic decline and the ageing of European population. How best to finance living conditions, housing of elderly and long-term care, have become a highly topical issues in recent years. In order to maintain a vital society in a vital town of inhabitants, it is necessary to develop new economic and social conditions and a new kind of facility management in European urban areas.

2 The Housing Crisis

Recent experience in several European countries (and also globally) has clearly and painfully shown that the housing market volatility has substantial impact on financial flows, supply chain flows and migration in the society, on wealth of households and on national and international financial systems, especially as collateral for mortgage financing. The macro implications of housing dynamics are more important today than ever, following one of the largest residential real estate boom and bust, as well as the subsequent recession which additionally disclosed the specific problems of ageing European and other developed societies and their dependence on the stable housing market. As it is reported by Guardian (February, 2014, see Fig. 1) there is more than 11 million empty housing units in cities and towns, many of them in the foreclosure and the bankruptcy procedures. There is 3.4 million empty houses in Spain, 2.4 million such residences in France and similar number in Italy. Therefore the share of construction sector in GDP which represents around 10 % of European economy is still in decline in South of Europe like in Spain. Activity of construction of new homes continues to decline in Spain and other Mediterranean countries. The average price of private housing units

Fig. 1 Number of empty properties in across Europe; (*Source* Guardian and SURS, February 23, 2014)

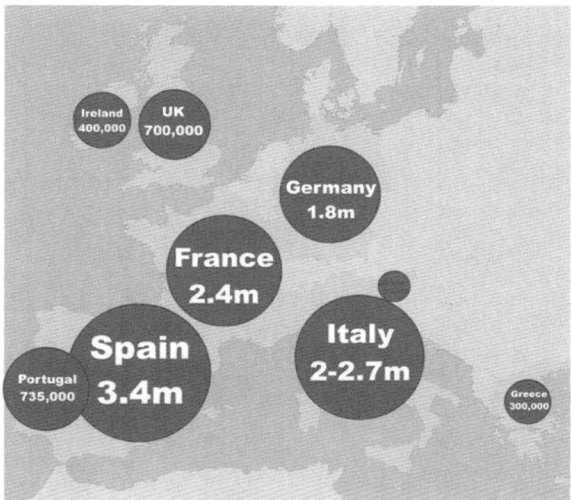

continues to decline and have declined already for 30 % since the beginning of the crisis. Meanwhile, the average price of urban land was in 2013 for 12 % lower than a year ago. Nevertheless the crises the foreign home ownership continues to increase. In the third quarter 2013 it amounted to 17 % of the total housing stock. In Spain the stock of new unsold housing units increased from 227,000 in 2005 to 811,000 in 2012 and only 10–20 % of them have been sold per year in the last 3 years. Following this dynamics on housing market these surpluses will not clear in another 10 years.

There is also trend of rising delinquencies in connection with questionable mortgages used for house purchase. Delinquency rate is 5 % for the new housing units, 8 % for the remodeled and rehabilitated housing units, and 31 % for financing the development and construction of new housing units. In the area of real estate management, the delinquency rate has also continued to increase and now has reached almost 34 % (Ministerio de Fomento 2013). These numbers show that flows of funds to new real estate developments has almost stopped. The major problem represents significant devastation of urban environment due to reduced maintenances—or in extreme cases even abandoned of buildings or complete urbanizations. The main question is how to increase the force of these flows on the pre-crises level and to get this underutilized built environment back into socially productive function. This challenge requires better understanding of the housing market dynamics. The availability of tools that enable policy makers to improve the liquidity of residential housing in Europe is especially important under impact of global crises (Fig. 2).

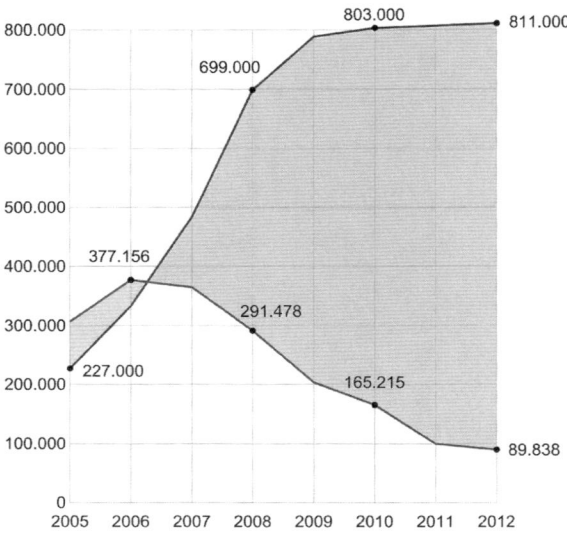

Fig. 2 Growing stock of new housing units in Spain (▬) and number of sold units (▬); (*Source* El PAIS, Nov 21, 2013, Ministerio de Fomento)

3 Existing Models, Improvements and Integration

To improve forecasting of supply and demand on the real estate market and to influence it, different models supporting decisions have already been developed, but they did not prevent the agents in the housing economy to behave in such a way that they would not generate the crises in 2008. The need is to further develop these models and integrate them into a complex decision support system addressing the issues of sustainable growth of cities in an ageing society and the housing market in general.

Six groups of models that were considered separately until now: (1) Gravity models; (2) Competing Risk Model/Multi-state transition model; (3) Housing equity and other financial models; (4) Urban land rent model and the mass appraisal; (5) Facility management and (6) Supply chain modelling should be integrated to support decisions towards Smart Urban Growth. Their integration is suggested like presented in Fig. 3.

3.1 The Gravity Models of Commuting and Migrations Structured by Age Cohorts

The Gravity Model is widely used in the analysis of migration flows in the USA and in other continents, but less well in Europe (Anderson 2011; Bertoli et al. 2013). Using a similar approach Ghosn et al. (2013) and Hunt (2006) addressed also the problems of ageing and the influence of certain indicators, also impact of the built

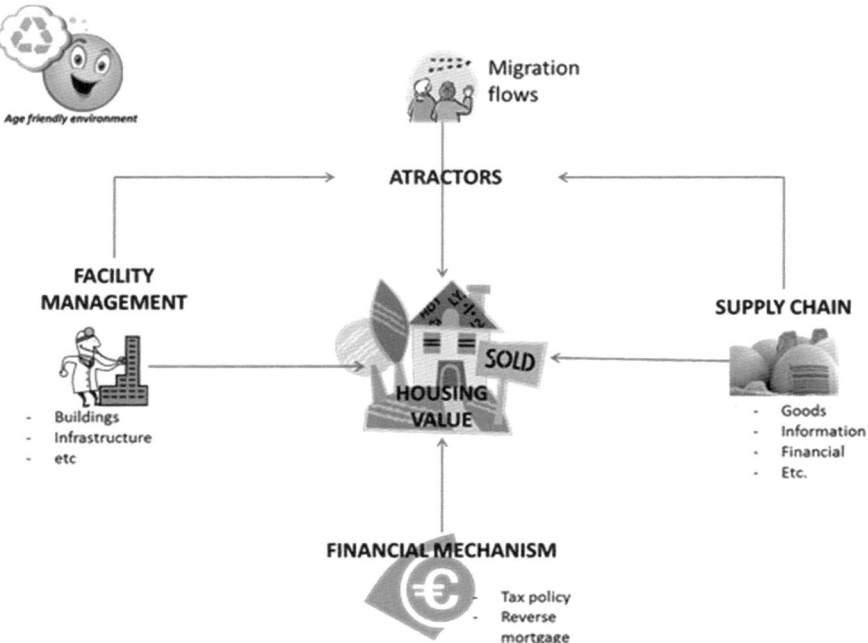

Fig. 3 Integration of forecasting and decision models in models of housing market regulation

environments on the mortality in cities. Therefore the regions need to start by the basic analysis of indicators that influence attractiveness and stickiness of regions in general, like we have analyzed it in the ESPON-ATTREG project (Russo et al. 2012). The detailed analysis of attractors including quality of facility management and supply chains available in the areas under consideration can be provided by in-depth analysis of these indicators (in the origin and destination of migration flows) and their relative value, studying intensity of migration flows:

$$I_{ij} = a^* P_i^{\alpha_1} P_j^{\alpha_2} d(t)_{ij}^{\beta^*} \prod_{s \in S'} C_{s,i}^{\kappa_s} C_{s,j}^{\lambda_s} \qquad (1)$$

In model (1), I_{ij} denotes intensity of flows, a^* is a constant of proportionality belonging to total analysed space (Europe), P_i is the population of origin i, P_j is the population of destination j, C denotes the ratio of analysed factors of attractiveness or stickiness s (ratio between the factor in the municipality and factor for the analysed space, for example all Europe), so s denotes the set of analysed socio and economic factors in the elementary spatial units respectively. α_1 and κ_s are measures of *emissivity* (in some applications also called measures of *stickiness*), while α_2 and λ_s are measures of *attractiveness*, β^* is a measure of *propensity to travel* (in some applications also called measure of *accessibility*) in distance function d.

Parameters $a^*, \alpha_1, \alpha_2, \beta^*, \kappa_s, \lambda_s$ could be evaluated by the regression analysis (Drobne and Bogataj 2012). Statistically significant estimates κ_s, λ_s of analysed factors were included as combined weights into the evaluation procedure evaluating the market potential of a certain area. The study is needed for each cohort of population separately to be able to estimate the ageing structure of buyers. For this purpose, spatial statistics could be explored based on GIS and geo-statistics, but also semi-structured interviews among individuals older than 50 years of age could participate to more precise elaboration of potential housing market and later compared with the results based on geo-statistical investigation to obtain more reliable parameters for the models. We shall start from the indicators evaluated in the ESPON-ATTREG project (Russo et al. 2012; Drobne and Bogataj 2011). The new indicators evaluated through geo-statistics and semi-structured interviews could be added to the metadata model.

3.2 Increasing Attractiveness by Proper Facility Management Through Categorization and Visualization

The inhabitants' need for housing, facility management, social services, health services, etc., changes during the lifetime; i.e. different age cohorts have different needs. Housing and services are important prerequisites for maintaining the citizens' independent living during their entire lifetime. MultiMap, which currently is a tool for strategic building portfolio analysis (Bjørberg et al. 2012), could facilitate mapping, planning and visualization of availability and needs of various age cohorts in an analyzed geographical area. facility management, facility services, health services, social services, etc. could be better managed and reported by MultiMap analysis of particular geographical areas. The demands for residential places differ on social and individual characteristic, also connected with the different age groups (Bjørberg and Temeljotov Salaj 2012). The ageing cohorts can be divided in 4° according to their functional capacities. On the basis of individual needs and quality of the built environment the state in the urbanisation area can be reported as attractor for individual cohorts. The supply of housing units on the market can be categorized on the way of categorization according to extended care dependency scale, facility management excellency and availability of logistic services. MultiMap which currently is a tool for strategic building portfolio analysis (Bjørberg et al. 2012) could facilitate visualization of various age cohorts in defined housing markets. Five-degree neighborhoods and urban development scale can be used and reported to attract the potential buyers of property in the analyzed housing area. Such information system which provide timely information regarding built environment and services including availability (costs) in competing European regions can increase or decrease attractiveness of these areas according to the age cohort and/or functional capacity of potential residents. The attractiveness is finally included in model (1) as a coefficient of stickiness $C_{s,i}^{\kappa_s}$ or a coefficient of attractiveness $C_{s,j}^{\lambda_s}$ and their powers κ_s, λ_s could be calculated.

3.3 Urban Networks Modelling and Supply Chain Control

According to the objective of Madrid International Plan of Action on Ageing (UN, 2002, http://undesadspd.org/Ageing/Resources) improvement in housing and environmental design to promote independent living by taking into account the needs of older persons, in particular those with disabilities is one of the main objective of UN programs. This objectives requires many actions like: (a) Ensure that new urban spaces are free of barriers to mobility and access; (b) Promote employment of technology and rehabilitation services designed to support independent living; (c) Meet the need for shared and multigenerational co-residence through the design of housing and public space; (d) Assist older persons in making their homes free of barriers to mobility and access. The impact of improvement of services (supply chain management, logistics) on attractiveness and therefore on migrations and commuting influence housing market (Bogataj and Bogataj 1988). The bases of this study are well developed model of supply chain management, where time delays evaluated in frequency domain and the net present value (NPV) approach has a special importance (Bogataj and Grubbström 2012; Kovačić and Bogataj 2013). While the attractiveness for service providers is measured through NPV, the level of services could be included in model (1) through coefficients of stickiness or coefficients of attractiveness and their powers could be calculated as in case in paragraph 2.2, but to evaluate the optimal functional region for regional supply networks the model of Drobne and Bogataj (2012) could be applied. The model can acquire data from plan4all or plan4business or similar data service platforms according to INSPIRE directives and return the data back to the multi-purpose databases for measuring attractiveness by (1) and consequently the housing market appraisal.

3.4 Multi-state Transition Model/Competing Risk Model

Physical environments that are age friendly can make the difference between independence and dependence for all individuals but are of particular importance for those growing older. Changes in the environment can lower the disability threshold, thus decreasing the number of disabled individuals in a given community (Kalachea and Kickbusch 1997). Therefore the model (1) developed individually for each group of functional capacity can be structured to forecast overall attractiveness and stickiness when disability thresholds, are determined. The structure is able to determine through multi-state transition model. Multi-state transition/competing risk models have been developed through the last quarter of the twentieth century, while some specifics have been reported in Christiansen and Denuit (2013). Each state in this model has own attractiveness or stickiness level described by (1).

3.5 Urban Land Rent Model and Appraisal of Housing Regarding Financial Mechanisms Available in Ageing Society

The housing market should be studied in two directions: leasing (rent) and ownership transactions. Urban land rent model as previously developed in bid rent theory by Alonso (1964) and further developed in many directions and dimensions can be a base for appraisal of rent and evaluation of the housing market. When having in mind accessibility to different services and facilities needed in society structured by functional capacity, we see that the price and demand for real estate change as the resultant of distances to this facilities and services. Analogous to the basic bid rent theory we can state that different land (housing unit) users will compete with one another for land close to these services and facilities. This is based upon the idea that users wish to maximize their quality of life, so they are willing to pay more for housing close to all kind of services and amenities, and also less for land (housing unit) further away from the optimal area. The attractors, described above through the gravity model increase the number of potential buyers or potential tenants and through increasing demand increase the market price. The potential buyers or tenants compete for the most accessible housing unit in dependence of their functional capacity and financial capacity. The amount they are willing to pay, named bid rent (the transaction value can be seen as its capitalization) depends also on their purchasing power, but the purchasing power depends on financial mechanisms available, text policies and maintenance costs on the level of community. These financial mechanisms should be developed in Europe at least on the level like developed in USA (mortgage, reverse mortgage, social programs of flexible reverse mortgage schemes). Today there are many different possibilities to develop such mechanisms due to low interest rate environment and possibilities of flexible dynamics of equity withdrawal—reverse mortgage products which are able to solve the today's problem of empty houses and reduce the delinquency rate, which now continue to increase in Europe and is over 30 % in Spain. Housing equity withdrawal (HEW, see Bogataj et al. 2012) most commonly takes the form of increasing the amount of debt secured on the house property. This loan can be used to increase the owner's purchasing power or to avoid his delinquency position, but the basic solution is in decision of ECB to accept HEW mortgage securities with insurance as the prime asset for the new loans. Such approach would enable more affordable housing market for the consumers and more stable housing market for investors. Such mechanism would enable clearing the market of the existing empty housing stock based on mortgage and reverse mortgage products. This mechanism would help to start the financial and migration flow for efficient utilization of societal resources tied up in European built environment.

4 Conclusion

The article gives the answer to the question, how comprehensive modelling of interactions between demographic change (ageing and migration patterns), housing dynamics (urbanization), real estate market on the bases of changing urban land rent differentials, taxation (local and state regulation) and mortgage credit provision (financial system) can support decisions with the aim of financial and housing crises prevention and sustainable growth of cities. The paper gives the general approach, but is also particularly focused on housing provision for cohorts 60+. We have seen how connection of existing models through gravity model, where better facility management and improved supply systems increase the parameters of attractiveness and stickiness in the model and therefore the migration flows, which increase demand of housing and therefore the value of bid rent. The improved financial mechanisms based on mortgage and reverse mortgages increase the purchasing power of those flows and additional increase the bid rent curve in dependence of interest rate. Models could be connected in a tool package supporting age-friendly policies where ageing cohorts are studied separately and in a changing age structure according to the Multi-state Transition Model. The models could obtain data from the databases built on standards adopted in Directive 2007/2/EC of the European Parliament and of the Council (2007) establishing an Infrastructure for Spatial Information in the European Community (INSPIRE, http://inspire.jrc.ec.europa.eu/). Integrated simulation packages could be available supporting decisions on urban, regional, and European level and could determine the proper feedback control. Mapping of the availability, affordability, and adaptability of the built environment and required services would provide better support to policy makers using MultiMap solutions.

References

Alonso W (1964) Location and land use. Harvard University Press, Cambridge
Anderson J (2011) The gravity model. Ann Rev Econ 3(1):133–160
Bertoli S, Fernández-Huertas Moraga J (2013) Multilateral resistance to migration. J Dev Econ 102:79–100
Bjørberg S, Temeljotov Salaj A (2012) Backlog of maintenance in public sector a huge challenge for FM. J Facility Manage Heft 5:8–22
Bjørberg S, Larssen AK, Listerud C (2012) Multimap—a tool for strategic analysis of building portfolios. IALCCE, Vienna
Bogataj L, Bogataj M (1988) Computer assisted control of urban growth through the valuation of the communal equipment and the land use value. Harvard Law School and Lincoln Institute of Land Policy, Cambridge, Massachusets
Bogataj D, Temeljotov Salaj A, Aver B (2012) Urban growth in ageing societies: delivering value to the community. Department of Construction Economics and Management, Cape Town, pp 437–446
Bogataj M, Grubbström RW (2012) On the representation of timing for different structures within MRP theory. IJPE 140(2):749–755

Christiansen MC, Denuit MM (2013) Worst-case actuarial calculations consistent with single- and multiple-decrement life tables. Insur Math Econ 52(1):1–5

de España G, Ministerio de Fomento (2013) Observatorio de vivenda y suelo, Buletin núm. 7

Drobne S, Bogataj M (2011) The accessibility and the flow of human resources between Slovenian regions at NUTS 3 and NUTS 5 levels. ESPON, ATTREG, Ljubljana

Drobne S, Bogataj M (2012) Evaluating functional regions. Croatian Operational Res Rev 3:14–26

Ghosn W, Kassie D, Jougla E, Rican S, Rey G (2013) Spatial interactions be-tween urban areas and cause-specific mortality differentials in France. Health Place 24:234–241

Hunt J (2006) Staunching emigration from East Germany: age and the determinants of migration. J Eur Econ Assoc 4(5):1014–1037

Kalache A, Kickbusch I (1997) A global strategy for healthy ageing. World Health 50:2

Kovačić D, Bogataj M (2013) Reverse logistics facility location using cyclical model of extended MRP theory. CEJOR 21(1):41–57

Russo A, Drobne S, Bogataj M et al (2012) ATTREG: the attractiveness of European regions and cities for residents and visitors. Ljubljana, ESPON, 2012

United Nations (2002) Madrid international plan of action on ageing. Available on http://undesadspd.org/Ageing/Resources, cited Feb 2012

The Role of the Entrepreneur in the New Technology-Based Firm (NTBF)

Juan Antonio Torrecilla García, Agnieszka Grazyna Skotnicka and Elvira Maeso-González

Abstract The aim of this article is to stress the role of entrepreneur from the point of view of the New Technology-based Firm (NTBF) perspective. The importance of this group of companies for the economic growth and their input into the economic development of regions, make the NTBFs an interesting issue to be analyzed. Although this group is relatively small but its potential to lead a substantial change of productive model of the country, is remarkable. The conceptual survey will allow us to contribute to formulating of a clarification of the entrepreneur's role at the New Technology-based Firm.

Keywords New technology-based firm · Entrepreneurship · Start-up · Entrepreneur · Technology

1 Introduction

Academic literature survey refers to the different factors and forces that cause the competitiveness of a nation in comparison to the others. Various authors come up with distinctive and relevant elements of the competitiveness of a country, such as: the capacity of its economy, the entrepreneurial initiative, the R&D investment, the culture, the developed knowledge, the human capital, the capacity to innovate, the intellectual capital, etc. Considering all factors that participate in greater or lesser

---

J.A. Torrecilla García (✉) · A.G. Skotnicka · E. Maeso-González
Universidad de Málaga, Málaga, Spain
e-mail: juantorrecilla@uma.es

A.G. Skotnicka
e-mail: ines@emotools.com

E. Maeso-González
e-mail: emaeso@uma.es

degree in the competitiveness processes of a country, we can set them in three collective concepts:

1. Capacity of raising external economic activity
2. Potential for internal economic activity generating by fostering the enterprises
3. Potential for internal economic activity generating by fostering the technology-based entrepreneurship

As far as the third collective aspect settled in the previous paragraph, the authors as Trenado and Huergo (2007) contemplate the public administrations have been designing different regional policies to enhance the New Technology-based Firms creation, as a result of NTBF potential awareness. They, however, clarify that policy design manifest errors and ambiguities because of lack of precise knowledge on what these organizations really are.

2 General Approach

The actual global crisis and its structural consequences impel us to deliberate on if the actual productive model is an adequate one, or if there is a need for a new economic model that implies an improvement of the productivity, the innovation and the sustainability, March et al. (2010).

Diaz et al. (2013), following the same general paradigm, raise the necessity of a change in the ways of production, that could allow not only a sustainable growth, but also that would contribute to certain stability between the economic cycles. To obtain the change in productive model, the creation of technology-based firms is considered as an important step. Many authors assume that the entrepreneurs of knowledge-intensive firms are major contributors to economic growth, development, and cycle change. March et al. (2010) keep up with this critical importance of entrepreneurs by suggesting the institutional support to the initiatives and industries of, so called, new economy, in focus to harness the NTBF creation.

According to this approach, Leon (2000) considers that the creation of the NTBFs is a step in the correct direction, due to its tendency to generate economic growth, to facilitate a superior innovating activity, to achieve the high quality employment, to produce more competitive goods and services and by its potential in the technology transfer (Quintana and Benavides 2012; Trenado and Huergo 2007).

Another excellent criterion is the one raised by Palacios et al. (2005), the importance of being the active part of a global economy, where the competitive forces are strong enough to select only products and services with real value added. Thus, the technology becomes the fundamental factor of the economic progress of any territory. However, other authors assume that it is a technology-based innovation, not technology, the competitiveness ingredient. They describe Spanish economy is characterized by the shortage of technology and technology-based innovations, with strong deficiencies in R&D investment and in a culture of innovation. According to the theorists, the productive capacity of Spanish economy will depend on innovation potential of companies already established, mainly the New Technology-based Firms.

Beraza and Rodriguez (2012) emphasize the globalization and the Knowledge-based society have transformed the links between technology, science and economy, and granting, this way, prominent role of the innovation in the employment, business opportunities, and in the economic growth and development. According to Díaz et al. (2013), the Member States of the European Union have reached an agreement to turn the European Common Market into a Knowledge-based economy and to consider the NTBFs as a fundamental part of this ecosystem.

Similar conclusions are drawn in March et al. (2010) work, granting the transition towards a new economy, with better productivity and competitiveness ratings. Thus, the New Technology-based Firms are enhancing key players of such change. In addition, the authors survey the "new economy" concept and describe the industries that compose it. These industries, so called emergent ones, are characterized by knowledge intensive aspects and enabling environment for the development and use of advanced technologies (high-tech). As emergent industries are recognized: Information and Communication Technologies (ICT), computer science services, microelectronics, biotechnology, new materials, pharmaceutics, electronics, alternative energies, nanotechnology, aeronautics, robotics, optoelectronic and photonics. The common elements that outline all of these industries are the following:

1. They are R&D intensive
2. Within continuous process of innovation
3. Industries with strong tendency to well trained and specialized workers hiring
4. They cause the multiplying effect to local economies
5. Multidisciplinary and multi applicability over the traditional industries

In this sense, Trenado and Huergo (2007) argue that the relevance of the New Technology-based Firms in the economy is clearly recognized through the scientific, economics and managerial, literature. However, Barajas and Ubierna (2011), emphasize the cardinal importance of these companies for the economic growth, in spite of being a very reduced group in comparison to the total set of enterprises.

Before turning to the implications of the role of the entrepreneur, we draw the conclusion, as stated by Salas (2010), that the entrepreneur is a key figure in the process of knowledge transfer and use. It is the nexus between the science and the market. It is the starting point for the technological advances that will lead to the product development and subsequent commercialization (Fig. 1).

The technical and scientific knowledge is usually complex to understand and apply outside the academic environment. This characteristic can constitute a serious obstacle to the commercial exploitation. The micro- and macroeconomic potential—that the technological advances and innovations provide to the development of new products, improvement of existing ones and its market entry- presents a challenge and an opportunity to be explored in due to accelerate the socioeconomic development.

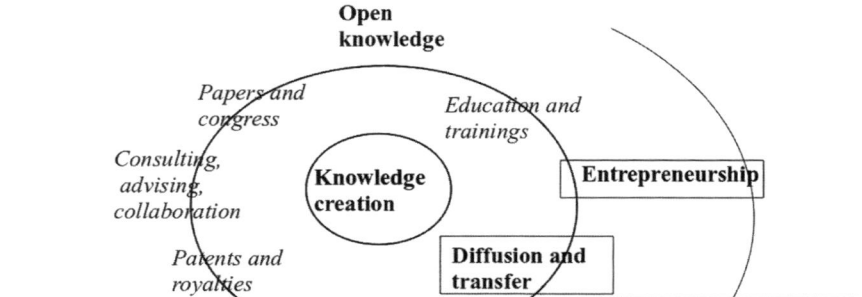

Fig. 1 Production, transfer and use of knowledge: entrepreneurship as a bridge. *Source:* self elaboration based on Salas (2010)

There are required persons capable to—by undertaking the entrepreneurial initiative—translate the language of science into business concepts, and in this way, to take the role of the nexus/bridge between both parts, generally lagging participants of the any communication. As it is well known, not always the scientific progress, or even knowledge advances, has an immediate impact on the market. That gap of R&D results' commercialization causes its lower economic productivity, possible lack of acceptance by the market, or requirements of additional innovations or improvements; reasons introduced by Fagerberg (2005) and cited by Souto (2013). This author places a historical example to illustrate the previous findings: Leonardo da Vinci and his sketches about the possibility men could fly. This innovative idea was not materialized until the new materials and modern means of mechanical traction were developed. Reason why there is a need for highly qualified people who serve as a link between the scientific knowledge generation and its commercial application. People who are able to design and produce goods and/or services that can be placed on the market. These qualified people able to interpret the knowledge, are precisely entrepreneurs, the key pieces of the economic dynamism.

In relation to the subject previously exposed, we define the NTBF's entrepreneur as *"any of company founder who achieves to create a New Technology-based Firm and who makes possible the economic growth, by means of the combined resources*

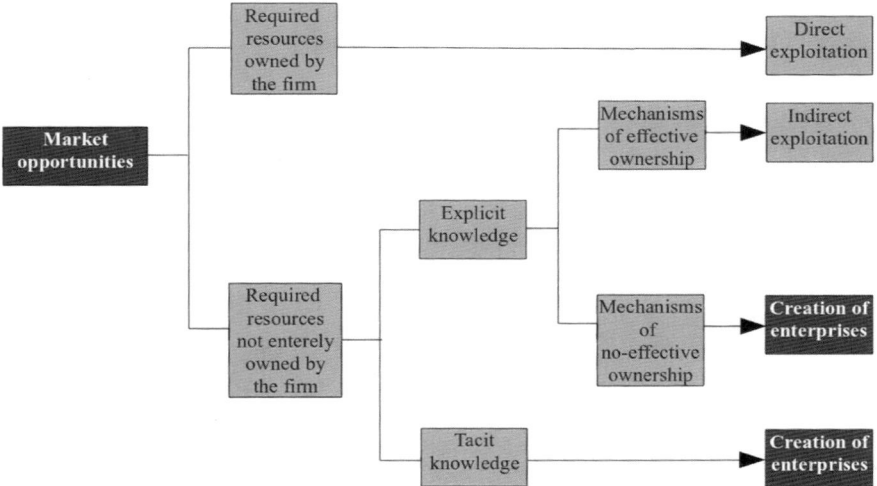

Fig. 2 Generation and ownership of the results of innovation. *Source:* Trenado and Huergo (2007) based on Álvarez and Barney (2004)

during the period from its birth to 42 months[1] (when companies consolidated on the market loose the denomination of entrepreneurial organization)"

But the scientists handle with several strategies to exploit the results of their R&D activity: for example the patents, characterized by a smaller investment in comparison to the NTBF creation. In this sense, Trenado and Huergo (2007) ask the following question: why then the New Technology-based Firms are constituted? To answer it, we have found some interesting aspects in Álvarez and Barney (2004) publications and we accomplish the sketch of the following figure (Fig. 2):

The creation of the NTBF, aimed to exploit the market opportunity for the new idea, depends on the combination of the following three elements:

(a) The level of control of all the resources required for the direct exploitation of generated knowledge.
(b) The commercialization of explicit knowledge is easier than the one of tacit knowledge.
(c) The effectiveness of the mechanisms of ownership and control over benefits generated within the innovation.

---

[1] The GEM report describes several stages of the entrepreneurial process. At this survey we can study the perspective that considers as an entrepreneurial activity those within the period of duration from the birth to 42 month.

As far as these three described factors concern, the entrepreneur will find him/herself "forced" to constitute a New Technology-based Firm, when:

(i) The entrepreneur doesn't have ownership or even control over the necessary resources and being the knowledge the tacit on, it's impossible to transmit commercially.
(ii) The second case has common ground with previous lack of control over the necessary resources required for direct exploitation, but although it is an explicit knowledge, where the mechanisms of appropriation of innovation results, are not effective. So it can cause the difficulties in innovation results benefits obtaining, and leads to NTBF creation.

At the majority of cases, these technology-based entrepreneurs have a technical and scientific profile. Storey and Tether (1998), in their insight on NTBF, review multiple studies on NTBF differential aspects and conclude that the entrepreneurs of this type of companies have some specific characteristics that make them different from the rest of companies founders, such as:

(a) Higher educational level; that confirms the previous assumption of the need of people able to decipher the scientific knowledge.
(b) Usually in possession of an extensive professional experience, predominating jobs for private companies and research centers.
(c) As a result of the previous one, the average age of undertaking is superior to the age of general companies' founders, between the 30 and 50 years.

However, in later studies, for example the one carried by Colombo and Delmastro (2001), cited by Fernandez and Hidalgo (2011), the another in-depth aspect has been observed, the entrepreneurs of Internet-based firms usually have an inferior age and an educational degree below the average companies founders of other industries. Another survey detects differential characteristics of NTBF entrepreneurs when they operate in different industries (Trenado and Huergo 2007). Complementing the role of entrepreneurs, we observe the lack of homogeneity of this entrepreneurial group, depending on the industry or technology they operate.

For Barajas and Ubierna (2011) the technological basis of the NTBF requires an extra effort focused on an exact explanation of the links between the business idea and the team of company founders. The emphasis is put on the requirement of certain degree of professional experience and higher education, not only to start the business but also to consolidate it on the market.

3 Conclusions

The difficulty for the general public to interpret correctly the potential of the knowledge, requires and demands a proactive approach from the entrepreneurs who are to be an emergent force of not only creation of companies, but also of the change of/for the society. The technology-based initiatives are supposed to became the vehicles of any possible substantial changes in people's life. So, as this change requires the transformation of knowledge (science, basic research, etc.) into the potential commercialization objects on the market, thus, the role of the entrepreneurs of the New Technology-based Firms lies in becoming the bridge between the knowledge generating environments of science and academia, and the market ruled by acts of satisfying the people's necessities.

Due to establish an empirical frame for the further study on the role of entrepreneurs, we provide the following definition of the NTBF's founder (Technology-based entrepreneur): *"any of company founder who achieves to create a New Technology-based Firm and who makes possible the economic growth, by means of the combined resources during the period from its birth to 42 months (when companies consolidated on the market loose the denomination of entrepreneurial organization)"*

References

Álvarez SA, Barney JB (2004) "Organizing rent generation and appropriation: toward a theory of the entrepreneurial firm." J Bus Venturing 19:621–635

Barajas MA, Ubierna A (2011) "Creación de empresas de base tecnológica, análisis de casos", Economía industrial, 382, pp 129–140. Ministerio de Industria, Energía y Turismo, Madrid

Beraza JM, Rodríguez A (2012) "Conceptualización de la spin-off universitaria, revisión de la literatura", Economía industrial, 384, pp 143–152. Madrid

Colombo MG, Delmastro M (2001) "Technology-Based Entrepreneurs: Does Internet Make a Difference?," Small Business Economics, 16(3):177–190, May, Springer

Díaz E, Souto JE, Tejeiro MR (2013) "Nuevas empresas de base tecnológica. Caracterización, necesidades y evolución en un periodo de crecimiento y en otro de ralentización y recesión económica (2004–2012)", Fundación madri+d y Netbiblo

Fagerberg, J. (2005) "Innovation: A guide to the Literature", In: Fagerberg J, Mowery D, Nelson, R (eds) The Oxford Handbook of Innovation, Oxford University Press, Oxford, 2005, pp 1–26

Fernández JC, Hidalgo A (2011) "Empresas tecnológicas creadas en España entre los años 2000 y 2010: perfil del emprendedor e importancia de la ayuda pública en su desarrollo" Dirección y organización: Revista de dirección, organización y administración de empresas, 45, pp 11–19

León G (2000) "La creación de empresas de base tecnológica desde el sistema público". Boletín SEBBM (Sociedad Española de Bioquímica y Biología Molecular) 128. Madrid

March I, Mora R, Yagüe RM (2010) "Las EBTs como motor de la nueva economía y revulsivo ante la crisis", XII Reunión de Economía Mundial de la Sociedad de Economía Mundial (SEM), Universidad de Santiago de Compostela

Palacios M, Del Val T, Casanueva C (2005) "Nuevas Empresas de base Tecnológica y Business Angels", Revista madrimasd, 31, Madrid

Quintana C, Benavides CA (2012) "Empresas de base tecnológicas", publicado en Jimenez, J.A. (Coord.), 2012, "Creación de empresas. Tipología de empresas y viabilidad estratégica", pp 63–84. Pirámide, Madrid

Salas V (2010) "Emprendedores y desarrollo económico", ciclo de Cátedra "La Caixa": Economía y Sociedad, del 12/04/2010 al 14/04/2010, CaixaForum Madrid

Souto JE (2013) "Innovación, emprendimiento y empresas de base tecnológica en España. Factores críticos e impacto sobre la competitividad de la economía", Fundación madri+d y Netbiblo

Storey DJ, Tether BS (1998) "New technology-based firms in the European Union: an introduction." Research Policy, 26, pp 933–946. Amsterdam

Trenado M, Huergo E (2007) "Nuevas empresas de base tecnológica: Una revisión de la literatura reciente", DT. 03, Departamento de Estudios, CDTI

Part IV
Quality and Product Management

Increasing Production and Minimizing Costs During Machining by Control of Tool's Wears and or Damages

Nivaldo Lemos Coppini and Ivair Alves dos Santos

Abstract Industries involved in machining processes need to increase their competitiveness to adequately meet the market demand, i.e. better quality of their final products and a more efficient machining process. There are some industries that prefer to work with the managers of machining tools in order to control the costs of the proceedings. However, the authors of this article have the opinion that it is essential to have professionals hired directly by the company to play the role of improvements introducer to the machining process. So, following this procedure, this paper proposes the use of the Six Sigma methodology to the management and stratification of the consumption and tool breakage during the machining process. Thus, the DMAIC method was used. By providing data collection during research in the areas of factory production, causes of tool breakage were identified. With this information were initiated actions seeking process improvements and also attempting to define new ideal conditions, both of machine tools, as tools. The result showed that it was possible to propose a model for controlling the tool breakage and, undoubtedly, contribute to reducing costs and increasing productivity.

Keywords DMAIC · Costs · Productivity · Tool wear · Tool damages

1 Introduction

Competitiveness is an important and constant factor for companies who intend to survive in the market. This is true for any productive sector and in particular is very true for the production sector by machining. The total cost of machining

---

N.L. Coppini (✉) · I.A. dos Santos (✉)
College of Mechanical Engineering, University of Taubaté—UNITAU,
Rua Daniel Danelli, S/N, Taubaté, São Paulo, Brazil
e-mail: nivaldocoppini@gmail.com

I.A. dos Santos
e-mail: santos_ivair@hotmail.com

corresponds to a part of the apportionment of the costs of the final product and decisively influences in the annual accounts of the company. According to Harry and Schroeder (2000), the researches show that companies adopting the Six Sigma methodology in their businesses have reduced their costs faster than their competitors who do not use it. Furthermore, achieve higher profit margins and greater customer satisfaction. According to Pande (2000), to succeed in reducing the analysis of the complexities of machining processes in the industry, DMAIC have been used, which is one of the SIX SIGMA tools. In large part, the success of companies adopting the DMAIC can increase their ability to reduce the complexity in the process of machining. The lack of a systematic management of tools breaks involving both quantity breaks as the cost of the tools and the cost to replace them, can create losses in terms of productivity and competitiveness in the labor market. Just as machining tools need to be developed for better performance and lower cost of production, companies also need to improve their way of managing these tools.

The aim of this study was to conduct a case study to evaluate the possible effectiveness of the implementation of the DMAIC method to introduce it in the management of premature failure of tools in an automobile company.

2 Review of Literature and Theoretical Foundation

2.1 DMAIC Method

According to Harry and Schroeder (2000), although the DMAIC method is sometimes represented in a sequential manner, the phases and steps do not always happen in a direct sequel.

According to Breyfogle et al. (2001), McAdam and Lafferty (2004) eight of the ten techniques among the most used tools in the DMAIC method are concentrated in step "Measure". Such occurrence is justified because at this stage the tools that measure the performance of machining processes are applied, which allows visualization of the current state of them for the target setting of improvements.

DMAIC is a tool consisting of lean manufacturing steps to be followed for its perfect application. They are: "Define, Measure, Analyze, Improve and Control." After these, the stages are defined.

- The **Define** Phase is to improve the understanding of the Project Team on the problem. This phase is to establish who are the internal or external clients to the target company from the method application and define customer needs and expectations. At this stage the Team is organized, establishing roles and responsibilities, goals and timelines and reviewing the process steps;
- The **Measure** Phase establishes the techniques and collects the data about the current performance allowing highlight the opportunities for improvement;
- The **Analyze** Phase is to visualize the opportunities for improvement based on the results of the analysis of the set of measured data. In this phase, the Project

Team should have a complete understanding of the factors that impact and should generate ideas on how to improve the process, project future improvements, make pilot plans, implement and validate improvements;

- The **Improve** Phase has the purpose of identify, test and implement a solution to the problem; in part or in whole. Identify creative solutions to eliminate the key root causes in order to fix and prevent process problems;
- The **Control** Phase is to institutionalize the improvements introduced to the system and be responsible for monitoring and ensuring the best performance achieved.

In accordance with Klefsjö et al. (2001), the benefits of SIX SIGMA are the main attractions that has motivated companies in the program. However, to evaluate each obtained benefit concretely with the implementation of SIX SIGMA, survey data are needed through researches with companies that apply the program or, that these organizations evidence their achievements through reliable information, such as balance sheets, letters to the shareholders or disclosure in the specialized press (Hoerl 2001).

It is reinforced by Antony and Banuelas (2002) that, if the main driver of the organization is maximizing the profit, it is necessary the creation of targeted projects to the business processes that provide attractive financial results, with a strong decrease in the index of definitive tailings, strong decrease in the rework rates and increase of the productivity. The authors add that in every SIX SIGMA project, the links between the research objectives and business strategy must be clearly identified.

Rodebaugh and Snee (2002) show explicitly the need to carefully examine the portfolio of Six Sigma projects to be initiated, in order to guarantee that the work proposals, that are in line with the strategic needs for improvement of the organization, are developed.

According to Carvalho (2002), the success of Six Sigma programs can not be explained only by the extensive use of statistical tools, but also for the harmonious integration of management by process and guidelines, keeping the focus on customers, in critical processes and in the company outcomes. Among other things, this new approach has made, unlike other quality programs, companies that use Six Sigma to disclose millions of chords gains from your projects.

2.2 Machining Costs

According to Wu and Ermer (1966), the importance of being able to select the optimal machining conditions has been recognized in the field of machining of metals. The basic mathematical model that has been used in the economic analysis is the model of unit cost, or the similar model of unit time.

These mathematical models adopt the criteria for determining the optimal cutting conditions: a model for determining the minimum cost and the other for maximum production. If, in a production sequence there is a "bottleneck" operation is recommended to work on cutting conditions for maximum production. However, it is more often than minimize costs, under the assumption that these operating conditions will eventually be possible to increase the profit.

The members of a team formed by the DMAIC method are responsible for finding ways to reduce costs in the process, introducing operational improvements aiming a greater yield of the processes.

2.3 Breakdowns, Burnouts and Life Tools

According to Diniz et al. (2013) the life of a cutting edge of a tool may be interrupted due to damage, or wear mechanisms.

The breakdown is an accidental process of destruction of the cutting edge of a tool. The breakdowns must be identified during the development process, and then should be taken arrangements to make it do not happen again. The major breakdowns are mechanical cracking, thermal cracks and micro cracks due to the sudden change or due to the intermittent cutting temperature variation, respectively; plastic deformation of the cutting wedge of the tool and breaking due to variations of the cutting forces (the input part with very high cutting speeds, presence of hard skins; variation of machining allowance). Any of these breakdowns can decree the end of the tool life in a sudden and unexpected way.

The wear, unlike the breakdowns, are a phenomenon of loss of tool material in the contact region with the workpiece due to different wear mechanisms, such as: mechanical abrasion (occurs throughout the range of speeds displayed to the tools, but are crescent with the increase of the cutting speed), adhesion (preferably occurs for lower speed ranges and consists of the formation of a fixation points of the workpiece material on the tool); false cutting edge (occurs for lower ranges cutting speeds and consists of fixation and growth of workpiece material on the surface of the tool output does not occur when the shear rate is higher). Unlike the breakdowns, the wears can be controlled by using machining parameters and especially by using carefully selected cutting speeds.

2.4 Machining Cells

According to Ohno (1997), the physical arrangement or layout is one of the factors directly related to the increase of productivity in companies, since it is a study of the spatial distribution or the relative positioning of the various elements that make up the job.

The physical arrangement called production cells is obtained when the machines or jobs are grouped so that each group (cell) manufactures the products from the beginning to the end of production (preparation, assembly and production). In the cell, products are produced in a given family which have some similarity in the production process.

The production cells depend mainly on the multifunctionality of operators, i.e., everyone should be able to operate the various machines that make up the cell. Some benefits of the physical arrangement in machining cells are:

- Better communication between the operators;
- Operators result more trained and capacitated with multifunctional activities;
- Elimination of intermediate stocks;
- Easier movement of goods;
- Lowest level of material in process;
- Reduction of the preparation time of the machines;
- Increase in productivity.

The use of production cells is directly connected to the concept just in time. It is worth noting that the work in the form of cells has the production machines used in a more dedicated way than in the traditional system. Thus, the concept of work cell demands a system of machine maintenance very efficient, as these are no longer shared. With machines more oriented by processes than by product or family of products, the new provisions of the machining cells are an alternative to increase the productivity as the concept of Lean Production. It is the replacement of the concept of individual productivity by overall productivity.

Shingo (1996) suggests that lean production is applicable to any factory, but the system must be adapted to the characteristics of each particular plant. Lean production is a conceptual framework that is based on principles that contribute to improving company performance: the total elimination of waste, continuous improvement, zero defects, production and just in time delivery (JIT); pull production; multifunctional teams; integrated functions, and decentralization of responsibilities (Sanches and Pérez 2001).

According to Sebastiani (2004), modern machines with CNC control, are ideal for a system of lean production and these can add a number of advantages, such as:

- Reduced machining time;
- Greater repeatability in the sequence of operations, making the predicted standard times safer;
- With the standard times safer, it has a greater accuracy in the calculation of costs, man-load control;
- Greater repeatability in the performance of tools. Because it is a machining with efforts, constant speeds, uniforms, and repetitive, it keeps the wear under control. This facilitates the control of the stock, the better development and suppliers test, a better control of the tools process by the wear, avoiding reworks and scraps;
- Reduced setup times making possible the production of small batches;
- Reduction in finished items in stock, so by enabling small batch production;
- Reduction in the time and frequency with which quality checks are performed.

2.5 Machining Lines: Concepts and Definitions

Assembly and machining lines are two basic categories of production lines, which differ from each other both by their characteristics as the models used for their analysis. Thus, while the assembly lines are characterized by high availability of machinery and equipment and intensive employment of staff, the machining lines in general are based on the intensive application of capital (machinery and equipment) and therefore its capacity is highly sensitive to the availability of physical resources. In the analysis of assembly lines are commonly used balancing models while probabilistic models of rows are often used to analyze the machining lines.

According to Askin and Standridge (1993), a transfer line is a "set of automatic machines and inspection stations arranged in series and connected by a common control system and material handling."

3 Methodology

This work is characterized as a case study developed on the plant floor of an industry in the automotive sector. The entire process was monitored and controlled by the researchers, who combined different forms of collecting data in the process. The instruments used were pen, paper and a fishbone diagram used to perform the brainstorming within the plant floor, providing hearing all operators involved in the machining of workpieces. The Minitab software was used to confection the graphics needed to assist in the methodology of SIX SIGMA/DMAIC.

The company selected for the study of this work operates in the sector of machining and assembly of automobile motors. This case study was developed to introduce improvements to the production process and adjust the machining process to the market reality.

The goods produced were developed according to the needs of the market, however, are manufactured in high volume, using a pull production. Due to the high cost of machining tools is necessary constant work to verify the damages and wears that generate premature end of life of tools.

This paper seeks to show according to the Six Sigma methodology (Harry and Schroeder 2000), a correct way to undertake efforts in the areas that raise the cost of production, since this cost includes tools, management tools, cutting fluids, labor, downtime and maintenance.

Although the target company of the research has a contract with another company to perform tools management, the latter does not focus to conduct surveys that demonstrate the appropriate use of the tools. For its part, in fact, the contract is primarily focused on business. The commitment is to keep tools beside the machine to perform the fast exchange work, focusing only on the loss caused by downtime on the machine with the lack of tools, not managing costs and the reasons of the end of life and their consequences on costs and productivity.

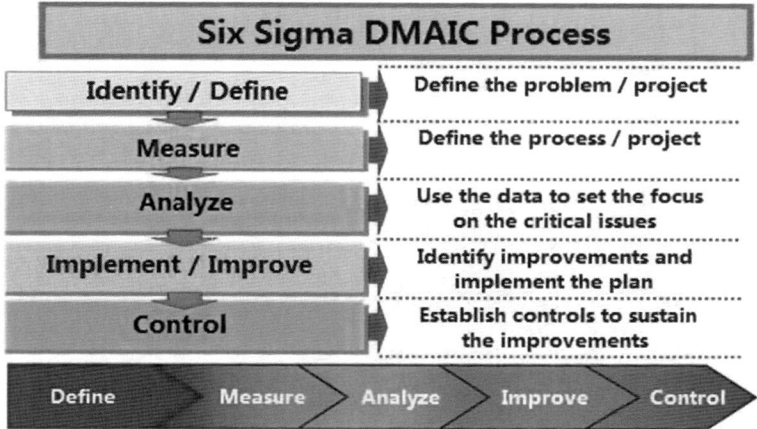

Fig. 1 DMAIC methodology used in the research

The survey of reasons for breaks was performed according to the DMAIC methodology, Harry and Schroeder (2000). Figure 1 illustrates the DMAIC sequence used during the application of the methodology.

Following the methodology, a survey was conducted to identify the end of life of tools, in which machines it occurred and what are the costs involved in each case. This procedure allowed us to map the events and, thus, define and analyze the results, determining the points where improvement efforts is necessary.

The survey was exhausting, however, it is impossible to present it in its entirety in this work, because there would not be enough space for that. Therefore, only one case more typical was presented and it is shown in Fig. 2. It was the case of T71 and T73 tools.

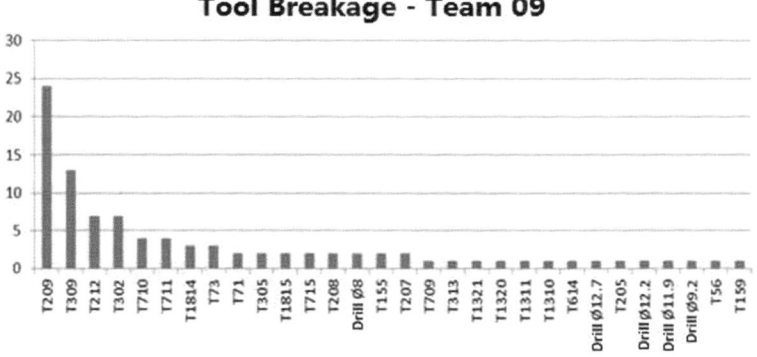

Fig. 2 Comparison among the tools for incidence of breakage of the cutting edge

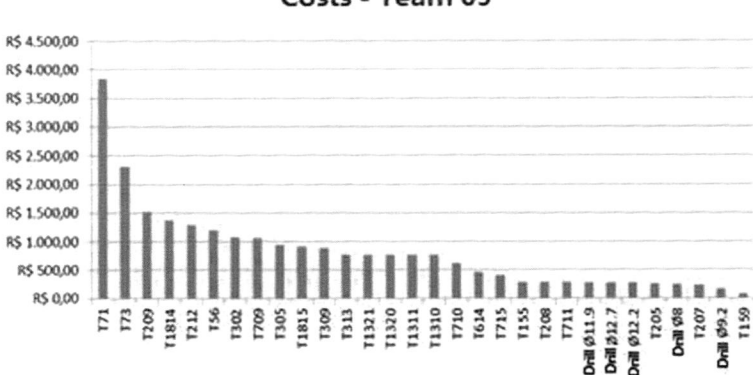

Fig. 3 Comparison among the tools considering the impact on the cost

In Figs. 2 and 3, it is observed that with the survey it was found that usually the tool that shows frequency of premature rupture it is not the one that leads the cost in this sector and Fig. 4 shows the cost of the highest value tool.

Following the methodology, after the survey, is necessary to devise a Fishbone Diagram and a Matrix of Cause/Effect. The purpose was to find the main causes of premature rupture of the cutting edges of the tools by the evaluation of the behavior of the tool and the characteristics of stability and performance of the machine that performed the machining. In this case, the T71 and T73 tools were used as example. Figure 5 illustrates the Fishbone Diagram. Figure 6 illustrates the matrix of cause/effect and Fig. 7, the Pareto's chart.

From the observation of Figs. 5 and 6 it is possible to detect and quantify the frequency of the main causes of breakage of tools. Then, with these data, we can construct the Pareto's chart relating to these causes to define the main actions to be taken, as illustrated in the graph of Fig. 7.

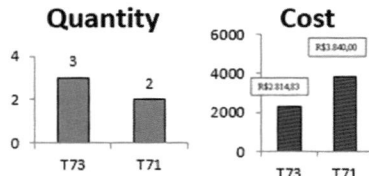

Fig. 4 Cost of the evaluated tools

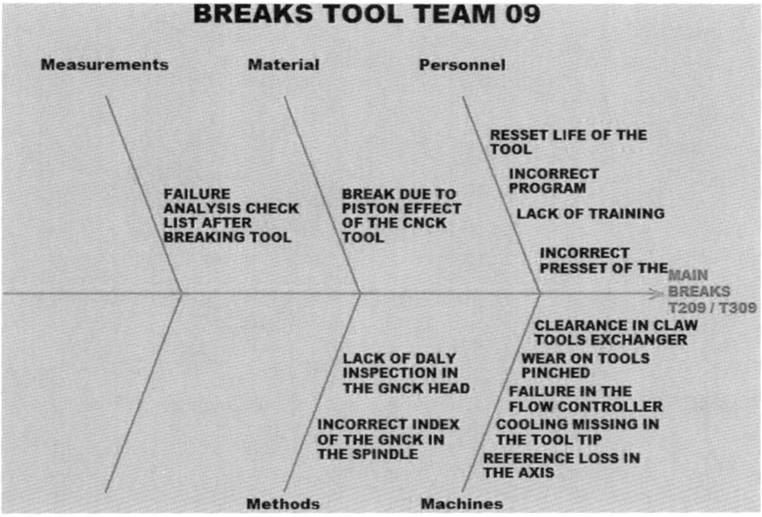

Fig. 5 Fishbone diagram for T71 and T73 tools

Fig. 6 Matrix of cause/effect for T71 and T73 tools

Rating of Importance to Customer		9			
Feature #		1	2	3	
Y / X's — Process Inputs		Quebra - T71 / T73			Total
1	Troca de ferramenta e falta de medição imediata.	1			9
2	OP#270 Perda de referência na máquina.	9			81
3	Aresta postiça (Emprastando material na ponta da ferramenta).	5			45
4	Resset de vida útil de ferramenta na OP#270.	5			45
5	Ajuste de programa incorreto, falta de treinamento.	5			45
6	Cilindro giratório do pallet	1			9
7	Sistema de limpeza do pallet.	1			9
8	Fixação do pallet.	1			9
9	Falta verificação da manutenção autonoma diária.	1			9
10	Falha no check air da Máquina 441 / 442.	9			81
11	OP#270 Falha no controlador de vazão.	5			45
12	OP#270 Máquina 442 vibração excessiva no fuso esférico.	5			45
13	OP#270 Máquina 441 desgaste na pinça do trocador da ferramenta.	5			45
14	OP#270 Folga na garra do trocador de ferramenta Máquina 441.	5			45
Total		522	0	0	

Fig. 7 Pareto's chart

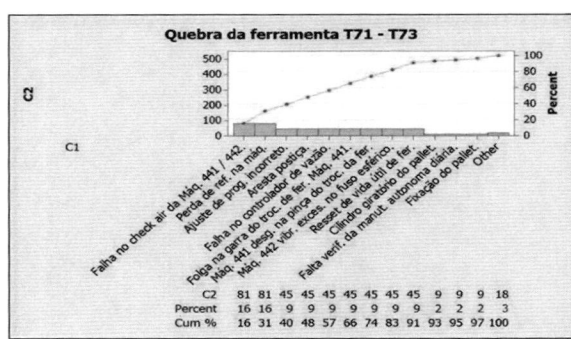

4 Discussions

The effectiveness of the SIX SIGMA principles was grounded in a systematic way using the DMAIC in the definition of the main causes of breakage of tools in the machining process. It was found that the main factors that caused the breakage of the cutting edges of the tools were related to the machine element: check air of the devices of workpiece clamping and loss of reference of the axes according to the matrix of Cause/Effect and Pareto Charts previously shown in Figs. 6 and 7, respectively. The results of applying this methodology allowed to demonstrate to the leadership where were the major improvement opportunities, thus enabling the maintenance focus to be directed to the necessary corrective actions. Opportunities to implement preventive maintenance directed to the true causes of the breaks were also observed. The mapping used to identify breaks and tool costs showed a significant change for the group of managers of the company and can be seen as follows:

- Assistance while using the DMAIC for managing of the main breaks of machining tools, integrating all production teams, managers and suppliers;
- Large reduction in breakage of tools, what reflects directly in the cost of machining;
- A systematic way of directing the maintenance to work for the major causes.

5 Conclusions

After this research, it was concluded that:

- The main reason for the premature end of life of tools was the breakage followed by issues of clamp workpieces devices;
- After eliminating the problem of breakage, the economy was R$15,126.49. The productivity, in this operation, had an increase of 35 %, calculated using the efficiency indicators of the machining line studied;

- The methodology allowed prioritizing the causes of broken tools in order to take effective actions, resulting in reduced costs and increased productivity, demonstrating a high level of confidence in the work performed.

References

Antony J, Banuelas R (2002) Key ingredients for the effective implementation of six sigma program. Measur Bus Excellence 6(4):20–27
Askin RG, Standridge CR (1993) Modeling and analysis of manufacturing systems. Wiley, New York
Breyfogle FW III, Cupello JM, Meadows B (2001) Managing six sigma: a practical guide to understanding, assessing, and implementing the strategy that yields bottom-line success. Wiley, New York
Carvalho MM (2002) Seleção de Projetos Seis Sigma. In: Rotondaro RG (Org) Seis Sigma: estratégia gerencial para melhoria do processo, produtos e serviços. Atlas, São Paulo
Diniz AE, Marcondes FC, Coppini NL (2013) Tecnologia da usinagem dos materiais. Artliber Editora Ltda, Campinas, SP, Brasil, 8ª Edição, pp 230–248
Harry M, Schroeder R (2000) Six sigma: the breakthrough management strategy revolutionizing the world's top corporations. Currency, New York
Hoerl R (2001) Six sigma Blak Belts: what do they need to know? J Qual Technol 33(4), 391–406
Klefsjö B, Wiklund N, Edegman RL (2001) Six sigma seen as a methodology for total quality management. Measur Bus Excellence 5(1):31–35
Mcadam R, Lafferty B (2004) A multilevel case study critique of six sigma: statistical control or strategic change? Int J Oper Prod Manage 24(5):530–549
Ohno T (1997) O sistema Toyota de produção: além da produção em larga escala. Bookman, Porto Alegre
Pande PS, Neuman R, Cavanagh RR (2000) The six sigma way: how GE, motorola and other top companies are honing their performance. New York, McGraw-Hill
Sánchez AM, Pérez MP (2001) Lean indicators and manufacturing strategies. Int J Oper Prod Manage 21(11):1433–1451
Sebastiani DJ (2004) Desenvolvimento e Utilização das Máquinas-Ferramenta com Comando Numérico Computadorizado (CNC) Aplicados no SENAI e nas Industrias Metal- Mecânica. Novo Hamburgo, Monografia, Agosto
Shingo S (1996) O sistema toyota de produção: do ponto de vista da engenharia de produção, 2nd edn. Porto Alegre, Bookman
Snee RD, Rodebaugh WF Jr (2002) The project selection process. Qual Prog 35:78–80
Wu SM, Ermer DS (1966) Maximum profit as the criterion in the determination of the optimum cutting conditions. J Eng Industry Trans ASME 88:435–442

Impact of 5S on Productivity, Quality, Organizational Climate and IS at Tecniaguas S.A.S

Paloma Martínez Sánchez, Natalia Rincón Ballesteros and Diana Fuentes Olaya

Abstract This paper describes an approach to assessing the impact of 5S on four study factors: namely quality, productivity, industrial security (IS) and organizational climate (OC) in the manufacturing processes carried out at SMEs in Bogota (Colombia). The approach is illustrated by a case study conducted at a manufacturing SME. The main purpose is to evaluate whether the 5S methodology could be considered as an effective tool to improve manufacturing processes at SMEs. A visual diagnosis is chosen to identify the areas that exhibit the largest amount of clutter and dirt. Once the location is identified, surveys, performance measurements and a risk landscape are conducted, focusing on particular study factors so as to understand the initial situation of the area. Subsequently, implementation of 5S is carried out and then three measurements are taken to monitor the performance of the study factors, thus finding out whether a trend can be observed during the measuring period. The results show the existence of a positive connection between the study factors and the implementation of the 5S methodology, since there is evidence of an increase in productivity (44 %), quality (44 %) and OC (52 %) and a decrease in the risks identified (90 %).

Keywords 5S · Productivity · Quality · Industrial security · Organizational climate

---

P. Martínez Sánchez
Facultad de Ingeniería, Programa de Ingeniería Industrial, Universidad El Bosque,
Av. Cra. 9 no. 131 A-02, Bogotá, Colombia
e-mail: martinezpaloma@unbosque.edu.co

N. Rincón Ballesteros (✉) · D. Fuentes Olaya
Grupo Gintecpro. Facultad de Ingeniería, Programa de Ingeniería Industrial,
Universidad El Bosque, Av. Cra. 9 no. 131 A-02, Bogotá, Colombia
e-mail: nrincon@unbosque.edu.co

D. Fuentes Olaya
e-mail: dfuenteso@unbosque.edu.co

© Springer International Publishing Switzerland 2015
P. Cortés et al. (eds.), *Enhancing Synergies in a Collaborative Environment*,
Lecture Notes in Management and Industrial Engineering,
DOI 10.1007/978-3-319-14078-0_28

1 Introduction

The 5S emerges in Japan and is created by Toyota in the 60s (Sánchez 2007) as part of the quality movement in Japan (Arrieta 1999). This methodology is recognized around the world because of its contribution to the improvement of processes focused on safety (Gapp et al. 2008; Tice et al. 2005; Jumar et al. 2007; HungLing 2011), working environments (Jumar et al. 2007; HungLing 2011; Cura 2003), productivity and quality (Moriones et al. 2010; Gapp et al. 2008; Tice et al. 2005; Osada 1991; Sacristán 2005), also yielding fast results (Moriones et al. 2010; Sacristán 2005; Ho 1999) as well as low implementation costs (Gapp et al. 2008).

5S methodology comes from five Japanese words: Seiri, Seiton, Seiso, Seiketsu and Shitsuke. The term was formalized by Takashi Osada in 1980 (Gapp et al. 2008; HungLing 2011). The word *Seiri*, refers to selecting and sorting the elements of the workspace into two main categories, namely essential and nonessential, in an effort to remove unused or rarely used elements that accumulate and promote untidiness (Cura 2003; Michalska and Szewieczek 2007). *Seiton* consist in establishing the adequate manner for locating and identifying the essential materials so that they can be easily accessible (Cura 2003; Ho 1999; Mateus 2011). *Seiso* seeks to maintain the workspace under clean conditions (Michalska and Szewieczek 2007) by having a regular schedule for removing dirt and dust (Osada 1991; Mateus 2011). Following the *Seiketsu* concept, everything should be easy to identify (Becker 2001) and with clearly visible labels for all operators (Riera 2010). *Shitsuke* consists in sustaining each of the 5Ss (Sacristán 2005), working permanently in accordance with the rules, agreements and commitments that were established to implement the methodology in the first place (HungLing 2011; Mateus 2011).

Even thought it is consider that 5S methodology is well known in the manufacturing sector, there is little evidence about its implementation (Sánchez 2007). In Colombia, the 5S methodology is being underutilized, especially in small a medium enterprises (Cura 2003).

The company chosen for the present study is called Tecniaguas S.A.S. This company focuses on the manufacturing and marketing of water treatment plants. Currently, the company operates in a dirty and untidy environment. This causes several problems such as unsafe working environments, longer times for finding tools as well as a critical loss of control regarding aspects like finished products, work in progress and reworks. All these inconveniences make it very difficult to know the actual level of productivity and quality.

2 Study Factors

This study evaluates the effects of the 5S methodology upon productivity, security, organizational climate and quality before and after its implementation. The study factors were defined as follows:

- *Productivity* pertains to how efficiently the resources of any business unit are being used. Thus, productivity can be defined as the relationship established between the amount of goods or services produced and the amount of resources used to produce them (Chase et al. 2009; Febrero 2000). Other authors refer to productivity as the efficiency in production, or simply the increase in quality caused by reducing rework (Syverson 2011; Deming 1989).
- *Quality* is defined as the extent to which a group of essential characteristics fulfill the consumer's needs or expectations. Likewise, in quality, the primary goals are customer satisfaction and processes-and-outcome improvement (Aenor 2000; UMH 2009). Quality also means doing things right from the beginning to the end of a process while satisfying the customer's expectations at the lowest cost possible (Nebrera 1999).
- *Industrial Security* is understood as a set of rules and principles that ensure the physical integrity of work as well as the proper use and maintenance of machines, equipment and tools of the company (Méndez 2009). Colombian regulations of industrial safety are defined by a Colombian safety decree as "a set of activities aimed at the identification and control of the causes of accidents". This concept was employed as a measurement tool in the Colombian Technical Guide (GTC 45).
- *Organizational Climate* refers to the way people perceive the surrounding environment in which organizations interact. These perceptions can be objective (i.e. related to organizational structures, policies or rules of the organization), or subjective (i.e. related to cordiality and support, which may affect the results of each individual) (Chase et al. 2009; Castillo et al. 2011).

3 Methodology

Diagnosis tests were applied using indicators, surveys and a risk landscape to establish the initial situation of the workshop regarding study factors of productivity, quality, industrial security and organizational climate. Once the 5S methodology was implemented in the workshop (June), the same diagnosis tests were performed after implementation in July, August and September. Afterwards, the measurement results taken before and after the implementation were compared in order to determine how the 5S affected each factor's performance and also to observe whether or not the variables followed a particular pattern.

4 Results and Findings

4.1 Productivity

Partial productivity indicators [see Eqs. (1–3)] were used to measure productivity.

$$\text{Human productivity} = \frac{\text{Total \# of cut pieces}}{\text{Total \# hours worked} - \text{man}} \quad (1)$$

$$\text{Capital productivity} = \frac{\text{Production base period}}{\text{Capital in terms of base period}} \quad (2)$$

$$\text{Multifactorial productivity} = \frac{\text{Net production}}{(\text{Labor} + \text{Capital})} \quad (3)$$

Figure 1 shows a decrease on capital and human productivity in July as a result of the beginning of a new project, which required some of the operator's time for understanding customer requirements correctly. However, since august, there is an upward trend. Multifactorial productivity exhibits a growing trend from June to September, which means that the resource allocation is appropriate.

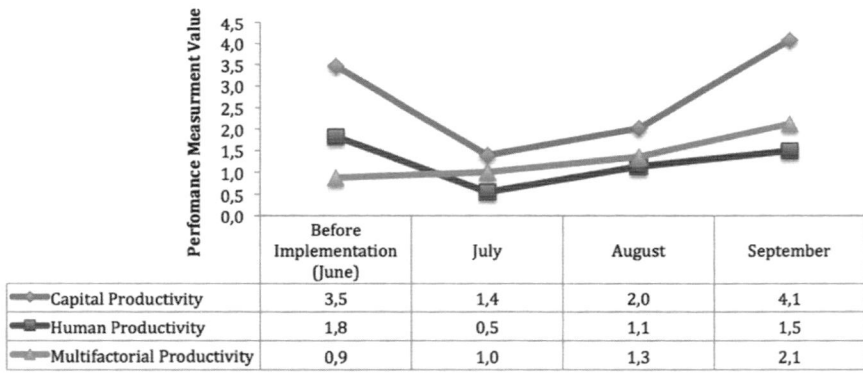

Fig. 1 Productivity performance measurement results

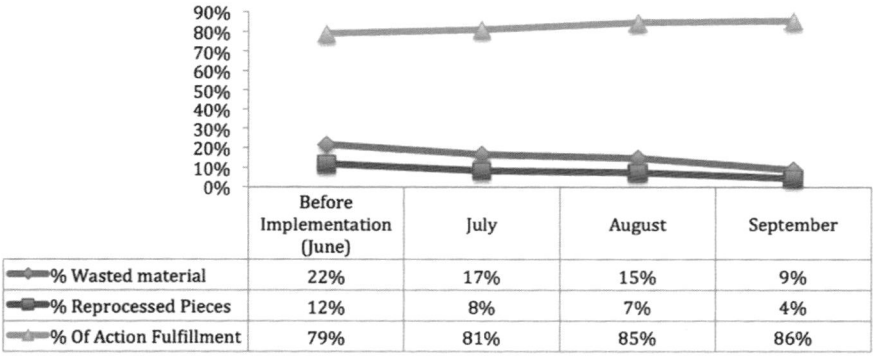

Fig. 2 Quality performance measurement results

4.2 Quality

Non-financial indicators [see Eqs. (4–6)] were used to measure the quality of the processes carried out in the workshop.

$$\text{Percentage of action fulfillment} = \frac{\text{Anomalies Corrected}}{\text{Anomalies Identified}} \times 100 \qquad (4)$$

$$\text{Percentage of wasted material} = \frac{\text{Total of reject cuts}}{\text{Total of processed cuts}} \times 100 \qquad (5)$$

$$\text{Percentage of reprocessing cuts} = \frac{\text{\# of reprocessed cuts}}{\text{Total of processed cuts}} \times 100 \qquad (6)$$

As shown in Fig. 2, the percentage of wasted material and reprocessed pieces decreased from 22 to 9 %; and from 12 to 4 %, respectively from June to September. Regarding the performance of action fulfillment, percentages increased from 79 to 86 %.

4.3 Industrial Safety

Because the company has no record of accidents and incidents, any risks the workers were exposed to were visually identified to reveal the current situation of the workshop with regard to industrial safety. Additionally, the Colombian Technical Guide (GTC 45) was employed as a measurement tool to help find and minimize these risks. Figure 3 shows a summary of results extracted from the risk matrices developed in June, July, August and September.

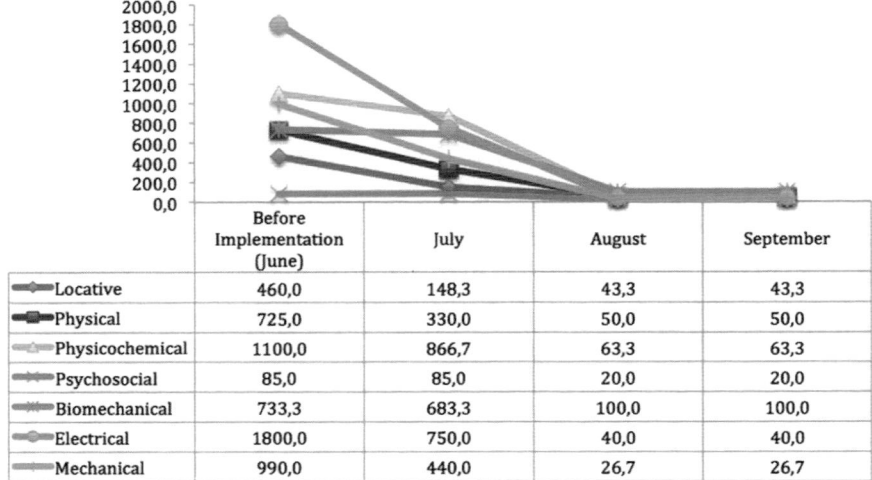

Fig. 3 Industrial safety performance measurement results

Figure 3 shows various types of risk reduction from June to September: namely Locative, Physical, Physicochemical, Psychosocial, Biomechanical, Electrical and Mechanical.

4.4 Operational Climate

The TECLA test was used to measure organizational climate because this test was developed within the Colombian culture. Moreover, TECLA is one of the models applied by students and consultants in the Colombian context. The questionnaire was answered by operators, supervisors and heads of the company so as to compare the viewpoints of each hierarchical level. In this test people were asked about particular aspects such as environmental conditions, communication, structure, motivation, cooperation, sense of belonging, labor relationships and leadership.

Figure 4 shows the improvement in the perception of operators and supervisors regarding the positive effect brought by the implementation of 5S. In the case of the board of directors, in September, there was a fall in organizational climate perception due to the board's constant absence, leading to a low assessment on communication, leadership and social relationships.

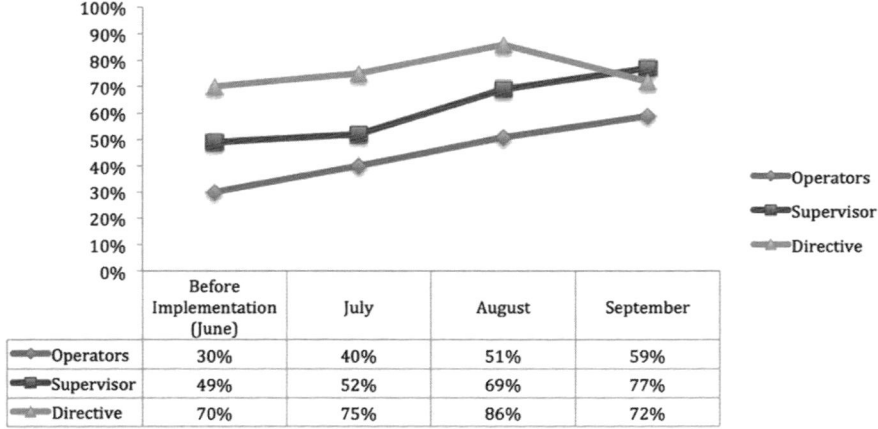

Fig. 4 Organizational climate performance results

5 Conclusions

The study factors' measurements corresponding to the third quarter (July, August and September) were taken once the 5S methodology had been implemented in the workshop. The impact of this methodology on each of the study factors is as follows:

- Partial productivity factors (i.e. partial capital productivity, multifactorial productivity) had a positive effect supported by an average percentage increase (44.45 %) despite the fact that human productivity declined when conducting a new project. This entailed spending some of the operator's time in understanding customer requirements correctly.
- Improvement is evidently reflected in quality indicators regarding the percentage of wasted material, reprocessed pieces and action fulfillment (59.09, 66.66 and 8.87 % respectively), which yielded an averaged improvement of 44.62 %.
- The organizational climate within the workshop achieved improvements in terms of perception of operators (96.6 %), supervisors (57.14 %) and the board of directors (2.86 %). It is worth noting that the board's perception in organizational climate exhibited an upward trend in July and August (7.14 and 22.86 % higher respectively) compared to June (before implementation). Moreover, in September, there was a constant absence of the board, which caused a fall in their organizational climate perception.
- Regarding industrial security of the workshop, the number of identified risks decreased by 90.4 %. Specifically, performance improved in various aspects, namely Locative (90.6 %), Physical (93.1 %), Physicochemical (94.2 %), Psychosocial (76.5 %), Biomechanical (86.4 %), Electrical (97.8 %), and

Mechanical (97.3 %). Workshop operators now understand the importance of using personal protection equipment and also know how to use it appropriately.
- All the factors that were evaluated in the short term showed an increase in their respective units, thus the results obtained in this research support the literature review, which mentions the positive impact the 5S methodology has on the quality, productivity, industrial security and organizational climate of any company.
- Unlike most 5S assessment studies, which focus only on the impact of 5S on single factors (as separate aspects); the present study offers a simultaneous assessment of four aspects that contribute to improving the overall company's performance. This suggests that the 5S methodology is an effective tool to solve a range of inconveniences within an organization in parallel.

6 Recommendations

It is recommended that additional case studies be performed to accomplish the following:

- Determination as to whether or not there is a tendency to reduce any sort of risk based on manufacturing activity
- Verification of organizational environmental factor tendencies, taking advantage of quantitative tools based on statistical data
- Validation of the relationship between the productivity and quality variables, and any changes according to the kind of manufacturing activity

References

AENOR (2000) Upcommons. Recovered 14 Aug 2012, from http://upcommons.upc.edu/pfc/bitstream/2099.1/3042/4/36146-4.pdf
Arrieta JG (1999) Las 5S, pilares de la fábrica visual. Revista Universidad EAFIT, pp 35–48
Becker JE (2001) Implementing 5S: to promote safety and housekeeping. J Prof Saf 46(8):29–31
Castillo L, Lengua C, Pérez P (2011) Caracterización psicométrica de un instrumento de clima organizacional en el sector educativo universitario Colombiano. Int J Psychol Res 4:40–47
Chase R, Jacobs R, Aquilano N (2009) Administración de operaciones: Producción y cadena de suministro, 12th edn. McGraw-Hill/Interamericana Editores S.A, Mexico
Cura HM (2003) UCEMA. Recovered 12 Feb 2012, from http://www.ucema.edu.ar/productividad/download/2003/Cura.pdf
Deming E (1989) Calidad, productividad y competitividad: la salida de la crisis. Diaz de Santos S. A, Madrid
Febrero E (2000) Valor trabajo, un indicador de productividad y competitividad: una aplicación empírica al caso español, 1970–1992. Ediciones de la Universidad de Castilla—La Mancha, España

Gapp R, Fisher R, Kobayashi K (2008) Implementing 5S within a Japanese context an integrated managemet system. Manage Decis 46(4):565–579

Ho SK (1999) 5s practice: the first step toward total quality management. Total Qual Manage Bus Excellence 10(3):345–356

HungLing C (2011) 5s implementation in Wang Cheng industry manufacturing factory in Taiwan. Master thesis, Wisconsin University, p 45

Jumar RS, Sudhahar C, Dickson J, Senthil V, Devadasan S (2007) Performance analysis of 5-S teams using quality circle financial accountant system. TQM Magaz 19(5):483–493

Mateus E (2011) 5S: Un metodo eficaz para el éxito en la organización y productividad empresarial. Metal Actual, pp 82–88

Méndez R (2009) www.seguridadindustrialapuntes.blogspot.com. Recovered 12, 2012, from http://seguridadindustrialapuntes.blogspot.com/2009/01/qu-es-la-seguridad-industrial-definicin_13.html

Michalska J, Szewieczek D (2007) The 5S methodology as a tool for improving the organization. J Achievements Mater Manufact Eng 24(2):211–214

Moriones A, Bello A, Merino J (2010) Use 5S in the manufacturing plants: contextual factor and impact on operating performance. Int J Qual Reliab Manage 27(2):217–230

Nebrera J (1999) INFOMED. 12 Jan 2012, from http://www.sld.cu/galerias/pdf/sitios/infodir/introduccion_a_la_calidad.pdf

Osada T (1991) The 5S: five keys to a total quality environment. Asian Productivity Organization, Tokyo

Riera A (2010) Universidad Politécnica Salesiana del Ecuador. Recovered 27 Apr 2012, from http://dspace.ups.edu.ec/bitstream/123456789/688/6/CAPITULO%20IV.pdf

Sacristán FR (2005) Las 5´: orden y limpeza en el puesto de trabajo. Fundación Confemetal, España

Sánchez R (2007) El proceso de las 5s en acción: La metodología japonesa para mejorar la calidad y la productividad de cualquier tipo de empresa. Gestión y estrategia, p 91

Syverson C (2011) What determinates productivity? J Econ Lit 49(2):326–365

Tice J, Ahouse L, Larson T (2005) Lean production and EMSs: aligning environmental management with business priorities. Environ Qual Manage 15(2):1–12

UMH (2009) Universidad Autónoma Del Estado De México. Retrieved 20 Aug 2012, from http://www.uaemex.mx/planeacion/docs/sgc/Algo%20acerca%20del%20concepto%20de%20calidad.pdf

Structuring a Portfolio for Selecting and Prioritizing Textile Products

Maurício Johnny Loos and Paulo Augusto Cauchick-Miguel

Abstract The objective of this paper is to demonstrate a proposal on product portfolio management for the development of new products in a textile company. The research methods address the application of real data from the studied company. This work is characterized both as conceptual and field-based research. It constructs a proposal based on existing theory and involves an empirical application of that proposal. By using the portfolio selection and prioritization proposal, the company was able to select new products for development in a structured way. As continuity for this work, the paper suggests the application of product selection and prioritization for other organizations.

Keywords Product development · Portfolio management · Project management

1 Introduction

Due to the competition among companies and the increasing demands regarding quality, cost and delivery, product development is critical for faster product production and the fulfillment of customer requirements. Companies require ongoing product development to remain competitive in the market (Rocha and Delamaro 2007). One problem faced by companies interested in improving their business performance is the variety of improvement methods and techniques; each method claims to provide the expected solutions for modern business problems (Poolton and Barclay 1998). Another problem is associated with the selection of projects for the development of new products. When companies generate multiple ideas for new products that are consistent with the current strategy and select projects with the highest probability of success, the performance of the new product should be

M.J. Loos · P.A. Cauchick-Miguel (✉)
Post-Graduate Program in Production Engineering, Federal University of Santa Catarina, Campus Universitário Trindade, Florianópolis, SC, Brazil
e-mail: paulo.cauchick@ufsc.br

acceptable (McNally et al. 2009). However, the portfolio of projects in an organization frequently includes projects that do not properly support the strategic intent, suffer from overlap and duplication, compete for limited resources, do not adequately share capabilities, or exceed the organization's capacity for change (Lockett et al. 2008). Cooper et al. (2000) emphasize two approaches for a company to succeed in the development of new products: "doing the right projects and doing projects correctly".

Portfolio management provides a detailed examination of the strategic dimensions that enable company executives to properly balance and prioritize projects, as well as establish mechanisms for the control and disposal of projects (Rabechini et al. 2005). However, the portfolio decision process is characterized by uncertain and fluctuating information, dynamic capabilities, multiple goals and strategic considerations, interdependence among projects, and multiple decision makers (Cooper et al. 1999; Killen et al. 2008). Cooper et al. (1999) suggest that companies suffer from a lack of effective portfolio management.

In this context, this paper explores product selection and prioritization as an initial phase of managing a portfolio of new product development. It aims at answering the following research question: how to structure the selection and prioritization of new product projects, aligned to organizational strategies to facilitate the development of new products can be developed? To conduct the proposal, important aspects of portfolio management were identified in the literature. Firstly, a preliminary assessment of how a company manages its product portfolio was carried out. Subsequently, a diagnostic of sales by product family and a classification of project types based on Griffin and Page (1996) were performed. Criteria for selecting products, which were obtained from the literature, were also considered. The proposal was applied using real historical data.

2 Research Methods

This paper adopts a methodological approach for analyzing a theoretical proposal with real historical data. It presents the characteristics of a conceptual and empirical study because the proposal is initially developed based on existing theory; its application is subsequently explored. Although the empirical part of the study presents the characteristics of action research, the proposal is not necessarily implemented, as suggested by Bryman (1989). To collect field data, several sources of evidence were employed: data primarily collected via a questionnaire from the Product Development Management Association (PDMA), participation by one of the authors in portfolio meetings, semi-structured and unstructured interviews with professionals involved in product development, and internal company documents. The rationale for selecting a company in the industrial sector can be justified by its representation in the Brazilian economy, which comprises 3.5 % of the Brazilian GDP (ABIT 2011). In addition, this sector increasingly contributes product innovations to the market. The following steps were considered in the development of this study:

Step I. Literature review: refers to the development of a theoretical foundation with regard to portfolio management and a brief description of the new product development (NPD) process.

Step II. Preliminary diagnostic: refers to an evaluation of how the company performs its portfolio management and utilizes a structured questionnaire that is based on a survey of portfolio management from the Product Development Management Association. The results of this step are detailed in related studies (Loos and Cauchick Miguel 2012); refer Table 2 in Appendix.

Step III. Second diagnostic: refers to the descriptive statistics of annual sales of the product family and the analysis of new product projects. The results of this step are detailed in Loos and Cauchick Miguel (2011); refer Table 2 in Appendix. This step is further divided into the following three substeps:

- Step III.1 Define revenues for family products: analysis of the distribution of the company's revenue by product families and its relationship to product development.
- Step III.2 Rate new product projects: analysis of how new products are classified by the company based on Griffin and Page (1996). This analysis considers the degree of novelty to the market and the company.
- Step III.3 Generate a list of products according to the business: identification of the product families that will have their products used for selection and prioritization.

Step IV. Develop portfolio management proposal (Fig. 1): refers to the selection and prioritization of new products, is divided into two substeps:

- Step IV.1 Definition and application of selection criteria: involves selecting the criteria that are relevant to the selection of new products and are linked to business strategies;
- Step IV.2 Prioritization: involves the prioritization of real products of the company and establishes the sequence in which they should be developed to ensure their suitability.

Step V. Comparison of the proposal with real company data from previous NPD projects: involves the evaluation of actual products and a comparison with the prioritization that was performed in the proposal.

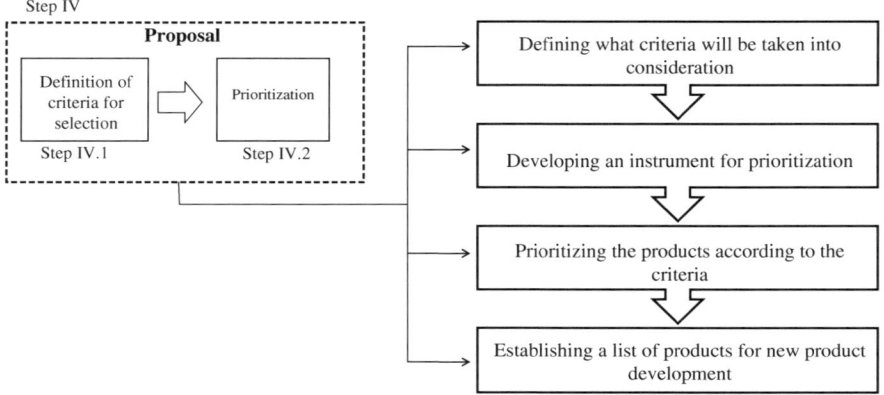

Fig. 1 Development of the proposal

2.1 Development of the Proposal

The following steps were performed in the development of the proposal:

- Define criteria: essential criteria for the selection and prioritization of product families based on the literature were selected. The experiences of company managers were also considered (commercial director, industrial manager, marketing manager, engineering manager, logistics manager, and controller). The criteria were discussed among executives at portfolio meetings (2-h duration) to identify organizational relevant criteria. The criteria from the literature were: impact on business strategy, stage of development ("active" or "standby"), technological difficulties, goals, contribution margin, and expected effectiveness. Criteria based on the vision and experience of executives were: source of ideas, class of projects ("platform" or "derivative"), sales volume, and importance of products ("common" or "specialty").
- Develop an instrument for prioritization: using the selection and prioritization criteria, a MS Excel spreadsheet was created to facilitate, sort, and prioritize products. The spreadsheet considered fields such as product identification, type of business, and impact on business strategy (refer Fig. 2 in Appendix).
- Prioritize the products according to the criteria: prioritization was performed by decision makers who apply all criteria in the assessment. The prioritization was conducted by members of a functional department (Product and Process Engineering—PPE) with the assistance of one of the authors. Part of the required data and information can be obtained from PPE; other data were obtained from the company planning team.

- Establish a list of products for new product development: after the previous steps, a list of products for development is obtained. This list indicates the sequence of design. A comparison was made with real historical data (i.e., with the prioritization performed by the company using the traditional method).

2.2 Product Prioritization

As mentioned previously, a spreadsheet depicting the selection and prioritization of new products was created. This worksheet includes important decision-making criteria that facilitate the selection of appropriate projects for development. However, the products and their respective families must be balanced (Cheng 2000). This balancing is critical because a company would not be able to survive if it considered only high-risk or long-term projects. For this reason, the criteria for the prioritization of products were differentiated among product families. Each product within a particular family should be prioritized based on the criteria described in the previous section. It is important that the prioritization of two families occur separately to prevent the exclusion of one family from new product development. The following criteria were considered in the prioritization: (i) "Extremely important" (strategic alignment, competitive position, revenues, margins and risks); (ii) "Very important" (brand position and the availability of resources); and (iii) "Important" (market share).

3 Main Outcomes

Using an initial diagnosis to identify opportunities for improvement, the company identified gaps in the management of its product portfolio and compared these disparities with survey responses of the PDMA (2010). By prioritizing the product families, the potential for more effective development of new products by improving the selection of new products linked to business strategies was identified.

After applying real historical data to the proposal, significant changes in the prioritization of products were achieved. These changes occurred due to certain factors; the most important factors consisted of the overestimated contribution margins and sales volumes. Using the portfolio selection proposal, the prioritization of products to be developed by the company can occur in a more structured manner through the selection of more suitable products, i.e., products that are more closely aligned with the company's strategy. Future applications of the proposed selection and prioritization of products is recommended in other organizations.

Table 1 lists the main findings of each previously outlined step.

Table 1 Outcomes of the steps

Step	Outcomes
Literature review	Theoretical basis
	Criteria for selecting new products
Preliminary diagnostic	Current practices of company product portfolio management based on a PDMA survey instrument
Second diagnostic	Descriptive statistics of the sale performance of 8 product families
	Identification of two families for application: 'bath A' and 'bath B'
	Categorization of products according to the degree of novelty to the market and the company
Portfolio management proposal	Application of selection criteria in a spreadsheet
	List of prioritized products
Comparison of the proposal and historic field data	Identify differences between the proposal and the field data
	Explanation of the differences

4 Conclusions

Several opportunities for improvement in product portfolio management were identified from the PDMA survey completed by the company. One improvement involved product selection and prioritization. Therefore, a proposal was developed. This research contributes an empirical finding to develop a proposal for the selection and prioritization of products. The proposal can effectively address the needs of a company in a developing country. By considering the limitations of the proposal, this study offers structure to decision making in a company with regard to product portfolio management. This preliminary proposal has the potential to contribute to a company's competitive advantage.

Acknowledgments The authors thank Brazilian agencies CAPES and CNPq for financial support.

Appendix

Table 2 shows an overview of previous work cited in the course of this work.

The worksheet header for selection and prioritization of new product development project can be seen in Fig. 2 (it is divided into two parts).

The fields of the worksheet are as follows:

Product number: product family identification.
Business: type of business (i.e., bath).

Table 2 Overview of the articles of authors

Article	Objective	Results
Loos and Cauchick Miguel (2011)	Making a diagnosis of the families of the company's products in order to have a better direction in implementing the proposal	Guiding points for proposal for selection and prioritization of the portfolio of new products in the textile company was developed
		Develop further in identifying how families of products the company behaved, assisting in the choice of families "Bath A" and "Bath B", for the simulation of the proposal
Loos and Cauchick Miguel (2012)	Making a diagnosis of the management practices of the product portfolio of the company searched in order to identify: (i) primary problems and risks in managing a portfolio of products; (ii) accuracy of investment projections and forecasts in terms of time, cost and expected revenue; (iii) challenges associated with managing resources; (iv) assessment and ability to project low end performance; (v) the use of management solutions portfolio and alternative instruments	Finding that the company has researched many similarities in their results compared respondents the PDMA, the main similarities observed: (i) have many projects for the resources available; (ii) decisions happen later or inefficiently; (iii) managing changing priorities and loss of seasonal windows present risks; (iv) integrate the voice of the customer in the PDP; (v) consider extremely important strategic alignment as a challenge and have the inability to see the demand for resources provided

Part 2

Product		Impact on Business Strategy			Stage		Origin		Difficulty Technology			Class	
N°	Business	High	Medium	Low	In which aspects?	Active	Standby	Company	Customer	High	Medium	Low	Platform	Derivative

Part 1

	Objetivo (Griffin e Page)					Sales Forecasting		Expected Duration			Importance		Priority
New to the firm	New to the world	Adding to the existing line	Improvements of existing products	Cost reduction	Repositioning	Average volume of monthly	Contribution margin expected %	6-12 meses	1-2 anos	>2 anos	Specialty	Common	1 2 3 4 5

Fig. 2 Worksheet header for product selection and prioritization

Impact on business strategy: "high", "medium" or "low" impact. It should consider aspects such as market share, added value, expansion of new markets, and how to handle competition.
Stage: indicates whether the product is "active" or "standby".
Origin: indicates whether the idea of the product originated from the company, customer demands, or suppliers.

Technological difficulty: "high", "medium" or "low" difficulty regarding the development of the product.

Class: indicates whether the product is "platform" or "derivative". Platform represents a solution for customers; both involve significant changes to the manufacturing process for the product (Cauchick Miguel 2008). Derivatives range from low-cost versions of an existing product or improvement of an existing production process (incremental changes in the product).

Objective: refers to the classification of the product structure according to Griffin and Page (1996), which considers the level of innovation; this classification includes "new to the world", "new to the firm", "adding to the existing line", "improvements and revisions of existing products", "repositioning", and "cost reduction".

Sale forecasting: estimates average volume of monthly sales and the percentage of expected contribution margin.

Expected duration: estimated lifetime of the product in the market, which is classified as 6–12 months, 1–2 years, and more than 2 years;

Importance: indicates whether the product is "common" or "specialty".

Priority: the product receives a priority number from 1 to 5 (1 denotes the highest priority), which defines their sequence in the product development process.

References

ABIT—Brazilian Association of Textile Industry (2011) Textile sector profile, São Paulo. Available at http://www.abit.org.br/site/navegacao.asp?id_menu=1&id_sub=4&idioma=PT. Accessed 7 July 2011 (in Portuguese)

Bryman A (1989) Research methods and organization studies. Unwin Hyman, London

Cauchick Miguel PA (2008) Portfolio management and new product development implementation: a case study in a manufacturing firm. Int J Qual Reliab Manage 25:10–23

Cheng LC (2000) Characterisation of new product development. In: Proceedings of the 2nd Brazilian Conference of Product Development Management, São Carlos, SP, Brazil

Cooper RG, Edgett SJ, Kleinschmidt EJ (1999) New product portfolio management: practices and performance. J Prod Innov Manage 16:333–351

Cooper RG, Edgett SJ, Kleinschmidt EJ (2000) New problems, new solutions: making portfolio management more effective. Res Tech Manage 43:18–33

Griffin A, Page A (1996) PDMA success measurement project: recommended measures for product development success and failure. J Prod Innov Manage 13:478–496

Killen CP, Hunt RA, Kleinschmidt EJ (2008) Project portfolio management for product innovation. Int J Qual Reliab Manage 25:24–38

Lockett M, Reyck B, Sloper A (2008) Managing project portfolios. Bus Strategy Rev 17:77–83

Loos MJ, Cauchick Miguel PA (2011) Analysis of the classification of projects for new products and sales in product development in a textile company. RACE: Rev Adm, Cont e Eco 10:185–214 (in Portuguese)

Loos MJ, Cauchick Miguel PA (2012) Product portfolio management practices based on the PDMA survey: a diagnostic in a textile company. Espacios 33:6–12

Mcnally R, Durmusoglu SS, Calantone RJ, Harmancioglu N (2009) Exploring new product portfolio management decisions: the role of managers' dispositional traits. Ind Mark Manage 38:127–143

PDMA (2010) Product Development and Management Association. In: Proceedings of 2nd annual planview product portfolio management benchmark study

Poolton J, Barclay I (1998) New product development from past research to future applications. Ind Mark Manage 27:197–212

Rabechini R Jr, Maximiano ACA, Martins VA (2005) Implementing portfolio management in an electronic data exchange company. Prod 15:416–433 (in Portuguese)

Rocha HM, Delamaro MC (2007) Product development process: using real options for assessments and to support the decision-making at decision gates. In: Proceedings of the ISPE international conference on concurrent engineering, vol 14, São José dos Campos. Available at http://urlib.net/dpi.inpe.br/ce@80/2007/01.07.18.13. Accessed 16 Nov 2011 (in Portuguese)

Methodology for Evaluating the Requirements of Customers in a Metalworking Company Automotive ISO/TS 16949 Certified

Gil Eduardo Guimarães, Roger Oliveira, Luiz Carlos Duarte, Milton Vieria and Daniel Galeazzi

Abstract This issue presents the process of implementation and management of customer-specific requirements aimed at the automotive industry, focusing stamping operations, machining, welding and surface treatment and certifications ISO 14001, ISO 9001 and ISO/TS 16949. Requirements management includes the implementation, monitoring and procedures, through quality management tools as indicators, audits, action plans, org charts, etc. This whole structure is required by ISO 9001 certification standards and ISO/TS 16949, quality standards required to supply parts and services to the automotive market. Initially it is shown the importance of quality in the product and in the productive process, enhanced by the opinion of the leading names of quality management. At the end it is shown the methodology and the results that will demonstrate the management strategy developed in the company, including but not limited to implementing, structuring and monitoring the processes until its full functioning till the point of being audited.

Keywords Requirements of customers · Management of automotive requirements · Focus and customer satisfaction

---

G.E. Guimarães (✉) · R. Oliveira · D. Galeazzi
DCEEng, Universidade Regional do Noroeste do Estado do Rio Grande do Sul, UNIJUÍ, Ijuí, Brazil
e-mail: gil.guimaraes@unijui.edu.br

R. Oliveira
e-mail: roger@hidroenergia.com.br

D. Galeazzi
e-mail: daniel.galeazzi@live.com

L.C. Duarte
AGIT, Universidade Regional do Noroeste do Estado do Rio Grande do Sul, UNIJUÍ, Ijuí, Brazil
e-mail: lduarte@unijui.edu.br

M. Vieria
Pós-graduação engenharia de produção, UNINOVE, São Paulo, Brazil
e-mail: mvieirajr@uninove.br

© Springer International Publishing Switzerland 2015
P. Cortés et al. (eds.), *Enhancing Synergies in a Collaborative Environment*,
Lecture Notes in Management and Industrial Engineering,
DOI 10.1007/978-3-319-14078-0_30

1 Introduction

The constant search for perfection leads the consumer society to increasingly demanding levels of quality. The automotive market has shown constant evolution showing great profits for automakers and their suppliers in recent years. It is a market moved by rigorous consumers and their needs to be interpreted and remedied by various competitors, which make it even more inflexible automakers with its supply chain.

Due to the large number and to the complexity of demands, many suppliers of parts and services to the automotive industry have difficulties in interpreting the needs of its customers and deploy them effectively in its organization to meet satisfactorily the costumer market. This issue presents tools and methods used to deploy and monitor these requirements into a supplier of parts to the automotive industry, by means of an easy management methodology in order to ensure its full operation and continuity, using tools of quality management within the models required in ISO/TS 16949 certification.

It will be shown, by means of the preparation of action plans for each of the items of the requirements demanded by the automakers, the ways for implementation of the requirements in areas liable to attend the actions bounded by deadline: How to establish the standards, describing them in the company's internal documents for specific areas; forms of elaboration of indicators, to measure the efficiency and effectiveness of the system; adequacy and maintenance through audits in the system.

2 Methodology

In implementing the requirements of customers in the enterprise, one must start by checking which apply to the company, requesting them to clients, if the latter is no longer valid, this scan can be performed via the customer's site or through direct or indirect contact with the client, after are analyzed and compared to the scope of the company and existing procedures.

In this case, the company owns mechanical metal parts production scope of automobile structure, with the printer operations, machining, welding, painting and tooling. It would apply, for example, requirements that aims the production of plastic products or applied in the field of aviation.

The step of reading and interpretation of the requirements requires a more in-depth knowledge of the company as: quality procedures, production process, overall organization and responsible for the various areas. This step is directed to persons in the area of Management System, because they have access to various departments, in addition to being responsible for certification documentation.

For a better management in the distribution of responsibilities and recovery of the deployment and operation of requirements, are defined action plans describing the points that are not served by the organization. In this plan are those responsible for restructuring/adequacy, implementation and evolution of the actions. We determined that as being the stage "P" (planning) of the PDCA cycle.

The responsible for the action is the person who occupies the highest hierarchical level of the affected area, so that this action can really be implemented efficiently and effectively; the leader can also indicate a subordinate to do the action. The date is then determined by this leader or responsible by action or by the client.

Documentation is important, mainly for certification audit of the company, which is held by a third party and requests the attendance of control and distribution according to ISO/TS 16949.

To meet the requirements, indicators should be established, preferably graphs that are reviewed the specified periods or according to customer requirements review. Giving closing the PDCA cycle have checked, acting with new preventive or corrective actions that feed the stock plan ("C" and "A" respectively of PDCA), this occurs through audits in the procedure and the requirement of the customer. Given the continuous improvement cycle established by the standard.

3 Analysis of Results

3.1 System of Analysis from Requirements

In the planning stage, we organize the management of the various requirements of the customers that the company should meet in a master list in this thread divide by listing are the first letters of requirements that mention the other manuals and procedures that need to be serviced.

The list shown below is an example that can be adapted according to the needs of the company, it can be controlled the dates and the various procedures to be met. We opted for the annual review of requirements, with direct query to the site or sending e-mail to the client requesting the changes in related documentation in the list (Table 1).

Besides the management it is important to educate and train those involved of how to query the lists and use the updated documents, for this set if a specific training to all supervisors and staff with some contact with clients within the enterprise.

The procedures are available in electronic media to facilitate access and control with regard to the use of obsolete documents, mentioned in the training access directories and the corresponding procedure.

3.2 Control and Distribution

In order to monitor the effectiveness and efficiency of the implementation of the requirements and entering the stage of systematic enforcement action plans were prepared, relating what, when, who is responsible for items that the Organization should meet (Table 2).

Table 1 Model of master list of specific requirements

LOGO EMPRESA			CONTROLE DOS REQUISITOS ESPECÍFICOS DOS CLIENTES		
Item	Cód.	Cliente	Normas/Procedimentos	Revisão	Local de Consulta
			Carta de aclaraciones y adicionales referentes a la ISO/TS 16949, aplicables a proveedores de general motors Argentina.	05/03/2012	
			1. GM Customer Specifics – ISO/TS 16949	Outubro/2010	
			1.1. GP 4 – Pre-Production/Pilot Material Shipping Procedures. (GM 1407)	Fevereiro/2005	
			1.2 Shipping Parts Identification Label Standard. (GM 1724)	Novembro/2000	
			1.5 GP-5 Supplier Quality Processes and Measurements Procedure. (GM 1746)	March/2011	
			1.6 GP-8 Continuous Improvement Procedure. (GM 1747)	15/Maio/2005	
			1.7 GP-10 Evaluation and Accreditation Test Facilities. (GM 1796)	Fevereiro/1990	
			1.9 GP 11 – Global Pre-Production Part Quality Process (PPQP)	01/01/09	
			1.10. Supplier Technology Information Global. (GM 1825)	03/12/2010	

Source The authors

Table 2 Customer requirements plan

LOGO EMPRESA					Plano de atendimento à Requisitos					
Cliente:		Norma:			Data recebimento da norma:	Revisão atual:		Total de itens:	18	Atendidas: 14
						Revisão anterior:		Não atendidas:	4	Indicador: 78%
N°	Item da Norma		Alterações:		Áreas afetadas:	Responsável:	Prazo:	Status	Data Conclusão:	Desvios:
01										
02										
03										

Source The authors

The plan of action drawn up is dynamic and automatically generates an indicator of customer requirements, but requires periodic review when an action is updated or a new deadline is requested.

3.3 Checking and Monitoring

After implemented actions is accomplished by the area control management system requirements, verification takes place through internal audits conducted 6 times to the year, divided by segment and also each semester during the certification audits.

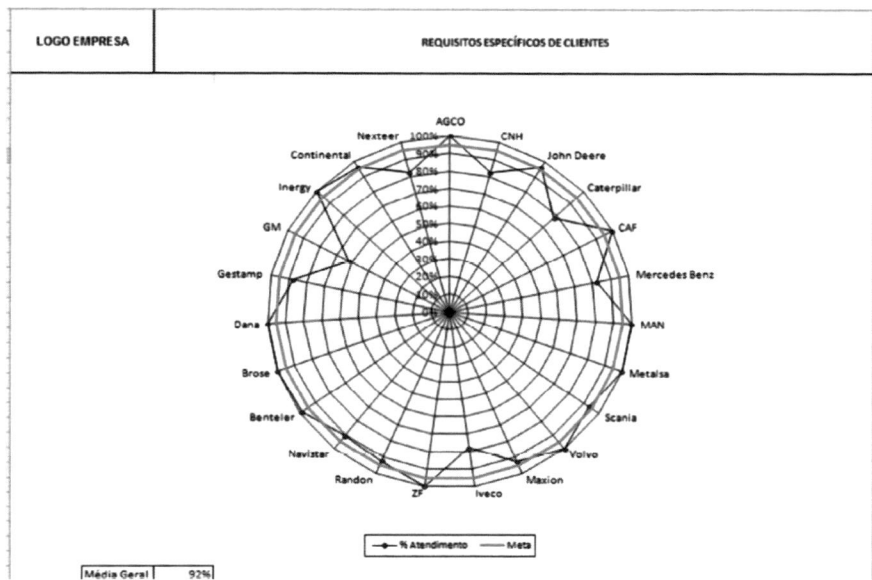

Fig. 1 Model radar indicator chart. *Source* the authors

For effective control, devised a graphic indicator in the form of radar, because it facilitates the interpretation and demonstrates the various results in compressed form (Fig. 1).

The bookmark is referenced in the quality manual of the company, situated at the top of the pyramid of documentation as a strategic point of the company ensuring the customer focus, this is available to everyone in the Organization, but is under the control of the system management area, after all the work you can check that the company serves approximately 92 % of customers' requirements, the others being 8 % constants to pending action plan.

The evaluation of the bookmark occurs twice a year and the goal is updated annually. In the second sentence of the same 2012 was 95 %, so the service was underwhelming and it was necessary to work with other areas to raise the indicator to your target.

4 Final Considerations

The search for excellence in customer service in the automotive business has been steady, increasing competitiveness, emerging companies in all parts of the world with products of competitive quality and price. These requirements causes companies to adopt innovative and practical management models that simplified way to compile as much information.

In this same line the various customers' needs need to be understood and passed on to all those who are involved in the supply of products, in the case of this work the control was accomplished by lists and in the course of job implementation tools that apply are action plans and indicators monitoring and audits, following the PDCA cycle.

The entire development was based on a bibliographical reference between comparative diversified authors and thinkers in the area of quality, the methods applied and the results were demonstrated in practice, in view of the customer and certification standards, enabling a wide range of practical knowledge in engineering.

The limiting work were the limited knowledge of some involved in the process, regarding the implementation or need to adapt to the customers' requirements, requiring considerable time to training. Also the difficulty in maintaining a requirement with the role switching and people within the sectors of the company.

For future challenges can be extensive combined horizontal and vertical structures: with identification of internal and external customers and outputs of each process, yet the quality function deployment product applied to processes such as stamping, welding, machining and painting, another theme is the management and implementation of process audits in companies certified and non-certified with gains and benefits generated for both.

References

Anderson C (2006) A Cauda Longa. Campus, Rio de Janeiro
Bhote KR (1993) Banas qualidade. Banas, São Paulo 1990–2012. In: O Cliente na Linha de Frente. Qualitymark, Rio de Janeiro
BSI Brasil (2013) Disponível em: http://www.bsibrasil.com.br/certificacao/sistemas_gestao/normas/iso_ts16949/. Acesso em 15 de abril de 2013
Campos VF (1991) Qualidade Total Padronização de Empresas. Belo Horizonte
Campos VF (1992) TQC Controle da Qualidade Total. 5. Ed. CIP, Belo Horizonte
Campos VF (2009) O Verdadeiro Poder. INDG, Belo Horizonte
Corrêa AA (2000) Avaliação de um Sistema Integrado de Gestão: Um Estudo na Indústria Automotiva. Porto Alegre: UFRGS, 2004.Trabalho de Conclusão (Mestrado Profissionalizante em Engenharia), Escola de Engenharia, Universidade Federal do Rio Grande do Sul
Crosby PB (1992) Qualidade é investimento. José Olympio, Rio de Janeiro
Esteves JC (2008) Gestão dos requisitos dos clientes no desenvolvimento de projetos, n° 44, São Paulo, 3 e 4 de novembro de 2008
Fenabrave. Disponível em. http://www3.fenabrave.org.br:8082/plus/modulos/listas/index.php?tac=indices-enumeros&idtipo=1&layout=indices-e-numeros. Acesso em 09 de junho de 2013
Filho HRP (2012) Qualidadeonline's blog. Disponível em. http://qualidadeonline.wordpress.com/2011/08/01/como-atender-aos-requisitos-especficos-dosclientes-do-setor-automotivo/. Acesso em 09 de fevereiro de 2012
G1. Disponível em: http://g1.globo.com/carros/noticia/2013/05/venda-de-veiculos-baterecorde-em-abril-com-alta-de-175-diz-fenabrave.html. Acesso em 21 de junho de 2013
Globo.com. Disponível em http://g1.globo.com/carros/noticia/2013/05/venda-de-veiculos-baterecorde-em-abril-com-alta-de-175-diz-fenabrave.html. Acesso em 09 de junho de 2013
Inmetro. Disponível em:http://www.inmetro.gov.br/gestao9000/dados_estat.asp?Chamador=INMETROCB25&tipo=INMETROEXT. Acesso em 21 de março de 2013

IRQA. Disponível em: http://www.lrqa.com.br/certificacao/qualidade/automotiva.asp. Acesso em 15 de abril de 2013
ISHIKAWA, Kaoru. Controle de Qualidade Total. Rio de Janeiro: Campus, 1991
Juran JM (1992) A Qualidade Desde o Projeto. Pioneira, São Paulo
Juran JM, Gryna, Frank M (1991) Controle da Qualidade. São Paulo, Makron
Liker JK, Meier D (2007) O Modelo Toyota. Bookman, São Paulo
Manual de Gestão Integrado, Bruning Tecnometal Ltda. Panambi, 2011
MARINS, Luiz. Projeto Cliente: Harbra, 2008. NBR ISO 9001:2008. Sistemas de Gestão da Qualidade—Requisitos. ABNT. Rio de Janeiro, 11 de Setembro de 2009
NBR ISO TS 16949:2010. Sistemas de Gestão da Qualidade—Requisitos particulares para aplicação da ABNT NBR ISO 9001:2008 para organização de produção automotiva e peças de reposição pertinentes. ABNT. Rio de Janeiro, 07 de dezembro de 2010
Revista Exame. Disponível em: http://exame.abril.com.br/negocios/album-de-fotos/as-20-maioresmontadoras-do-brasil. Acesso em 10 de junho de 2013
Revista Isto É. Disponível em: http://www.istoedinheiro.com.br/noticias/114256_MAIS+6+MONTADORAS+CHINESAS+CHEGARAO+AO+BRASIL+NO+SEGUNDO+SEMESTRE. Acesso em 10 de junho de 2013
Schmidt FC (2011) Mapeamento por Processos e Indicadores de Desempenho de Uma Empresa Metal-Mecânica do Setor Automotivo com Base nas Normas ISO 9001, ISO/TS 16949, ISO14001, OHSAS 18001 e ISO 26000. Panambi, UNIJUI, 2011. Trabalho de Conclusão de Curso (Engenharia Mecânica), Universidade Regional do Noroeste do Estado do Rio Grande do Sul
Seven Empresarial Consultoria. Disponível em http://sevenempresarial.blogspot.com.br/. Acesso em 27 de maio de 2013
Universo Chevrolet. Disponível em: http://www.chevrolet.com.br/Universo-Chevrolet/sobrea-gm/a-companhia.html. Acesso em 21 de junho de 2013
Wikipédia. Disponível em: http://pt.wikipedia.org/. Acesso em 27 de maio de 2013

Improving Resilience and Performance of Organizations Using Environment, Health and Safety Management Systems. An Empirical Study in a Multinational Company

Barbara Seixas de Siqueira, Mercedes Grijalvo and Gustavo Morales-Alonso

Abstract This article presents and analyzes the experience of a European multinational company working in diverse infrastructure sectors on the deployment and enhancement of a new internal Environment, Health and Safety (EHS) management system. As the company works in various sectors and had a much decentralized approach to EHS management, there was a need for centralization and integration of its practices to a system that could allow the headquarters to assess, control and properly administer its processes and its people under one single management system. The new system aimed to improve the performance of the company's sites and also has shown as an effective tool to build up corporate resilience. During the deployment of the system, various industrial units presented very critical issues concerning bad EHS administration and commitment, which enabled organization to tempt its capacity to realignment of EHS management to corporate strategy on those sites. The case of one specific unit is showed here in order to illustrate the situation prior to and subsequent to the execution of an orientated action plan and implementation of the new EHS management system on site.

Keywords Operational resilience · Integrated management systems · Environmental · Occupational health and safety management system · EHS assessment and performance · Roadmap

---

B. Seixas de Siqueira · M. Grijalvo (✉) · G. Morales-Alonso
Dpto. de Ingeniería de Organización, Administración de Empresas y Estadística,
Universidad Politécnica de Madrid, C/José Abascal, no 2, 28028 Madrid, Spain
e-mail: mercedes.grijalvo@upm.es

B. Seixas de Siqueira
e-mail: barbara.s.siqueira@gmail.com

G. Morales-Alonso
e-mail: gustavo.morales@upm.es

© Springer International Publishing Switzerland 2015
P. Cortés et al. (eds.), *Enhancing Synergies in a Collaborative Environment*,
Lecture Notes in Management and Industrial Engineering,
DOI 10.1007/978-3-319-14078-0_31

1 Introduction. EHS Management in Multinational Groups: Decentralization Versus Integration

With the intent to enhance market share and financial performance, many organizations go for mergers, acquisitions, horizontal or vertical expansions. They are an increasingly important mean for achieving corporate diversity and growth (Malekzadeh 1988). Although the main focus is given to the strategic and financial fit between organizations during those expansion processes, the integration of various organizational systems, such as technology and management systems are extremely important for a wealthy continuance of the business.

The management of Environment, Health and Safety (EHS) systems represents a significant piece of those integration practices, given that EHS issues can account from 5 to 30 % of the deal value of industrial mergers and acquisitions, and that they are almost always among the most difficult issues to deal with (Pilko 2007).

Bringing new EHS cultures and systems to the organizations give the companies the asset of having more dynamic and versatile environments; however, it also brings different methods of managing, assessing and controlling the group EHS performance. Proper governance looks for understanding the target company's EHS vision, integrating policies, EHS culture, EHS recruitment and evidently EHS management systems (Pilko 2007).

This article presents and analyzes the experience of a European multinational company working in diverse infrastructure sectors on the deployment and enhancement of a new internal Environment, Health and Safety (EHS) management system.

The company studied, as many others multinational companies, holds different business divisions, usually coming from previous acquisitions, and offers its clients a diversified portfolio of products and services. The various sectors have individual EHS systems managed according to its necessities and culture, following a much decentralized approach. Consequently, evaluating the EHS performance of its different businesses became a hard task to the headquarters as the assessments brought to light deficient and conflicting results. In order to unify and harmonize the EHS processes under one single management system, the company deployed its own EHS program and reference document.

The new system launched not only had the objective of standardizing and integrating the procedures within the organization but also two other more: first to improve the EHS performance and assess equally the EHS maturity level of its units, being able to produce more reliable outcomes and secondly to support and guide the industrial units on attaining environmental and safety international standards such as ISO 14001 and OHSAS 18001. In addition the new internal EHS management system has shown as an effective tool to build up corporate resilience (Ignadiadis and Nandhakumar 2007; Park et al. 2011).

During the deployment of the system, various industrial units presented very critical issues concerning bad EHS administration and commitment, which enabled organization to experiment the new management system and tempt its capacity to improve the EHS performance on those sites. The case of one specific unit is showed here in order to illustrate the situation prior to and subsequent to the execution of an orientated action plan and implementation of the new EHS management system on site. The authors conduct a qualitative research in order to set some of the key factors to take into consideration (Yin 2009).

2 Corporate Resilience. Standardization Versus Adaption of Management Systems

Studies at the macroeconomic and microeconomic levels in various European countries and around the world have demonstrated the clear benefits of standards and standardization to the wider economy. The ISO 9000 standards were pioneers in the development of management systems in companies other than financial and economics. Their success has made them have been imitated in other areas of management and another "ad hoc" models and certification schemes development as the ISO 14000 or OHSAS 18000 with which they share both its virtues and its limitations largely related to their pragmatic general application and not integrated within the organization, (Zeng et al. 2011), as the case study shows.

The organization had two different problems regarding the management systems: first EHS processes were managed individually, each division of the company held its own system and managed EHS aspects according to their principles and culture and secondly, the systems were not all integrated, meaning that generally environment and safety were managed separately according to different criteria.

In an attempt to address both issues the company designed an EHS roadmap. It was fully developed inside the organization as a management system assessment tool which combines both environmental and safety systems in one integrated and single tool and which serves the management of sites (industrial units, offices, research centers, etc.) to progress in the various EHS topics with a maturity matrix. Each of the topics is broken down into initiatives, which are developed into several themes. The evaluation of each theme, based on the iterative four-step management principle based on Deming's cycle, is provided on a five levels implementation, being 1 the lowest and 5 the highest. The model does not indicate in what order and how processes should be implemented, it only define and characterize them. The scores obtained allow the units to know their EHS maturity level showing them their strengths, weaknesses and points of improvement.

This tiered evaluation scheme not only promotes the development of an improvement disciplined approach but allows efforts to formalize and implement both systems have visibility and recognition within and outside the organization.

Being benchmarked and continually updated with ISO 14001 and OHSAS 18001 standards, the system is thus planned in a way that it could lead organization's sites to attain those certifications. Certifications for small sites are potentially economically unviable, since they implicate in high costs for the units, financial and human resources (Downs 2003).

2.1 Design of the Assessment Tool as a Maturity Model

Regarding the scoring basis, to reach a certain level, criteria of all previous levels should be reached; however, as the method resembles more as an assessment system rather than an audit system, it is not required that all criteria are verified with exactness, leaving to competent assessors the verification of existing processes. This evaluation scheme allows each organization to mark their own pace in the process of improvement based on its goals and priorities what makes easier the necessary cultural change (Camisón et al. 2007).

As the idea of the tool is to encourage the sites to improve first their weaknesses, the initiative's score will be the lowest score of a theme and the final score for a topic is the average of the initiatives. This mechanism adopted avoids falling into temptation, to address the less problematic actions to quickly demonstrate improved performances (Ivey and Carey 1991).

2.2 The Top Management Commitment as a Concept of Sponsorship

Each EHS theme has a sponsor inside the organization selected by job title who must regularly report to the site's managers on the control of these themes. In this sense, the concept of sponsorship brings the community the responsibility and ownership of specific themes, including EHS management to their routine jobs (ISO 2004).

The application of the tool thus addresses two aspects what have been shown keys in the implementation of improvement models in recent years: the management commitment and the alignment of the system goals with the overall strategy of the company.

2.3 The Implementation of the Tool as an Assessment Program

The roadmap tool defines the basis the organization's standards for environment, health and safety and evaluates the units' EHS maturity level. Assessments aim at supporting the continuous improvement of EHS performance by measuring gaps between actual situation and reference levels/targets on different key EHS themes, defining recommendations and implementing improvement plans.

The group headquarters sets the general guidelines for the assessments (as frequency, associated costs, reporting instructions and the accreditation process). It also sets minimum annual objectives for units based on overall scores for each topic and/or individual scores on specific themes identified as key priority areas. The sectors set their own objectives but cannot target lower performance than group overall objectives.

The system works in a cycle where the self-assessments are validated by formal assessments which in turn, enable the sites to structure themselves to go for certifications. The results of those assessments support the setting of new targets and the identification of improvements to the system.

3 Results

Six years after the deployment of the new integrated management system, the scores of formal assessments started to be used by the corporate headquarters to closely observe the performance of its main industrial sites and set up new objectives, targets and provide its sectors with enough resources to ensure the continuity of the program. Therefore, the results of formal assessments were directly issued to the corporate EHS team strengthening communication and interaction between the EHS communities at all organizational levels. The cross interaction between EHS communities and the close surveillance of top managers reinforced the EHS management practices among the all the community and also the employee's involvement and acceptance of organizational policies. Management communication can also influence the employees towards the organizational goals by communicating them the importance of their involvement and unique role in the organization (Smidts et al. 2001).

3.1 EHS Performance

It is still difficult to evaluate the efficiency of the new EHS system. As some segments still embrace more than one management system, the results, could be attributed to any or to the combination of them. Some real improvements on EHS aspects were observed during last decade in industrial sites, though.

A score of 3.5 was established as being the minimum assessment score to be achieved in all topics by all its sites. Since the corporate headquarters started to assess the outcomes from the site assessments, it was observed a positive trend in all EHS topics with few eventual drops. In 2006 the organization had less than 50 % of its reporting sites scoring above 3.5 in all EHS topics, while in 2012 this number rose to more than 55 %, reaching almost 70 % for Environment.

The evolution on EHS roadmap scores was followed by negative trends in some key performance indicators as it could be observed from the Injury Frequency Rate (IFR) indicator, calculated as the number of occupational injuries suffered by employees with one day lost or more related to a common exposure of 1,000,000 (a million) hours. From 2006 to 2012 the indicator was decreased by more than 75 %. The negative trend was also observed for environmental indicators as energy consumption, water consumption and greenhouse gases emissions. There was also a substantial increase in the number of industrial sites certified by international standards. All the main units are already certified and the organization is now looking for having its medium sized units qualified on top.

First class EHS performance is vital for the health of an organization of this nature and in many kinds of industry is considered as "license to operate". Poor EHS performance was an indication that the organization held a great deal of inefficiencies on its value-chain; as waste of scarce resources, pollution, fines paid to public authorities, high indices of absenteeism, low productivity and huge costs associated to medical treatments and indemnities.

The progress on EHS management led the organization to cost-cuttings, reduced absenteeism and turnover of employees. It also improved the organization's overall reputation, which contributed not only to more efficient operations but also to retain clients and gain new contracts.

3.2 Study Case

A study case was developed and is showed here in order to analyze and discuss the effectiveness of the EHS management system deployed internally. During the years when the new method was being deployed, some industrial units started to show serious issues concerning increased number of accidents with employees and contractors.

Those distressing events enabled organization to attempt the new assessment system and assess its capability of improving the EHS performance on site. In 2006, some industrial sites were elected to receive an EHS coaching program. They figured among the poorest sites regarding safety, holding the highest indices of accidents. Table 1 shows the bad results of the unit analyzed here at that point.

A customized safety coaching program was so launched to this unit. The action plan created by the corporate top-management team together with the sector's president and local EHS team started to be implemented in the facility, after that, the new EHS assessment system was enforced and closely watched by the top management during the subsequent years. The outcomes of the assessment done in 2012 illustrate a completely different situation on the site, it even pointed to this facility to be example of EHS excellence inside organization (Table 2).

The overall positive results in EHS performance allowed the company in 2012 the license to expand the industrial facility creating hundreds of new jobs and bringing more prosperity for the local community. Today the site has been working with best practices.

Table 1 Results from EHS assessment in the analyzed industrial unit, year 2006

	Management	Environment	Health and safety	Global
Results obtained from EHS assessment	1.3	3.5	2.3	2.4

Table 2 Results from EHS assessment in the analyzed industrial unit, year 2012

	Management	Environment	Health and safety	Global
Results obtained from EHS assessment	4.6	5.0	4.7	4.8

4 Conclusions

The new management system facilitated the creation of one structure that allowed the harmonization of EHS procedures and management principles to efficiently deliver the organization's objectives. The method comprised the main management principles of international standards, but all together it was shaped to the organization's necessities and culture. In general, it was observed a change of mind regarding EHS aspects among the community, reflected on progressive higher assessment scores, improvement of KPI's, increased number of certifications and decrease in absenteeism.

The method delivered the ownership of EHS process all the community and encouraged them towards excellence increasing their environmental and safety awareness. In this way it embraced people among one only objective, the enhancement of their unit/business EHS performance. In fact all developed systems facilitate and enhance the active participation of all levels of the organization and cannot be understood without the participation of all (Camisón et al. 2007).

Nevertheless, the organization still needs to raise the integration of all segments under a unique EHS management system. This harmonization of processes is primordial to support the corporate headquarters to access the EHS performance of its units and track the achievement EHS objectives and targets.

Moreover, dealing with different information platforms generate conflicting results, increase time, lead to mistakes and rework. The transfer of management techniques is not a linear process of diffusion, but a process of adaptation by those involved in the context where you want to implement, and the new approaches taken by each of them are different both from the standpoint of to the organization of work (Ates and Bititci 2011).

The integrated approach helped the organization to enhance the effectiveness of various EHS processes as managing employees' needs, pushing managers towards EHS commitment, monitoring competitors' activities, encouraging best practices among sectors and the cross communication (Hamel and Valikangas 2003).

In addition, EHS management practices are expected to have an increasingly important role in organizational performance, since they can provide a good measure of organization's environmental and social performance, constituting already two pillars of a sustainable development management system (Giagnorio 2005).

References

Ates A, Bititci U (2011) Change process: a key enabler to building resilient SMEs. Int J Prod Res 49(18):5601–5618

Camisón C, Cruz S, Gonzalez T (2007) Gestión de la calidad: conceptos, enfoques, modelos y sistemas. Pearson Educación, Madrid

Downs DE (2003) Safety and environmental management system assessment: a tool for evaluating and improving SH & E Performance. Prof Saf 48(11):31–38

Giagnorio ML (2005) Towards corporate sustainability indicators. Forumware Int 2:22–31

Hamel G, Valikangas L (2003) En busca de la reiliencia. Harward Bus Rev 81(9):33–39

Ignadiadis I, Nandhakumar J (2007) The impact of enterprise systems on organizational resilience. J Inf Technol 22:33–46

ISO (2004) ISO 14004. Environmental management systems—requirements. International Organization for Standardization

Ivey M, Carey J (1991) The ecstasy and the agony. Bus week, Oct 21

Malekzadeh AN (1988) Acculturation in mergers and acquisitions. Acad Manage Rev 13(1):79–90

Park J, Seager TP, Rao PSC (2011) Lessons in risk- versus resilience-based design and management. Integr Environ Assess Manage 7(3):396–399

Pilko G (2007) EHS: a new paradigm for mergers and acquisitions. Directors and Boards 31(3):48–49

Smidts A, Pruyn H, Riel C (2001) The impact of employee communication and perceived external prestige on organizational identification. Acad Manage J 44(5):1051–1062

Yin R (2009) Case study research: design and methods, 4th edn. Sage Publications, Thousand Oaks

Zeng SX, Xie XM, Tam CM, Shen LY (2011) An empirical examination of benefits from implementing integrated management systems (IMS). Total Qual Manage Bus Excellence 222):173–186

Part V
Knowledge and Project Management

Effectiveness of Construction Safety Programme Elements

Antonio López-Arquillos, Juan Carlos Rubio-Romero,
Jesús Carrillo-Castrillo and Manuel Suarez-Cebador

Abstract Occupational health and safety levels in construction continue to be an international cause of concern because of high accident rates in the sector. In order to reduce negative incident rates in the industry, construction companies implement safety programmes that include different elements. In current research, an expert panel of 8 members was selected with the aim to quantify the effectiveness of the essential safety program elements using staticized groups methodology. Results showed that effectiveness of Job hazard analyses and communication, and upper management support as safety program elements, obtained the best scores by the experts. These results can be used to prioritize the most effective safety elements for construction projects in order to control construction safety risk.

Keywords Construction · Occupational · Safety · Programme · Elements

1 Introduction

Occupational health and safety levels in construction continue to be an international cause of concern. In Europe construction sector had the worst incident rates per worker (Eurostat 2011) and in the rest of the world, the rates are not better.

A. López-Arquillos (✉) · J.C. Rubio-Romero · M. Suarez-Cebador
Dpto. de Economía y Administración de Empresas, Universidad de Málaga,
C/Dr. Ortiz Ramos s/n, 29071 Málaga, Spain
e-mail: investigacioncatedra@gmail.com

J.C. Rubio-Romero
e-mail: juro@uma.es

M. Suarez-Cebador
e-mail: suarez_c@uma.es

J. Carrillo-Castrillo
Dpto. de Organización Industrial y Gestión de Empresas II, E.T.S. de Ingeniería,
C. de los Descubrimientos, s/n. Pabellón Pza. de América, 41092 Seville, Spain
e-mail: jcarcas@gmail.com

These rates could be a signal that the management of the construction safety is not been implemented properly in construction projects.

European contractors are legally required to develop and to implement their health and safety plans (Directive 92/57/EEC), but this obligation is not always carried out properly. Consequence of a poor development and implementation of the health and safety plans is an increase in the risk levels. Correct identification of hazards in the construction place is a key factor of successful construction safety management (Carter and Smith 2006) but this identification cannot be reduced to the construction site and workers' influence. Contractors have a big influence in the construction hazards because they are in the middle of the production chain between designers and workers so their view and opinion are decisive in the construction safety conditions.

In order to reduce negative incident rates in the industry, construction companies implement safety programmes that include different elements. Previous research had studied and discusses the elements of an effective safety programs. Sawacha et al. (1999) studied the relative impact of various safety programme elements using a factor analysis. They quantified relationship between each safety programme element safety performance. Rajendran and Gambatese (2009) using the Delphi method created a Sustainable Construction Safety and Health (SCSH) rating system. Other authors discuss the formation on an effective safety program (Hinze 1997; Gibb 1995).

2 Methodology

An expert panel of 8 members was selected in order to quantify the effectiveness of the essential safety program elements using staticized groups methodology. Hallowell and Gambatese (2009) created a list with the essential elements of safety prevention strategies based on the review of the existing literature. From hundred of safety prevention strategies only 12 were mentioned as essential in four or more of the publications reviewed. Descriptions of these elements are included in Table 1.

3 Results

Data matrix resulted from the experts opinion showed the effectiveness of each specific preventive program element for each civil construction activity studied. Mean values of each element program effectiveness were calculated. According to the experts opinion the two most effective elements were job hazard analyses and communication (Mean = 7.56) and upper management support (Mean = 7.13), following by safety and health and orientation training (Mean = 7.05) and safety manager on site (Mean = 7.00). On the other hand, written and comprehensive safety and health plan (Mean = 6.52) and substance abuse programs (Mean = 5.97) obtained the lowest values (Table 2).

Table 1 Description of the safety program elements

Safety program element	Description
Upper management support (E1)	Participation and commitment of upper management involves the explicit consideration of worker safety and health as a primary goal of the firm. Upper management must demonstrate commitment by participating in regular safety meetings, serving on committees, and providing funding for other safety and health program elements
Employee involvement safety and evaluation (E2)	Employee involvement and evaluation is a means of including all employees in the formulation and execution of other programs elements. Involvement in safety and health activities may include activities such as performing job hazards analyses, participating in toolbox talks, or performing inspections. Evaluation of employees' safety performance involves considering safety metrics during regular employee performance evaluations. This may include the consideration of incident frequency, inspection results and near misses
Substance abuse programs (E3)	Substance abuse programs target the identification and prevention of substance abuse of the workforce. Testing is a crucial component of this safety program. Methods of testing and consequences of failure may differ from one firm to another. However, repeated violations are typically grounds for dismissal of the employee. Testing may occur on regular or random basis and always employees involved with an incident that involves a medical case or lost work-time injury or fatality
Written and comprehensive safety and health plan (E4)	A safety and health plan serves as the foundation for an effective safety and health program. The plan must include documentation of project-specific safety and health objectives, goals, and methods for achieving success
Project-specific training and regular safety meeting (E5)	This element involves the establishment and communication of project-specific safety goals, plans, and policies before start of the project. Safety training may include reviewing project-specific or task specific hazard communication, methods of safe works behavior, company, policies, safety and health goals, etc. This element also involves regular meetings such as toolbox talks to reinforce and refresh safety and health training
Subcontractor selection and management (E6)	This element involves the consideration of safety and health performance during the selection of subcontractors. That is, only subcontractors with demonstrated ability to work safety should be considered during the bidding or negotiation process. Once a contract is awarded, the subcontractor must be required to comply with the minimum requirements of the general contractor's safety and health program

(continued)

Table 1 (continued)

Safety program element	Description
Job hazard analyses and communication (E7)	Job hazard analysis may be initiated by reviewing the activities associated with a construction process and identifying potential hazardous exposures that may lead to an injury. Other sources as OSHA logs, violation reports, accident investigation reports, interviews with laborers, or simply intuition may be used to identify hazards
Record keeping and accident analyses (E8)	Record keeping and accident analyses involve documenting and reporting the specifics of all accidents including information such as time, location, work-site conditions, or cause. The element also includes the analyses of accident data to reveal trends, points of weakness in the firm's safety programs, or poor execution of program elements
Emergency response planning (E9)	Emergency response planning involves the creation of a plan in the case of a serious incident such as a fatality or incident involving multiple serious injuries. Planning for emergencies can define the difference between an accident and a catastrophic event. Such a plan may be requires by the owner or insurance carrier
Safety and health committees (E10)	A committee made up of supervisors, laborers, representatives of key subcontractors, owner representatives, OHSA consultants, etc. may be formed with the sole purpose of addressing safety and health on the worksite. Such a committee must hold regular (e.g. weekly or biweekly) meeting to address safety and health by performing inspection, discussing job hazard analyses, or directing safety meetings and training
Safety manager on site (E11)	Simply, this safety program involves the employment of a safety and health professional (i.e., an individual with formal construction safety and health experience and/or education). This individual's primary responsibility is to perform and direct the implementation of safety and health program elements and serve as resource for employees
Safety and health and orientation training (E12)	Orientation of all new hires may be the most important safety training. Even skilled and experienced workers should be provided with a firm-specific safety and health orientation and training. Such training and orientation informs new hire of company safety goals, policies, program resources, etc. This element involves the firm-specific, but not necessarily project-specific, orientation, and training of all new hires (or existing employees if a safety and health programs is new to the firm)

Table 2 Ranking of the safety program elements

Ranking	Safety program element	Effectiveness score
1	Job hazard analyses and communication	7.56
2	Upper management support	7.13
3	Safety and health and orientation training	7.05
4	Safety manager on site	7.00
5	Project-specific training and regular safety meeting	6.90
6	Subcontractor selection and management	6.89
7	Safety and health committees	6.88
8	Employee involvement safety and evaluation	6.82
9	Record keeping and accident analyses	6.80
10	Emergency response planning	6.78
11	Written and comprehensive safety and health plan	6.52
12	Substance abuse programs	5.97

4 Conclusions

Effectiveness of Job hazard analyses and communication, and upper management support as safety program elements, obtained the best scores by the experts. These results can be used to prioritize the most effective safety elements for construction projects in order to control construction safety risk. This strategic information can be especially useful when the resources in the projects are small and limited, because it can be used to select the most effective measures.

The findings obtained in the current study must be considered since the beginning of the design phase of the project, to the end of the works at the construction site, and adequate preventive measures must be improved as the experts recommend.

References

Carter G, Smith SD (2006) Safety hazard identification on construction projects. J Constr Eng Manage 132(2):197–205

Eurostat (2011) European statistics on accidents at work. http://epp.eurostat.ec.europa.eu/tgm/table.do?tab=table&init=1&plugin=1&language=en&pcode=tps00042

Gibb A (1995) Effective implementation of a safety strategy during the construction phase of major, complex construction projects. NICMAR J Constr Manage Natl Inst Constr Manage Res India 10(2):116–126

Hallowell MR, Gambatese JA (2009) Activity-based safety risk quantification for concrete formwork construction. J Constr Eng Manage 135(10):990–998

Hinze J (1997) Construction safety. Prentice-Hall, Upper Saddle River

Rajendran S, Gambatese JA (2009) Development and initial validation of sustainable construction safety and health rating system. J Constr Eng Manage 135(10):1067–1075

Sawacha E, Naoum S, Fong D (1999) Factors affecting safety performance on construction sites. Int J Project Manage 17(5):309–315

Operational Issues for the Hybrid Wind-Diesel Systems: Lessons Learnt from the San Cristobal Wind Project

Yu Hu, Mercedes Grijalvo Martín, María Jesús Sánchez and Pablo Solana

Abstract Hybrid wind-diesel power systems have a great potential in providing energy supply to remote communities. Compared with the traditional diesel systems, hybrid power plants can offer many advantages such as additional capacity, being more environmentally friendly, and potential reduction of cost. The O&M of a hybrid power project requires comprehensive knowledge from both technical and managerial points of view. This study focuses on one of the largest existing hybrid wind-diesel power system, the San Cristobal Wind Project. Performance analysis and computer simulation are conducted to illustrate the most representative operational issues. We demonstrate that the wind uncertainty, control strategies, energy storage, and the wind turbine power curve have a significant impact on the performance of the system.

Keywords Hybrid wind-diesel power plant · Operations and maintenance · Feasibility study · Wind power and project management

---

Y. Hu (✉) · M.G. Martín
Unidad Administración de Empresas. Escuela Técnica Superior Ingenieros Industriales, Universidad Politécnica de Madrid, José Gutiérrez Abascal 2, 28006 Madrid, Spain
e-mail: yu.hu@alumnos.upm.es

M.G. Martín
e-mail: mercedes.grijalvo@upm.es

M.J. Sánchez
Dpto. de Organización de Empresas. Escuela Técnica Superior Ingenieros Industriales, Universidad Politécnica de Madrid, José Gutiérrez Abascal 2, 28006 Madrid, Spain
e-mail: mjsan@etsii.upm.es

P. Solana
ATI Consultores, Madrid, Spain
e-mail: psolana@aticonsult.es

1 Introduction

Nowadays, wind power is becoming a widely accepted solution for providing alternative power supply to remote or isolated areas. Many applications and demonstration projects with wind power have been established around the world. Being different than the traditional power systems, the O&M of the new hybrid wind-diesel energy system requires comprehensive knowledge from both technical and managerial points of view.

In order to demonstrate some key issues concerned about the management of a hybrid wind-diesel power project, this study focuses on one of the largest existing wind-diesel energy systems, the San Cristobal Wind Project.

Before year 2007, three 650 kW diesel generators were used to provide energy supply to the nearly 6 thousands inhabitants in the San Cristobal Islands, with an average electricity demand of approximately 900 kW, and peak energy demand at around 1700 kW (Kornbluth et al. 2009). A total of three 800 kW synchronous type wind turbines were installed in Oct 2007. The minimum wind speed for the wind turbine starting generating energy is 3.5 m/s and it reaches the maximum capacity with a 12 m/s wind speed (EOLICSA 2013).

2 Financial Investment and Tariff

The investment for constructing the San Cristobal Wind Project is approximately $9,840,000. The funding of the construction cost is raised from different parties, as shown in Table 1.

The project cost for the San Cristobal Wind Project is higher than other similar wind energy projects in remote areas. According to Hinokura, the high cost of construction is mainly due to the logistical and environmental challenges in the Galapagos Islands, lack of infrastructure, and the diesel displacement objectives of the project (Hinokuma 2008).

Table 1 Funding for the San Cristobal project

$4,850,000	Cash donation provided by Global 3e, a charitable fund established by e8 members. The majority of the donation was provided by AEP as project leader, with support from RWE, HQ, and ENEL
$625,000	Cash donation provided by RWE
$930,000	Matching grant donation provided by United Nations Foundation (UNF)
$3,195,000	Capital subsidy provided by Ecuadorian Government (FERUM)
$240,000	2004 Designated income tax payments by Ecuadorian contributors through municipality of San Cristobal
$9,840,000	Total funding

Source http://www.eolicsa.com.ec/

According to the official website of the project (http://www.eolicsa.com.ec/), the San Cristobal Wind Project will receive a fixed tariff of $0.1282 per kWh for a twelve-year period, with no adjustment for inflation. This tariff is less than the current price paid for diesel generation on San Cristobal, estimated at $0.1585 per kWh. Meanwhile, the project will not impact on the electricity price paid by the consumers (approximately $0.10 per kW h) with the help of the energy subsidy by the Ecuadorian government.

3 Feasibility Study

The feasibility study was carried out in 2005. Different expected results have been found in the published documents (Baring-Gould and Corbus 2006). According to "e8 The Galapagos San Cristobal Wind and Solar Projects" (2008), the wind project is expected to significantly reduce San Cristobal's consumption of diesel fuel. Expected annual performance of the hybrid wind-diesel power system is modeled using the HYBRID2 computer software. The wind turbines are expected to deliver more than 4.12 GW h energy to the grid, covering 52 % of the total energy consumption in the year 2008. And in 2028, the wind energy will reach about 6.62 GW h (covering 42 % of the total demand) due to the increase in the power demand.

However, more modest estimations has been found in other documents. According to the "Project Design Document of the San Cristobal Wind Project submitted to United Nations Framework Convention on Climate Change (UN-FCCC)" in 2007, the project is expected to produce 3.32 GW h during the first year of operation (assuming 52 % kWh annual diesel displacement and 96.5 % wind turbine availability) This estimation claims an 37 % diesel displacement in the year 2026, with annual wind generation of 4.43 GWh.

It is worth mentioning that, since the investments involved are mainly donation-based, no report about cash flow payback and profitability is found.

4 Actual Performance

After 6-years operation, EOLICSA has published the report of "Galápagos San Cristobal Wind Project: Highlighting 6 years of Operations". A general overestimation is found in the wind power penetration level. Table 2 compares the actual and estimated energy delivered by the San Cristobal Wind Project. For the five complete accounting years from 2008 to 2012, the wind turbines generate an average of 84 % (from 64 to 96 %) of the estimated energy. Meanwhile, the wind turbine availability is 93 %, also lower than the expectation of 96.5 % in the feasibility study.

Table 2 Real and estimated energy generation in the first period 2007–2013

Year	Estimated electricity supplied to the grid	Real energy generated from wind turbines	Real/estimated wind energy production rate
2007	3,316,759	790,398 (Oct–Dec)	Not applicable
2008	3,387,240	2,682,461	79.2 %
2009	3,471,068	3,204,436	92.3 %
2010	3,555,581	3,434,854	96.6 %
2011	3,640,671	3,344,625	91.9 %
2012	3,747,344	2,398,372	64.0 %
2013	3,816,214	2,453,916 (Jan–Sep)	Not applicable

The reason of this under performance has not been disclosed by the report. More analysis must be done in detail in order to correct the over-estimation with more data.

Large volatility of the wind energy production along the time horizon is shown in the report (EOLICSA 2013). In 2010, approximately 3.4 GWh wind power is produced. Compared with the 2.4 GWh wind power production in 2012, the difference is 1 GWh. Meanwhile, large difference is found in the monthly performance of the project. A clear decrease of penetration happened periodically from Jan to Apr. The average penetration level for the month from Jan to Apr is 13.2 %, and for the other months is 42.5 %.

This finding is in accordance with the wind character within the year. The wind speed is generally lower from Jan to Apr. The average monthly wind speed from Jan to Apr is 4.4 m/s, while the average monthly speed in the other months is 6.7 m/s. The average wind speed data also show that an unusual low-wind year occurred in 2008, which is in accordance with the relative low production in this year. These findings suggest that the long term wind power uncertainty has an important impact on the project performance, which should be noticed by the researchers.

Clear positive correlation has been found between the monthly wind energy penetration level and the monthly average wind speed, shown in Fig. 1. When the monthly average wind speed is lower than 5 m/s, the wind penetration level is generally less than 20 %. On the other hand, if the wind speed level is higher than 6 m/s, the wind energy penetration level would be expected to be larger than 25 %.

The periodical wind speed change and relationship between wind speed and wind power production do not explain the low penetration in 2012. In 2012, only 2.4 MWh energy had been produced by the wind resource, even lower than that in 2008, in which an unusual low wind year occurred. According to the report, a notable event occurred in May 2012, "which resulted in the unavailability of 80 days of one wind turbines and required 2 specialists from the Manufacturer to be on site". This event could explain part of the unusual low production in 2012. This event has significantly dropped the wind energy production in the months of May and June, compared to other months. As a result, more accurate analysis on rare events should be included for other hybrid wind-diesel power plants.

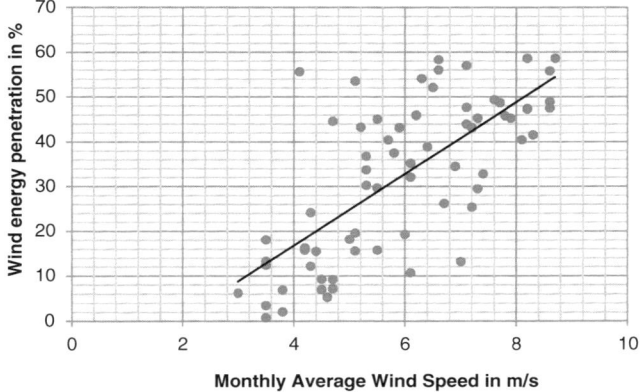

Fig. 1 Wind energy penetration level compared with the monthly average wind speed

Because no energy storage systems are designed to work along with the hybrid system, the operational mode for the power system is based on a diesel always-on criteria, supervisory control is active in order to keep a minimum diesel load. Meanwhile it is necessary to maintain at least one diesel unit operating at 25 % of its rated capacity for technical requirements according to the EOLICSA report.

One of the drawbacks of this control strategy is the decreasing fuel efficiency, since the diesel engine is not working at nominal power load. Due to the wind uncertainty, more reserved power should be considered in the case of sudden drop in wind speed.

According to Jargstorf (2008), during the first 3 months the wind energy penetration level was 45 %. However the fuel consumption of the diesel engine increased from 0.33 to 0.4 L per kWh, which means that the diesel generation efficiency went down from 11.45 to 9.31 kWh per gallon. Therefore, the effective fuel saving was only 33 % (with a nominal of 45 %) in these months. Advanced power dispatch strategy considering both the wind uncertainty and the diesel efficiency is therefore necessary for future developments.

5 Performance Simulation

Seeking for a detailed analysis about how the related factors will impact on the performance, we use the software HOMER (Hybrid Optimization Model for Electric Renewables) developed by NREL (National Renewable Energy Laboratory, USA), to simulate the performance of the San Cristobal Wind Project.

The monthly average wind speed data and the monthly energy demand data from the year 2011 is used to generate the hourly wind speed and demand series. The demand data are adapted based on the work of Hinokuma (2008). The wind turbine character and wind speed response are also simulated. The operating strategy is adapted as reported by the EOLICSA document (2008).

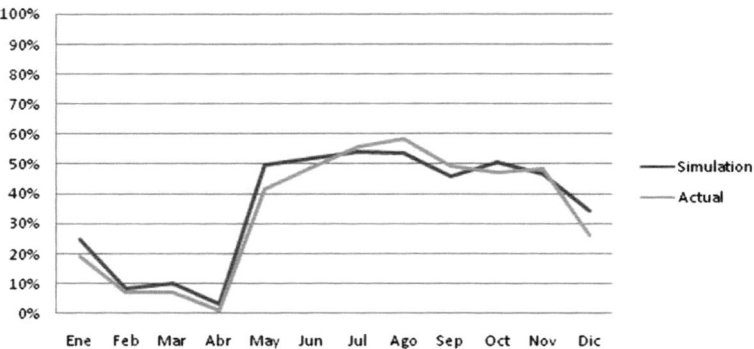

Fig. 2 Simulation result of the wind penetration level in year 2011

The simulation result is shown in Fig. 2, indicating that HOMER reaches a similar result as the real wind penetration level in year 2011. The average absolute error for the simulated penetration level is 3.9 %.

In order to validate the simulation with the assumed daily demand, we apply the same setting to the wind speed data and monthly demand data in year 2010. The average absolute error for the simulated penetration level is 3.0 %.

The simulation and validation process has shown that the simulation is of significance for managing and optimizing the wind energy delivered by the San Cristobal Wind Project.

From the operational experience, according to Villagómez (2013), the wind farm of the San Cristobal Wind Project is oversized with the current diesel power plant. The potential improvement based on the existing system depends on the available wind energy excess. We simulate the wind energy excess using HOMER for year 2011, shown in Fig. 3. Due to the seasonality of the wind speed in the San Cristobal

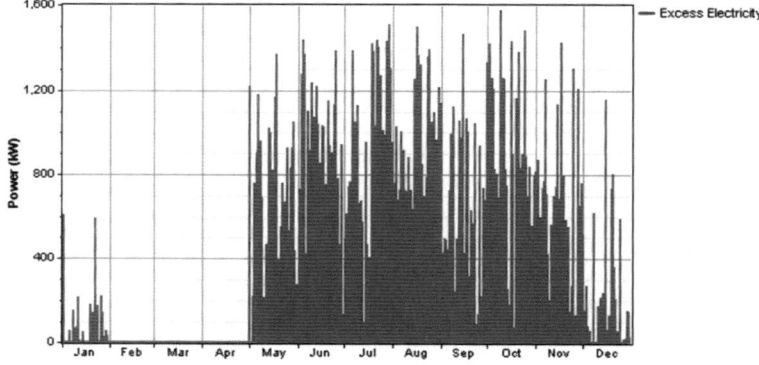

Fig. 3 Excess power from wind energy simulation with HOMER. Example year 2011

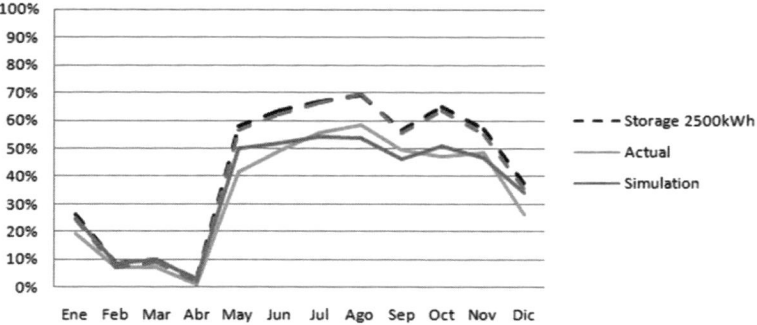

Fig. 4 Simulation of system performance with 2500 kW h battery storage

island, the excees energy also shows a significant seasonality. Most of the unused wind energy is concentrated during the months of high wind speed from May to Nov.

One possible improvement comes from the energy storage system. The application of energy storage offers the energy producers the ability to shift the energy load (Denholm and Margolis 2007). We simulate also the performance by including additional 2500 kWh battery storage to the isolated system. The penetration level can be improved approximately in 12 % in the high wind season, as shown in Fig. 4. It is not a surprising fact that the wind energy penetration in the low wind seasons is hardly to be improved based on the current wind turbines, due to the fact that the potential wind energy excess available at the low wind season is significantly lower than in the high wind seasons. Wind turbines designed especially for low speed wind conditions should be considered in the following development of the next phase of the San Cristobal Wind Project or similar projects.

6 Conclusion

According to the EOLICSA report (2013), this project is successful in the sense that a positive cash flow has been created for covering the costs of operation and maintenance and other related duties since it started working. However, the cash flow is still not enough to recover the high initial investment. A feasibility study must be carried out with more intensive work from profit/cost point of view.

As a short summary, the following aspects should be learnt as a lesson from the San Cristobal Wind Project.

- Feasibility study should be revised in order to avoid biased estimation.
- Wind turbines for low speed wind conditions should be considered to adapt for the low wind season.
- The uncertainty regarding to the long term wind resource uncertainty should be taken into consideration.

- Modeling of operational issues such as system failure should be considered in the feasibility study.
- More research should be devoted to the application of energy storage to improve the efficiency of the hybrid system.
- Novel dispatching strategies and control modes for diesel engines is necessary to optimize the diesel efficiency.

References

Baring-Gould E, Corbus D (2006) Modeling results of wind-diesel retrofits options for Santa Cruz, Galápagos. NREL, Ecuador

Denholm P, Margolis RM (2007) Evaluating the limits of solar photo voltaics (PV) in electric power systems utilizing energy storage and other enabling technologies. Energy Policy 35 (9):4424–4433

EOLICSA (2008) The Galapagos San Cristobal wind and solar projects (Online). Available: http://www.globalelectricity.org/upload/File/e8_san_cristobal_wind_and_solar_projects_publication_final.pdf. Accessed 01 Mar 2014

EOLICSA (2013) Galapagos San Cristóbal Island wind project 2007–2013 (Online). Available:http://www.globalelectricity.org/upload/File/Projects/SanCristobalGalapagosWindProject/Galapagos-Wind-Project_Eng_fin.pdf. Accessed 01 Mar 2014

Hinokuma R (2008) Optimization of a small wind-based hybrid electrical grid on the island of San Cristobal in the Galapagos. ProQuest, Ecuador

Jargstorf B (2008) Comparison between the operating San Cristóbal and planned Baltra/Santa Cruz. International Wind/Diesel Workshop, USA

Kornbluth K, Hinokuma R, Johnson E, McCaffrey Z (2009) Optimizing wind energy for a small hybrid wind/diesel grid in the Galapagos Islands. University of California, Davis Energy Efficiency Centre

Villagómez JC (2013) Energy demand analysis and management according to availability of renewable energy sources in Galapagos Islands. The first LACCEI international symposium on mega and micro sustainable energy projects, Mexico

Points of Convergence Between Knowledge Management and Agile Methodologies in Information Technology (IT)

Eliane Borges Vaz and Maria do Carmo Duarte Freitas

Abstract Software companies operate in an environment where there is global competition and demand for innovation is increasing. In these environments, the knowledge generated daily is primarily priority raw material for the development of programs, which will be delivered to increasingly demanding customers. The difficulty in systemizing the process of knowledge creation in order to retain it and improve it, even with the intense changes in staff, in technology and in the defined organizational objectives, is the major challenge for the organizations. The methodologies used in the software development process can influence directly the way knowledge is shared by the team and the customer and in adding value to the final product. The research is relevant due to link agile practices and knowledge management for competitive survival of IT companies. The methodology used was based on exploratory research to establish the theoretical bases of the relevant topics. From this theoretical foundation it was possible to establish the points of convergence between the pillars of Knowledge Management and Agile Methodologies.

Keywords Agile · Knowledge management · Organizational culture

---

E.B. Vaz (✉)
Programa de Pós-Graduação em Engenharia de Produção,
Universidade Federal do Paraná, R. Hildebrando de Araújo,
189 Ap. 501. Jardim Botânico, Curitiba, Paraná, Brazil
e-mail: vaz.elianeborges@gmail.com

M. do Carmo Duarte Freitas
Programa de Pós-Graduação em Engenharia de Produção, Universidade
Federal do Paraná, Avenida Coronel Francisco Heráclito dos Santos,
210, Jardim das Américas, Curitiba, Paraná, Brazil
e-mail: mcf@ufpr.br

1 Introduction

Software development projects are fertile fields for the creation and sharing of knowledge. Sveiby (1998) argues that the concept of knowledge is related to factors such as information, awareness, knowledge, cognition, wisdom, perception, science, experience, expertise, insight, competence, ability and learning.

When considering software artifacts as accumulated knowledge belonging to the stakeholders of an organization, the process of software development itself should be oriented to knowledge and must ultimately reflect the know-how of the organization (Dakhli and Chouikha 2009). Software development requires various forms of explicit and tacit knowledge. This knowledge is dynamic and involves technology, organizational culture and adaptation needs in its practice of software development (Aurum et al. 2003).

After the relationship of the information with the values, ideas, emotions, motivations and commitments, it can be internalized and become part of tacit knowledge for those involved in the process (Cassapo 2010). The knowledge is composed of elements in a structured way. It is intuitive, difficult to express and verbalise and it is part of the human being, with all its complexity and unpredictability (Silva 2004). The creation of new knowledge requires a model of management. Santos et al. (2001) adds that an organizational environment must be favorable to creative behavior and where learning should be continuous and collective.

So the question is: which knowledge management practices are related to agile methodologies for software development? Considering the relevance of the themes for the IT companies survival, an exploratory survey was conducted in the literature to minimize the gap found mapping this relationship.

2 Knowledge Management

The many authors studied agree that Knowledge Management can be defined as the conversion of tacit to explicit knowledge. The knowledge management can be classified into three types: economic school, organizational school and strategic school (Torres 2006).

In the economic school, achieving knowledge generates income through the accounting and management of intellectual capital. There is a greater concern with management than with the development of knowledge that is seen as an object that can be acquired, stored and distributed. In the organizational school, knowledge is socially acquired and shared through organizational structures. In the strategic school, it starts to have a competitive strategic dimension (Torres 2006).

This article explores the approach of the organizational school where the focus is on groups of people with the same interest, where the process is sustained by the interaction between people, within the Information Technology (IT) environment. In Awazu (2004), Knowledge Management is also treated in two distinct approaches:

either as a technological or humanistic initiative. In the first one, the focus is on the use of information technology through systems that encode and create knowledge networks for organizational management of this. On the other hand, the humanistic approach has a people-centric perspective, where the focus is on managing individuals and teams to enable the creation and sharing of knowledge.

Considering the interaction between individuals, the Model Socialization, Externalization, Combination, Internalization (SECI) of Nonaka and Takeuchi (1997), features four modes of knowledge creation. Socialization is defined by sharing experiences, tacit knowledge being transmitted in a simplified way and turning into new tacit knowledge. Externalization is understood as the creation of new concepts, which will be the basis for new knowledge, extracting explicit knowledge from tacit knowledge. The combination systematizes various concepts into a knowledge system, formalizing a path for learning. Finally, the internalization is an inverse end path which turns the explicit knowledge again into tacit knowledge by reproducing experiments that have been documented by other performers (Nonaka and Takeuchi 1997). Kikoski and Kikoski (2008) state that tacit and explicit knowledge are not separate or distinct from one another but work in different areas in human intelligence and change explicit to tacit and again to explicit.

3 Knowledge Management and IT Agile

Utilizing the theoretical foundation on Knowledge Management established here, the purpose of this article is to explore the organizational environment related to software projects and their management methodologies to identify convergence points between the Agile Project Management, known by professionals like Agile, a term adopted in this study, and practices of knowledge management in their humanistic approach.

The traditional method of project management is characterized as an evolutionary process of creation and use of techniques, tools, processes and systems to assist in the search for the expected results of the project (Sone 2008). In the IT environment, development and market operations occur in parallel. Requirements are never finished until its delivery and, subsequently, changes are inevitable and cannot be rationalized (DeCarlo 2004). This was an enabling environment for the origin of agile methodologies to project management.

Agile is a set of principles, practices and values that help overcome difficult challenges that cannot be attained by traditional methods which are more rigid (Highsmith 2004). It aggregates the evolution of the methodologies for managing and executing software projects, incorporates a considerable amount of influence of iterative and evolutionary development methodologies, empirical process control, game theory, lean production and learned lessons from the development teams (Judy 2009). Agile methodologies emphasize the contact between client, team and collaboration between individuals, which allows the creation of an environment where knowledge sharing should be constant.

Studies indicate that, culturally, a significant challenge is identified among members of organizations to the externalization of their knowledge due to fears of errors, inaccuracies, loss of exclusivity and lack of recognition. To create an environment of trust that inspires and motivates organizational learning, the adoption of knowledge management systems, with the provision of adequate infrastructure, the use of various technologies and approaches and developing strategies that include information and rating of knowledge assets contributors (Davenport and Prusak 1998) is required.

Agile teams generally promote an open organizational environment that respects the individual, showing care with the actions and opinions of other members of the team, and a better understanding of their own opinions and profile in relation to the whole. The practice of transparency and honesty from the developer towards managerial levels shall be encouraged. Problems, critical issues, holdups and special points of interest should be discussed openly and immediately (Whitworth and Biddle 2007).

Xiaohua et al. (2008) support the principles of Agile to argue that agile methodologies emphasize the adaptation to changes and collaboration between developers and the customer, who must participate throughout the development process. Based on the principles of Agile, practices were categorized into a set of six interdependent values, namely: individuals, teams, value, customers, uncertainties and context. These values form an interdependent whole, each of which is important in itself and the six together form a system of core values for project management, especially the more complex ones. The interdependence that has as its foundation: the members of a team of a project do not form a group of disconnected individuals, but are part of an interdependent whole. This is also reflected in the relationship between team and client sponsors (DOI 2005).

The "Manifesto for Agile Software Development" summarizes the outcome of the meeting of seventeen members of the global software development community to discuss the good practices adopted by each of them in developing. Their words, as seen below:

- Individuals and interactions over process and tools;
- Working software over comprehensive documentation;
- Customer collaboration over contract negotiation;
- Responding to change over following a plan.

In other words: that is, while there is value in the items on the right, the items on the left are more valuable (Agile 2001).

At this meeting has been set, established and registered (DOI 2005) that:

- the increase of the return of investment is achieved by making the continuous flow of value the focus of the project;
- the most satisfactory results are delivered to customers who are engaged and involved within the project through frequent interaction;
- changes and uncertainties are expected and managed through iterations, anticipation and adaptation;

- creativity and innovation are desired and individuals are recognized as the ultimate source of value and they create an environment where they can make a difference;
- a better performance is achieved by creating an environment where the team shares the responsibility for the results and the efficiency, and
- the effectiveness and reliability are higher by creating strategies, processes and specific practices, considering the context.

Levy and Hazzan (2009) discuss some of the practices related to Agile Team, Profile, Daily Meetings, Measurement, Customer Collaboration and Pair Programming for aspects of Knowledge Management, and this analysis can be expanded to include the "Seven dimensions of Knowledge Management" established by Terra (2001) which must be observed by managers in order to create a enabling environment for individuals to pursue an organizational creative role, namely: strategic factors and the role of top management, organizational culture and values, organizational structure, human resources, information systems, measuring outcomes, learning with the environment.

From this analysis, it is concluded that:

- Team: this practice means that all staff, including all those involved, should look for face to face communication whenever possible. Initially, even the work space should be friendly and facilitate the communication. Then, all team members must participate in the presentation of the product to the customer, listening to their requests and actively participating of the planning. And finally, there is integration between the development team and managers. The organizational structure and the human resources are part of the dimensions that directly influence these practices, since they determine more flexible hierarchical structures, multidisciplinary and motivated teams.
- Profiles: the distribution of responsibilities in the form of profiles enables better management of them. It translates into involvement of all team members in all parts of the software development, contributing with their point of view. Profiles also establish the connection with the dimension of organizational culture and values, since the Agile aims at continuous improvement of the team, and preaches acceptance of the mistakes that are a reflection of the learning process.
- Daily meetings: held with the whole team, every day, for approximately 10–15 min. The dimensions of culture and organizational values and organizational structure are essential for these meetings fulfill its purpose with the effective participation of those involved.
- Measures: it is useful to accompany the process of software development, aiming for a higher quality product, generating measures that will be used by stakeholders for decision making, also sending a message that the process is being monitored in a transparent manner for all involved. The influence of the dimension of results measurement is clear for both the tangible information about the projects as the intangible assets generated in the course of these to be seen and measured.

- Customer collaboration: the whole team should have access to the client during the development process. This direct communication increases the chance of software requirements to be more easily understood and adapted prior to any necessary changes. The dimension of learning about the environment stipulates that the internal and external environments are source of learning and that includes strong partnerships with the client.
- Pair programming: this practice means that all code is developed by two members of the team, who share the same workstation, in an interactive process for the development of the tasks. Each member has a specific responsibility within the task, even if working in pairs. With this, all team members become familiar with all parts of the software and improve their own understanding, reinforcing a culture of knowledge sharing. Here also the dimension of the learning environment is present, since knowledge is being shared among pairs, creating a friendly environment for the practice of this process.

The collaboration with the client is linked to periodic review, where doubts are resolved, activities adjustments are made and new knowledge is acquired. The direct involvement of all participants in the project also facilitates the process of acquiring knowledge and their own development through the transfer of collective knowledge for the individual and the frequent deliveries of software.

The response to changes is related to the positioning of the team, which aims to constant learning and application of knowledge generated from learned lessons, the direct involvement of all participants and the pursuit of continuous improvement that is reflected in the construction of a simplified code. Levy and Hazzan (2009) discuss some of the Agile practices in relation to aspects of Knowledge Management, mentioning among them, the monitoring of the software development process through measures that will be used by stakeholders for decision making and as direction to any changes.

The software running is a reflection of the continuous development of knowledge and its use to satisfy the customer. The process of knowledge retention is directly linked to organizational guidelines establishing how organizational memory is created and maintained. Between the agile practices analyzed by Levy and Hazzan (2009) is the distribution of responsibilities in the form of profiles that enables better management of these.

The rotation of responsibilities established as a practice of Agile brings benefits to the organization. This is one of the characteristics of learning organizations and contributes to a highly innovative performance, besides the benefits of sharing tacit knowledge and the development of redundant knowledge and an array of needs for them. The result is the increased efficiency of the team, faster problem solving, and the understanding of the more macro organizational aspects and hence the exploitation of new knowledge (Fægri 2009).

4 Conclusion

Management of organizational knowledge is essential for companies to not depend on specific people for its survival and financial health. To overturn the barriers in the use of organizational knowledge in software processes it is necessary that the organization disseminates its importance in its culture, as well as the gain it can bring to all involved. It is necessary to manage much more than the information itself, but the people and the interaction between them.

The organizational culture is essential for the construction of this environment. It is important to remember that even an error should be considered a rich learning opportunity. External relations such as the contact with customers, suppliers, competitors, partners and support institutions, should be valued and it is an effective way of acquiring and identifying processes and innovative products.

This work aimed to minimize the gap found mapping the relations between the two themes, based on the constructed knowledge through research in the literature, linking the identified practices. This goal was achieved by establishing points of conceptual convergence between the two issues.

Finally, it was concluded that the practices of Agile provide a favorable environment for knowledge management. The organization should establish policies and guidelines that promote continuous learning, encouraging, directly or indirectly, courses, training, improvement of academic training and the knowledge development activities, and providing an environment of respect and trust, allowing individuals to exchange information, actions that will be reflected in both the individual and organizational.

As subjects of future studies, it is suggested the identification of mechanisms for using Agile in order to systematize the practice of Knowledge Management. Another point showed is that as the organizational culture and the particular characteristics of the individuals involved in the process are essential to success, an anthropological deepening focusing on basic profile of Agile practitioners, searching for ways of more efficient promotion of continuous learning is also seen as an opportunity for further studies.

References

Agile I (2001) Agile manifesto. http://agilemanifesto.org/. Cited 16 May 2011
Aurum RJ, Wohlin C, Handzic M (2003) Managing software engineering knowledge. Springer, New York
Awazu Y (2004) Knowledge management in distributed environments: roles of informal network players. In: Proceedings of the 37th Hawaii international conference on system sciences—IEEE
Cassapo F (2010) As duas leis fundamentais da gestão do conhecimento. http://www.redeinovacao.org.br/LeiturasRecomendadas/As%20duas%20leis%20fundamentais%20da%20Gest%C3%A3o%20do%20Conhecimento.pdf. Cited 22 March 2010

Dakhli SBD, Chouikha MB (2009) The knowledge-gap reduction in software engineering. IEEE Computer Society

Davenport T, Prusak L (1998) Working knowledge: how organizations manage what they know. Harvard Business School Press, Boston

DeCarlo D (2004) Extreme project management: using leadership, principles, and tools to deliver value in the face of volatility. Jossey-Bass, San Francisco

DOI Declaration of Interdependence (2005) http://pmdoi.org/. Cited 18 May 2011

Fægri TE (2009) Improving general knowledge in agile software organizations: experiences with job rotation in customer support. In: Agile conference/09. IEEE Computer Society

Highsmith J (2004) Agile project management. Adison-Wesley, Boston

Judy KH (2009) Agile principles and ethical conduct. In: Proceedings of the 42nd Hawaii international conference on system sciences—2009, IEEE Computer Society

Kikoski CK, Kikoski JF (2008) The inquiring organization-tacit knowledge, conversation and knowledge creation. Skills for 21st-Century Organizations, Wilk

Levy M, Hazzan O (2009) Knowledge management in practice: the case of agile software development, ICSE'09 workshop. Vancouver, Canada

Nonaka I, Takeuchi H (1997) Criação de Conhecimento na Empresa – Como as empresas japonesas geram a dinâmica da inovação. Campus, Rio de Janeiro

Santos ARD, Pacheco FF, Pereira HJ, Bastos JR, Paulo A (2001) Gestão do Conhecimento - Uma experiência para o sucesso empresarial. Editora Universitária Champagnat, Curitiba

Silva HFN (2004) Criação e Compartilhamento de Conhecimento em Comunidades de Prática: Uma Proposta Metodológica. UFSC, Florianópolis

Sone SP (2008) Mapping agile project management practices to project management challenges for software development. Faculty of Argosy University/Washington DC College of Business, Doctor of Business Administration, June 2008

Sveiby KA (1998) Nova Riqueza das Organizações: gerenciando e avaliando patrimônios de conhecimento. Campus, Rio de Janeiro

TERRA JCC (2001) Gestão do Conhecimento: o grande desafio empresarial. São Paulo: Negócio Editora

Torres AHS (2006) Captura e Disseminação de Conhecimento em Projetos de Software. UnB, Brasília

Whitworth E, Biddle R (2007) The social nature of agile teams. IEEE Computer Society

Xiaohua W, Zhi W, Ming Z (2008) The relationship between developers and customers in agile methodology. In: International conference on computer science and information technology, IEEE Computer Society

Recent Advances in Patent Analysis Network

Javier Gavilanes-Trapote, Rosa Río-Belver, Ernesto Cilleruelo and Jaso Larruscain

Abstract The databases of patents are considerable, with many authors, as a source of information very valuable within the innovation process. One of the most important methods in patent analysis is based on the citations. The basic concept of patent citation analysis is that there exists a technological linkage between two patents if a patent cites the other. The networks codifying the cited-citing relationship between patents are useful for visualizing the overall status of a given technology and helps the experts in the identification of the technological implications using analysis network techniques. The potential offered by the measuring citations for planning and assessing of policies from Science and Technology is immense. The aim of this paper is to describe the utilities and limitations of the analysis network of patents as well as recent advances.

Keywords Patent citation · Patent citation network · Patent classification · Technological knowledge flow · Citation frequency

---

J. Gavilanes-Trapote · R. Río-Belver · J. Larruscain
Foresight, Technology and Management (FTM) Group, Department of Industrial Engineering, University of the Basque Country UPV/EHU, Calle Nieves Cano, 12, 01006 Vitoria, Spain
e-mail: javier.gavilanes@ehu.es

R. Río-Belver
e-mail: rosamaria.rio@ehu.es

J. Larruscain
e-mail: jaso.larruscain@ehu.es

E. Cilleruelo (✉)
Foresight, Technology and Management (FTM) Group, Department of Industrial Engineering, University of the Basque Country UPV/EHU, Alameda Urquijo s/n, 48030 Bilbao, Spain
e-mail: ernesto.cilleruelo@ehu.es

1 Introduction

Many studies have revealed that the patents are more than 90 % of the latest technical information in the world and 80 % of the information in the patents have not been published in any other form (Zha and Chen 2010). In this sense, considering the patent as a substitute of technology, the analysis of the patents has been considered as an analytic tool for forecasting technology.

The first analysis of the patents consisted of counting the patents and making a comparison by nation, company or technological fields (Wartburg and Teichert 2005). The last decade has triggered the use of citations from patents and from non-patent literature. The basic concept of the analysis of the patent citations is that there exists a technological linkage between two patents if one patent cites the other or a knowledge flow if the citation is between one patent and a scientific article.

Besides the simple frequency counts there exists other indicators like technological cumulativeness, citation impact, generality of a patent, technology strength or technology cycle time.

The potential offered by the measuring citations for planning and evaluation of the policies from Science and Technology is immense. Mainly their utilities can be grouped into three: the measurement of knowledge flows, the measurement of patent quality and the strategic behavior of companies (Podolny et al. 1996).

In respect to the measurement of the quality of the patents, the citation analysis is one of the methods most used (Narin 1994). Harhoff et al. (1999) discovered that the American and German patents with high economic value tend to be cited more often. Various studies found a relationship between citation frequency of patents and their market value (Nagaoka 2005). In recent literature we can find numerous studies that try to measure the value of the patents (Breitzman and Narin 2001; Hall et al. 2005; Cheng et al. 2010). However the real value of a patent and its relation to innovation is a complex subject (Meyer 2000; Wang 2007).

The citations as indicators of technology flow have been utilized in numerous studies, for example to identify the core technologies in the telecommunication sector (Lee et al. 2009), discover the knowledge flows among traditional and emerging industries (Han and Park 2006) or identify core and emerging technologies in Taiwan (Cho and Shih 2011).

A form of representation of the technology flows is through the patent networks that allow to visualize the results and simplify the understanding. These networks are based in the co-citation of the documents to discover the bonds of technological knowledge that bind.

The aim of this paper is to describe the contributions and limitations of the patent network analysis as well as recent advances.

2 Patent Citation Network

The analysis of co-citation is one of the most important methods of patent analysis. It goes back to the classic work of Small (1973) where a new form of document coupling called co-citation is defined as the frequency with which two documents are cited together.

Recently this analysis has been used to study the knowledge structure of various technological fields including nanotechnology (Huang et al. 2004; Kostoff et al. 2006). Wallace et al. (2009) took the research further and used the co-citation network to detect clusters based on the topology of the citation-weighted network (Blondel et al. 2008).

Sung et al. (2010) adopted a microscopic approach to measure and evaluate the level of technological convergence through the analysis of patent citation. Yoon and Kim (2011) used SAO-based semantic patent network to identify the rapidly evolving technological trends for R&D planning.

Betweenness centrality is an important measure to take into account in the citation network of patents. It is equal to the number of shortest paths from all vertices to all others that pass through that patent. Betweenness centrality is a more useful measure (than just connectivity) of both the load and importance of a patent.

Recent studies have shown that most of the citations are linked with a small group of patents with a high betweennes centrality. Take into account that measure allowed to detect the key patents in the knowledge diffusion process in a network.

3 Latest Advances

The patents are grouped by technology by way of two classification systems: used the United States Patent and Trademark Office (USPTO) and the International Patent Classification (IPC) developed by the World Intellectual Property Organization (WIPO). These classifications are very hierarchical and can commit errors, hinder the detection of the technological spillovers and not discover emerging technologies.

For this reason, some author has created their own patent classification by the co-citation of the IPCs of each patent (Kay et al. 2012; Schoen 2011). This new classification tends to be heterarchy continuously adapting to the technological changes and being more sensitive to the classification errors.

Kay et al. (2012) have been a global patent map developed using the data of the European Patent Office (EPO) between 2000 and 2006. In Fig. 1 the map is represented and through the special positioning of the nodes or categories and the relationship and distance among them. The different colours represent each of the 35 new technological areas and the nodes are the 466 new sectors or technological categories that form the new classification of patents. For proper representation, the correct number of digits was varied in the IPC hierarchy in order to optimize the size of the categories.

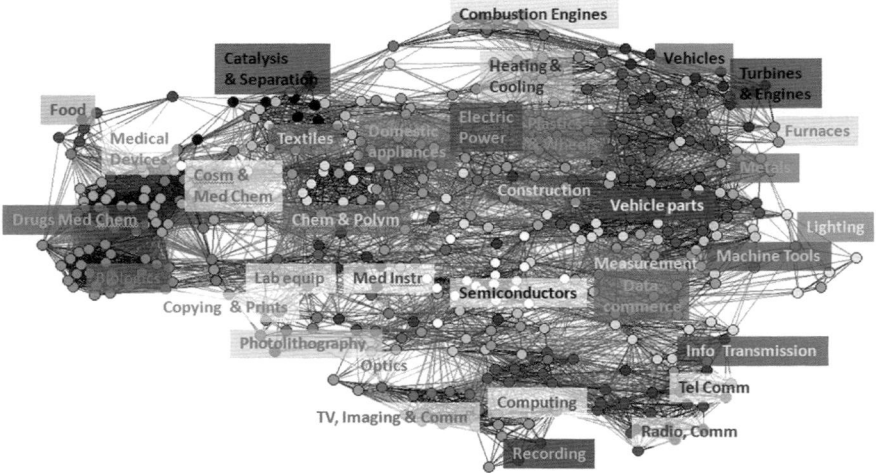

Fig. 1 Full patent map of 466 technology categories and 35 technological areas (Kay et al. 2012)

The subsequent studies such as Leydesdorff et al. (2014) intended to systematize the process more and propose normalizing the categories through the cosine as a similarity measure. Even more some maps can be overlapped in the Google Map for geographic visualization (Leydesdorff and Bornmann 2012).

The patent citation networks or maps based only on the citations have limitations from the richness of information perspective and visualization to represent only static analysis. To overcome these limitations Lee et al. (2011) used the formal concept analysis (FCA) and they modified it for the new algorithm taking into account the time periods and the changes of keywords amongst patents. In this way they were able to design dynamic patent network (Fig. 2).

Figure 2 represents "radial dynamic patent network" for laser technology in lithography for the fabrication of semiconductors between the years 1984 and 2009.

Unlike a conventional patent map that network explains and visualizes the detailed technological changes along a timeline, this permitted the analyst to better understand the technological overview.

Each color of the nodes corresponds with a different technology semiconductor manufacturing. The nodes and arcs differ from one another in the dynamic patent network according to the number of patents in a concept and types of changes of keyword. Finally they identified five types of technology groups and its evolution time.

The technology fusion has recently become the way to achieve innovation by creating new inventions with the convergence of diverse technologies (Jin et al. 2011; Kodama 1986). The technological limits have been blurred because the major inventions do not permit only one technological field but several (Hacklin et al. 2009).

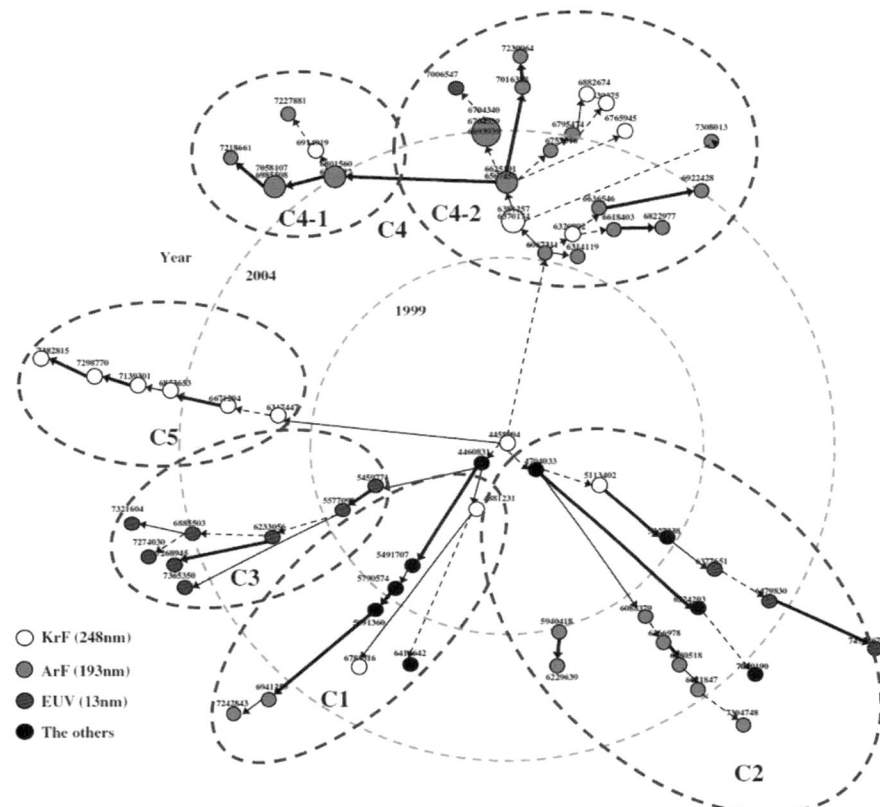

Fig. 2 Radial dynamic patent network for laser technology in lithography (Lee et al. 2011)

This is the reason they require the analysis not only one technology group but all of them. Therefore Ko et al. (2014) presented a method to analyzing dynamic trends of technology fusion industry-wide by the measurement of knowledge flows from patents (Fig. 3).

Figure 3 graphically represents every step of the investigation process. It mainly consist of three steps: constructing a knowledge flow matrix by extracting the classification codes and citation information from patent data; generating a knowledge flow map from and industry by associating the classification codes with industrial sector; and constructing a technology fusion map using assessment indicators for analyzing industry-wide knowledge flows.

This work attempts to develope a systematic method on how to analyze the interdisciplinary trends about different technologies. These technological convergences are analyzed from a global perspective industry-wide measuring knowledge spillover.

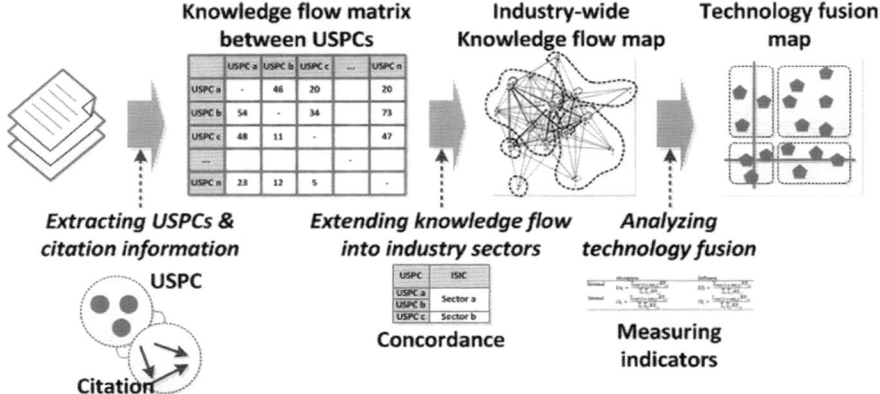

Fig. 3 Overall procedure of this research (Ko et al. 2014)

4 Discussion and Conclusions

In the last decade there has been a skyrocket in the use of patent citations as an innovation indicator. The analysis have become more complex and have evolved from a simple count of citations to determining the quality of a patent or the knowledge flow between a group of patents to complex algorithms taking into account the coreness and intermediarity of the technological sectors (Park and Yoon 2014) or the time periods and the keywords amongst patents (Lee et al. 2011).

Even so, there are still biases and limitations that introduce an important quantity of noise in the results. Bacchiocchi and Montobbio (2010) discovered that the different legal rules and procedure of patent examination and approval in the national offices produced a clear bias. Even more, it has to be taken into account the self-citation bias and the tendency to cite patents from the same nationality (Jaffe and Trajtenberg 1999).

On the other hand, it has to be remembered that the latest patents have less opportunity to be cited from other subsequent patents, because the citation analysis may not function well to reflect the most recent technological tendencies. In the case of the Korean and Japanese patents the problem is bigger because the citations are largely omitted and are not mandatory in its inclusion (Yoon and Kim 2011).

There are other methods not based on citations to determine the similarity between patents such as text mining (Kostoff et al. 2006), keywords analysis (Huang et al. 2004), and the word co-ocurrence analysis (Garechana et al. 2012). Even so, the unexpected scientific discoveries, the changes in the patent laws or the habits of the patent examiners continue to be factors that influence the development of technology and are not part of the patent analysis.

In general, one can not say that patent citation analysis is the final solution to direct the policies of R+D but observe that most traditional methods of analysis of the patent paths and the new methods that are converging (Fontana et al. 2009).

References

Bacchiocchi E, Montobbio F (2010) International knowledge diffusion and home-bias effect: do USPTO and EPO patent citations tell the same story? Scand J Econ 112(3):441–470

Blondel V, Guillaume JL, Lambiotte R et al (2008) Fast unfolding of communities in large networks. J Stat Mech Theory Exp P10008

Breitzman AF, Narin F (2001) Method and apparatus for choosing a stock portfolio, based on patent indicators. United States Patent 6175824

Cheng YH, Kuan FY, Chuang SC et al (2010) Profitability decided by patent quality? An empirical study of the US semiconductor industry. Scientometrics 82(1):175–183

Cho TS, Shih HY (2011) Patent citation network analysis of core and emerging technologies in Taiwan: 1997–2008. Scientometrics 89(3):795–811

Fontana R, Nuvolari A, Verspagen M (2009) Mapping technological trajectories as patent citation networks. An application to data communication standards. Econ Innov New Technol 18:311–336

Garechana G, Rio-Belver R, Cilleruelo E, Gavilanes-Trapote J (2012) Visualizing the scientific landscape using maps of science. In: Industrial engineering: innovative networks, part 2. Springer, London, pp. 103–112. ISBN: 978-1-4471-2320-0

Hacklin F, Marxt C, Fahrni F (2009) Coevolutionary cycles of convergence: an extrapolation from the ICT industry. Technol Forecast Soc Change 76(6):723–736

Hall BH, Jaffe A, Trajtenberg M (2005) Market value and patent citations. Rand J Econ 36(1):16–38

Han YJ, Park Y (2006) Patent network analysis of inter-industrial knowledge flows: the case of Korea between traditional and emerging industries. World Patent Inf 28(3):235–247

Harhoff D, Narin F, Scherer FM et al (1999) Citation frequency and the value of patented inventions. Rev Econ Stat 81(3):511–515

Huang Z, Chen H, Chen ZK et al (2004) International nanotechnology development in 2003: country, institution, and technology field analysis based on USPTO patent database. J Nanopart Res 6:325–354

Jaffe AB, Trajtenberg M (1999) International knowledge flows: evidence from patent citations. Econ Innov New Technol 8:105–136

Jin JH, Park SC, Pyon CU (2011) Finding research trend of convergence technology based on Korean R&D network. Expert Syst Appl 38(12):15159–15171

Kay L, Newman N, Youtie J et al (2012) Patent overlay mapping: visualizing technological distance. Arxiv preprint arXiv:1208.4380

Ko N, Yoon J, Seo W (2014) Analyzing interdisciplinarity of technology fusion using knowledge flows of patents. Expert Syst Appl 41:1955–1963

Kodama F (1986) Japanese innovation in mechatronics technology. Sci Publ Policy 13(1):44–51

Kostoff R, Stump J, Johnson D et al (2006) The structure and infrastructure of the global nanotechnology literature. J Nanopart Res 8:301–321

Lee H, Kim C, Cho H et al (2009) An ANP-based technology network for identification of core technologies: a case of telecommunication technologies. Expert Syst Appl 36(1):894–908

Lee C, Jeon J, Park Y (2011) Monitoring trends of technological changes based on the dynamic patent lattice: a modified formal concept analysis approach. Technol Forecast Soc Chang 78:690–702

Leydesdorff L, Bornmann L (2012) Mapping (USPTO) patent data using overlays to Google Maps. J Am Soc Inform Sci Technol 63(7):1442–1458

Leydesdorff L, Kushnir D, Rafols I (2014) Interactive overlay maps for US patent (USPTO) data based on International Patent Classification (IPC). Scientometrics 98:1583–1599

Meyer M (2000) What is special about patent citations? Differences between scientific and patent citations. Scientometrics 49(2):93–123

Nagaoka S (2005) Patent quality, cumulative innovation and market value: evidence from Japanese firm level panel data. IIR working paper, WP#05-06. Institute of Innovation Research, Hitotsubashi University, Tokyo

Narin F (1994) Patent bibliometrics. Scientometrics 30(1):147–155

Park H, Yoon J (2014) Assessing coreness and intermediarity of technology sectors using patent co-classification analysis: the case of Korean national R&D. Scientometrics 98:853–890

Podolny JM, Stuart TE, Hannan M (1996) Networks, knowledge and niches: competition in the worldwide semiconductor industry, 1984–1991. Am J Sociol 102(3):659–689

Schoen A (2011) A global map of technology. Paper presented at the IPTS patent Workshop, Seville, Spain, 13–14 June 2012

Small H (1973) Co-citation in the scientific literature: a new measure of the relationship between two documents. J Am Soc Inf Sci 24(4):265–269

Sung K, Kim T, Kong H (2010) Microscopic approach to evaluating technological convergence using patent citation analysis. In: U-and E-Service, Science and Technology, pp 188–194

Wallace M, Gingras Y, Duhon R (2009) A new approach for detecting scientific specialties from raw cocitation networks. J Am Soc Inform Sci Technol 60(2):240–246

Wang SJ (2007) Factors to evaluate a patent in addition to citations. Scientometrics 71(3):509–522

Wartburg T, Teichert KR (2005) Inventive progress measured by multi-stage patent citation analysis. Res Policy 34(10):1591–1607

Yoon J, Kim K (2011) Identifying rapidly evolving technological trends for R&D planning using SAO-based semantic patent networks. Scientometrics 88(1):213–228

Zha X, Chen M (2010) Study on early warning of competitive technical intelligence based on the patent map. J Comput 5(2):274–281

Information Enclosing Knowledge Networks: A Study of Social Relations

L. Sáiz-Bárcena, J.I. Díez-Pérez, M.A. Manzanedo del Campo and R. Del Olmo Martínez

Abstract We design and test a multiagent-based knowledge-exchange network, the results of which will guide corporate decision-making in this field. Its novelty resides in the application of multiagents to the field of knowledge management. We selected this methodology because it is valid for the purpose of the study, because it generates realistic and practical conclusions and, moreover, because it serves to negotiate the difficulties that these aspects of a qualitative and social nature can in reality imply for the firm. We designed the knowledge network through simulation with multiagents, studying some of their most important parameters. In doing so, we arrived at a set of conclusions, prominent among which are that by using the network, we may easily identify experts, and people that either share or conceal knowledge. The network also helps us locate people with leadership capabilities and those that remain isolated, as well as the formation of subgroups that segregate themselves from the original network and that can set up their own independent relational structures, unsupportive of knowledge exchange.

Keywords Knowledge sharing · Knowledge network · Multi-agent · Simulation

---

L. Sáiz-Bárcena (✉) · J.I. Díez-Pérez
Dpto. de Ingeniería Civil. Escuela Politécnica Superior,
Universidad de Burgos, Av/Cantabria S/N, 09006 Burgos, Spain
e-mail: lsaiz@ubu.es

J.I. Díez-Pérez
e-mail: ignacorreo@hotmail.com

M.A. Manzanedo del Campo · R. Del Olmo Martínez
Dpto. de Ingeniería Civil. Escuela Politécnica Superior,
Universidad de Burgos, C/Villadiego S/N, 09001 Burgos, Spain
e-mail: mmanz@ubu.es

R. Del Olmo Martínez
e-mail: rdelolmo@ubu.es

1 Introduction

Management of the firm based on the knowledge of its workers, together with the organizational culture and the influence of social relations between people carrying out different tasks, have a significant effect on knowledge-exchange within the firm (Gutiérrez and Flores 2011; Sáiz-Bárcena et al. 2013, 2014). Such a large-scale flow of social relations and the stream of exchanges that it stimulates have given rise to popular social networks, on which basis knowledge networks have emerged, over recent years (Shin and Park 2010). Their study and analysis in the firm can assist in diagnosing the strengths and weaknesses of knowledge and locating people that transmit and receive knowledge, the leaders at mastering and sharing knowledge (Pedraja-Rejas and Rodríguez-Ponce 2008a, b; Mládková 2012), and the obstacles that make exchange difficult. The objective of this article is to design and to test a multiagent-based knowledge-exchange network, to arrive at results that can guide decisions and actions on knowledge exchange in a business setting. Its novelty resides in the application of multiagents to the field of knowledge management. We selected this methodology as valid for the purpose of the study, because it obtains realistic and practical conclusions and, moreover, because it serves to negotiate the difficulties that these aspects of a qualitative and social nature imply for the firm in reality.

To do so, in the first section, we analyze the design of the knowledge network through simulation with multiagents. In the following section, we study and interpret some of the most important network parameters, such as input degree and output degree and modularity. By doing so, we arrive at a set of conclusions, among the most relevant of which is that the network greatly facilitates the location of knowledge experts, people that share knowledge and those that conceal it. The network also helps us locate people with leadership capabilities and those that remain isolated, as well as the formation of subgroups that segregate themselves from the original network and that can set up their own independent relational structures, unsupportive of knowledge exchange.

2 The Design of a Knowledge Network Through Simulation

The method applied is a simulation with multi-agents. It is a program which uses Netlogo software. The agents move randomly in all directions. The program takes control of various parameters: such as to fix the distance of movement, number of agents, predisposition to social relationships, display the directional links or ties established between agents.

The control of these parameters may modify the density of the network due to the number of links. This simulation has been done with 34 agents. After 200 cycles of time, the agents convert into nodes. The nodes simulate people, and the links simulate social relationship between agents to exchange and share knowledge. The network obtained can be visualized by Gephi. The links can be directed or undirected, depending on what and how you want to analyze and to study the network. If you want to analyze the relations, you have to study the undirected links, but in this article directed networks are used because it is interesting to study the direction of the exchange of knowledge, for locating the people who share and receive the most.

There are two sorts of simulation, using two different sorts of software. The first simulation uses multi-agents, for the creation of the network with nodes and links. In the second, the network is assimilated to the network of a company where there are relations sharing knowledge, and the network is studied to reach very interesting conclusions. The network of a company can be more or less dense due to the relations between employees, and it can detect social problems, segregation, isolation, or be used to find people who have the most and the fewest links, because they have exchanged or shared knowledge. With the network, you can locate key employees, who can prevent other workers from running the risk of being isolated.

From a sociological perspective, when the nodes of a network represent people (or organizations) and the links between them their relations, we can refer to it as a social network (Barber et al. 2011; Arcila and Said-Hung 2012). Such a structure, for the purposes of this study, helps to design and to generate the knowledge map of the firm through the use of specific software (Cobo et al. 2012; Irani et al. 2009), improving corporate performance (Lee et al. 2011). One application of information technology to knowledge networks is that it facilitates and contributes to knowledge exchange in the firm or between firms (Martin-Rios 2012). So, by identifying the network and by analyzing the links, we can almost immediately locate the people that exchange, contribute and conceal knowledge and the extent to which they do so.

The networks present different configurations, which may depend on the links that are defined by the exchange of knowledge or the established social relations, among others. Their representation is in the form of nodes and relations or links (Watthananon and Mingkhwan 2012; McLinden 2013). They are therefore tools that mean we can detect and study social relations arising from the exchange of knowledge, either because of the software they employ or because they are linked with certain information technologies that facilitate knowledge processes (Vaccaro et al. 2010).

Multiagent-based network simulation serves to study social and engineering situations (Terán et al. 2010; Izquierdo et al. 2008) in a business setting. Netlogo 5.0 software is appropriate, in our case, for the simulation of these social relations, in order to depict the network. After its study, we can arrive at practical conclusions, by using Gephi software (Fig. 1) (Wu 2001; Wang et al. 2009; Caballero et al. 2011; Rodríguez et al. 2011), and obtain data for subsequent statistical studies (Thiele et al. 2012).

3 Relevant Parameters and Study of the Network

The analysis of network parameters is very important (Yu et al. 2013; McLinden 2013), given that they represent all the information on the nodes, the links or the relations, the dysfunctions, and if applicable, the need for other nodes and actors

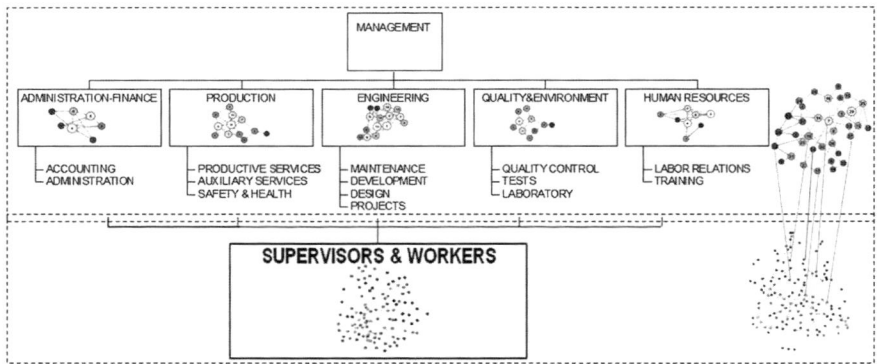

Fig. 1 Netlogo simulation of the network visually represented with Gephi software

(Shin and Park 2010) and even the power of the relevant nodes (Simpson et al. 2011). In our case, a knowledge network with many links is conducive to exchange between people, although it can also lead to the creation of subgroups, which diminish social cohesion in the firm (Janhonen and Johanson 2011).

3.1 Input Degree and Output Degree

Input degree represents the nodes that receive more knowledge and shows those agents that other members of the group have supported and taught. So, workers that have established more relations for learning purposes may be located (Fig. 2).

The output degree of the nodes shows the actors that contribute most knowledge to the group (Fig. 2). The case may arise in which the person who shares most knowledge is not the authentic expert; in consequence, we should analyze whether the expert acts as a transmitter of knowledge, as the one that exchanges most knowledge, or whether the expert shares it, on the majority of occasions, with the same person. That particular circumstance is important, as it would indicate that the knowledge was transmitted hierarchically, from the expert to a subordinate and then onto others. The firm should therefore investigate the location of the expert and the resultant links, because the expert may be protecting and concealing knowledge, with the devastating consequences that such an action would entail for the network.

On a positive side, the output degree of the network also serves to show new experts or leaders, the identification of whom would otherwise be more difficult.

Network degree represents, at the same time, the aforementioned parameters—output and input degree. These are nodes of great importance for the firm, as they are agents and people that show many relations and deploy very high levels of knowledge exchange activity. These people demonstrate leadership skills within the social group

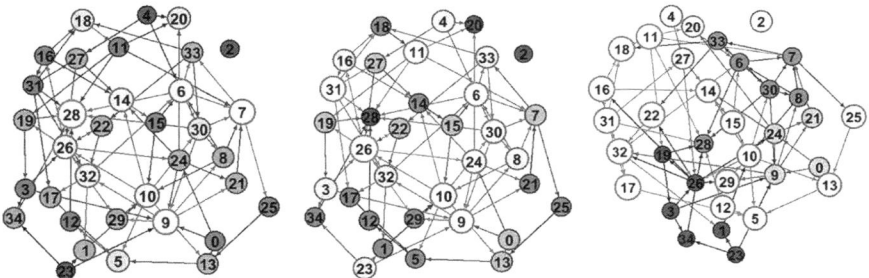

Fig. 2 Representation of input degree and output degree and modularity

or subgroup, with high influence over other group members, favoring and provoking the exchange of knowledge between them all (Pedraja-Rejas et al. 2006).

3.2 Network Modularity and Group Formation

The study of modularity serves to analyze the groups that form in the firm when its members exchange knowledge. These groups can imply a problem for the diffusion and exchange of knowledge, because subgroups and small factions can form that limit knowledge exchange (Fig. 2).

Modularity is the term given to the nodes with links to similar nodes (Banos 2012) and, it also locates those nodes that are capable of establishing relations with other different ones. Thus, we may visualize isolated nodes and analyze the cause and possible solutions, at the same time as detecting problems or negligence that relate to a person or to management decisions.

Likewise, if the firm is hierarchically structured into departments, various groups of networks that interrelate between each other may form, where, in turn, some node appears that stands out in the intra and inter-departmental relations.

It is precisely the relations between the different departments and the different people that compose each department that can help to determine who takes the decisions, making problem-solving more flexible and maintaining fluid links with other departments.

4 Conclusions and Discussion

The objective of this study has been achieved, as it has been demonstrated that the multi-agent based simulation and social network analysis software allows us to study sociological circumstances that arise in the context of knowledge management, in particular its exchange and sharing. A powerful tool becomes available by relating both multiagent and network applications, with which to test some aspects of knowledge management, the application of which is practically impossible in the day-to-day reality of the firm. Thus, we can generate conclusions to guide future actions, in accordance with the results of the simulation. It also allows researchers and business directors to understand and to interpret some of the social relations that arise in the exchange of knowledge and to avoid certain circumstances that hinder knowledge management. This study, in addition to the novelty of its contribution, has served to recapitulate the utility of having a network associated with the knowledge map of the firm, as in this way, it becomes possible to locate who possesses the knowledge, who is the expert, who shares it and, even, who conceals it. All these responses are of vital importance for the survival and the competitiveness of the firm, as it makes the flow of knowledge between employees more agile and efficient, facilitating their learning and avoiding situations of alarm in the face of the exclusive possession of relevant knowledge in the hands of a single person. As demonstrated, relevant software can represent social problems and generate a visual display of the network.

Other interesting aspects, of great importance for the firm, are that it can detect people with leadership capabilities through the knowledge network as well as people who remain isolated or who have fewer links. Likewise, it may also detect the formation of subgroups segregated from the original network that can constitute their own relational structure independent from the latter.

The detection of leaders brings highly trained workers forward who were previously ignored, as they were not known and, on the contrary, other overvalued employees that demonstrate their isolation or lack of relations, and in consequence, lack of knowledge exchange with their subordinates. Finally, the position that the expert occupies in the network determines the degree to which they share their knowledge that, as demonstrated, may be broad or narrow. In the latter case, a narrow band would indicate a desire for the knowledge to remain tacit. Thus, the network map also warns of this and other types of barriers or obstacles for the exchange of knowledge in the firm.

References

Arcila C, Said-Hung E (2012) Factores que inciden en la variación de seguidores en los usuarios top20 más vistos en Twitter en América Latina y Medio Oriente. Interciencia 37(12):875–882

Banos A (2012) Network effects in Schelling's model of segregation: new evidence from agent-based simulation. Environ Plann B-Plann Desi 39(2):393–405

Barber MJ, Fischer MM, Scherngell T (2011) The community structure of research and development cooperation in Europe: evidence from a social network perspective. Geographical Analysis 43(4):415–432

Caballero A, Botía J, Gómez-Skarmeta A (2011) Using cognitive agents in social simulations. Eng Appl Artif Intell 24(7):1098–1109

Cobo MJ, López-Herrera AG, Herrera-Viedma E et al (2012) SciMAT: a new science mapping analysis software tool. J Am Soc Inform Sci Technol 63(8):1609–1630

Gutiérrez L, Flores M (2011) Un concepto sobre las redes de conocimiento entre organizaciones. Revista de Ciencias Sociales 17(3):473–485

Irani Z, Sharif AM, Love PED (2009) Mapping knowledge management and organizational learning in support of organizational memory. Int J Prod Econ 122(1):200–215

Izquierdo L, Galán JM, Santos I et al (2008) Modelado de sistemas complejos mediante simulación basada en agentes y mediante dinámica de sistemas. EMPIRIA, Revista de Metodología de Ciencia Sociales 16:85–112

Janhonen M, Johanson J-E (2011) Role of knowledge conversion and social networks in team performance. Int J Inf Manage 31(3):217–225

Lee WL, Liu CH, Wu YH (2011) How knowledge cooperation networks impact knowledge creation and sharing: a multi-countries analysis. Afr J Bus Manage 5(31):12283–12290

Martin-Rios C (2012) Why do firms seek to share human resource management knowledge? The importance of inter-firm networks. J Bus Res 67(2):190–199

McLinden D (2013) Concept maps as network data: analysis of a concept map using the methods of social network analysis. Eval Program Plan 36(1):40–48

Mládková L (2012) Leadership in management of knowledge workers. Procedia Soc Behav Sci 41:243–250

Pedraja-Rejas L, Rodríguez-Ponce E (2008a) Estilos de liderazgo, gestión del conocimiento y diseño de la estrategia: un estudio empírico en pequeñas y medianas empresas. Interciencia 33(9):651–657

Pedraja-Rejas L, Rodríguez-Ponce E (2008b) Estudio comparativo de la influencia del estilo de liderazgo y la congruencia de valores en la eficacia de empresas privadas e instituciones públicas. Interciencia 33(1):8–13

Pedraja-Rejas L, Rodríguez-Ponce E, Rodríguez-Ponce J (2006) Liderazgo y decisiones estratégicas: una perspectiva integradora. Interciencia 31(8):577–582

Rodríguez S, de Paz Y, Bajo J et al (2011) Social-based planning model for multiagent systems. Expert Syst Appl 38(10):13005–13023

Sáiz-Bárcena L, Díez Pérez JI, Manzanedo del Campo MA et al (2013) Intercambio del conocimiento en la empresa. Aprendiendo de la experiencia. Interciencia 38(8):570–576

Sáiz-Bárcena L, Díez Pérez JI, Manzanedo del Campo MA et al (2014) Predisposition of workers to share knowledge: an empirical study. In: Lecture Notes in Management and Industrial Engineering, 2:77−83

Shin J, Park Y (2010) Evolutionary optimization of a technological knowledge network. Technovation 30(11–12):612–626

Simpson B, Markovsky B, Steketee M (2011) Network knowledge and the use of power. Soc Netw 33(2):172–176

Terán O, Quintero N, Ablán M et al (2010) Simulación social multiagente: Caso reserva forestal de Caparo. Venezuela. Interciencia 35(9):696–703

Thiele JC, Kurth W, Grimm V (2012) RNETLOGO: an R package for running and exploring individual-based models implemented in NETLOGO. Methods Ecol Evol 3(3):480–483

Vaccaro A, Parente R, Veloso FM (2010) Knowledge management tools, Inter-organizational relationships, innovation and firm performance. Technol Forecast Soc Chang 77(7):1076–1089

Wang J, Gwebu K, Shanker M et al (2009) An application of agent-based simulation to knowledge sharing. Decis Support Syst 46(2):532–541

Watthananon J, Mingkhwan A (2012) Optimizing knowledge management using knowledge map. Procedia Eng 32:1169–1177

Wu DJ (2001) Software agents for knowledge management: coordination in multi-agent supply chains and auctions. Expert Syst Appl 20(1):51–64

Yu Y, Hao J-X, Dong X-Y et al (2013) A multilevel model for effects of social capital and knowledge sharing in knowledge-intensive work teams. Int J Inf Manage 33(5):780–790

Part VI
Service Systems

Cost System Under Uncertainty: A Case Study in the Imaging Area of a Hospital

Victor Jiménez, Carla Duarte and Paulo Afonso

Abstract An activity-based costing (ABC) system is based on the premise that activities consume resources and cost objects (e.g. products and services) consume activities. The development of effective costing systems requires a good design of processes and activities. But some business and production systems can be characterized by high levels of uncertainty. In the particular case of hospitals, the amount of resources consumed for the same type of examination by two different patients can be completely different, which may hinder costs and to bias the process of budget planning. This paper presents an ABC costing model for the imaging area of a hospital, including the uncertainty of the processes. A Monte Carlo simulation was made and interesting results were obtained. The probabilistic ABC costing model is a tool that can be used to improve decision making.

Keywords Hospitals cost management · Activity based costing · Uncertainty and risk management · Monte Carlo simulation

---

V. Jiménez (✉) · C. Duarte · P. Afonso
Departamento de Produção e Sistemas, Escola de Engenheiras, Universidade do Minho, 4800-058 Braga, Portugal
e-mail: victor.jimenez@correounivalle.edu.co

C. Duarte
e-mail: carlatrindade@dps.uminho.pt

P. Afonso
e-mail: psafonso@dps.uminho.pt

V. Jiménez
Departamento de Contabilidad y Finanzas, Facultad de Ciencias de la Administración, Universidad del Valle, Cali, Colombia

1 Introduction

Activity-Based Costing (ABC) can be used to report precise and relevant information to health care professionals. In this particular industry there is an unavoidable need to refresh cost models and costing techniques in order to allow managers and clinicians to obtain more relevant, precise and transparent information on costs. Current financing systems for hospitals are not flexible enough and they are based on the controversial concept of Diagnosis Related Groups (DRG). ABC could be considered as a more effective alternative to manage costs in hospitals by supporting the decision making process in many levels, such as decisions of make or by, expand or reduce services based on the patients' needs and services efficiency (i.e. costs versus reimbursement schemes). Nevertheless, ABC models have some limitations.

Namely, as other costing systems, ABC models are typically deterministic even if some important variables could have a stochastic pattern. This is the case of the imaging service in a hospital. The amount of resources consumed by two patients for the same type of exam can be completely different, which may hinder an effective process of budgeting and cost management. This paper presents an ABC model that includes uncertainty in some parameters. A Monte Carlo simulation model for including the probability distribution of the uncertain variables was used. This model was applied in the imaging service of a Portuguese medium-sized hospital where a total of more than 200 different types of exams have been conducted. During the period under study (the year 2011) they were made in the imaging service More than 150,000 exams corresponding to approximately 100,000 patients. In this research project, real data collected in the hospital was used to test and validate the proposed model. This new model to deal with uncertainty results in an extended ABC model (which includes uncertainty) that offers additional and very valuable information for budgeting and cost management in the imaging service and can be used for other hospital services and departments as well as in different types of organizations. Thus, this article presents an ABC model that includes the uncertainty in some parameters. A Monte Carlo simulation model was used to include the probability distribution of the uncertain variables.

2 ABC in Hospitals and Monte Carlo Simulation

Today, health care organizations face one of its most important challenges: to decrease the total expenditure without affecting the quality of the services provided to patients. The key to the success to accomplish the so expected financial viability relies on the development of relevant and strict cost information to support strategy, price adequacy and management decision (Demeere et al. 2009).

Current financing system for hospitals is relatively rigid and is based on Diagnosis Related Groups (DRG), as well as more strict financing rules. Regarding the situation faced by health care sector, it is recognized that there is an urgent need to evaluate all priorities of investment and expenditure. In addition, we also have the recognition from external entities that cost accounting systems must be reformulated in order to give reliable information for the funding and pricing process, without underline the lack of usage of advanced cost accounting practices (such as Activity-Based Costing (ABC)) in health care organizations. ABC could be considered as a more effective alternative to manage hospitals by supporting the decision-making process in many levels, such as the decision to expand or contract services based on the patients' needs and their profitability.

Cost management in health care organizations should support the effort to accomplish positive financial results and higher efficiency without adversely affect the quality of health care services provided to the patients, supply information to optimize resources and ensure quality and continuous improvement (Baker and Boyd 1997). The traditional cost accounting systems applied to health care services have not accomplished these objectives once they fail to emphasised the role given to the activities performed in order to meet the patients' needs and because their departmental analysis does not reflect the process work flows that go through different departments in the organization (Lawson 2005).

An ABC system can be used to balance the health care delivery perspective and the financial dimension that makes the sustainability of the organization possible, as well as to report accurate and relevant information to health care professionals, as this information should support health care services, achieve financial accounting and drive higher performance in the organization (Ross 2004). Several studies have been conducted to show the applicability of ABC in hospitals (e.g. Laurila et al. 2000; Ross 2004; Baxendale and Dornbusch 2000). Even though there is a great recognition of the benefits and value of the ABC model, the implementation of this method is not yet to be generalized (Kaplan and Anderson 2007). Indeed, the development of an ABC system requires extensive data collection and analysis.

Yet, in health care there is a progressive need to refresh the techniques for the cost accounting system in order to allow managers and clinicians to obtain more accurate information so that it may be possible to analyse the profitability of each department, service and other relevant cost objects, supporting future investments.

Considering that work requirements are now more complex, more uncertain, and changing, control systems cannot be static and formal (Davila et al. 2009). Even ABC data are often estimated due to cost and time constraints, which leads to inherent imprecisions and uncertainty (Nachtmann and Needy 2003).

Different techniques can be used for describing the uncertainty of input parameters. For example, fuzzy method uses membership functions for describing an uncertain parameter while the stochastic methods use probability density function. The similarity of them is that all of them try to quantify the effect of input parameters on model's outputs. These methods can be classified as probabilistic approach or possibilistic approach, Hybrid Possibilistic-probabilistic and robust optimization. One of the methods used to understand and attempt to manage uncertainty has to do with the Monte Carlo Simulation.

The Monte Carlo simulation method can be included in the experimental branch of mathematics that deals with experiments on random numbers based on statistical probabilities (Hammersley and Handscomb 1964). Essentially, the analyst describes, in a probabilistic sense, the variables that determine the value by their respective probability distributions. By repeating this process thousands of times, the analyst gets many possible values for the object variable. From these values we can determine a mean value and a set of statistics that provide a quantification of uncertainty. For example some authors have approached the subject of uncertainty in costing systems using either the Monte Carlo Simulation or other techniques such as Fuzzy methods (e.g. Nachtmann and Needy 2001, 2003; Chansaad and Rattanamanee 2012; Rivero and Emblemsvåg 2007). In all cases, clear evidence include managing uncertainty within costing systems is shown.

3 Case Study and Proposed Model

In this research project it was important to study the different processes and the differences among the different types of exams. Furthermore, health services have an additional complexity. Indeed, each service may depend on the state of health of the patient and that is not anticipated by health care providers in general and hospitals in particular. For the design of the costing model, firstly, the exams were identified and grouped in families' exams. Secondly, they were identified the activities, followed by resource identification. In a third stage, they were identified the cost drivers, followed by the calculation of the rate of allocation. Fourthly, we proceeded with the identification of probability distributions. Fifthly, the ABC model was designed. Finally, through simulation and analysis of results, costs were allocated to the exams.

Thus, after the identification of families' exams, the activities, resources and cost drivers were identified by this order and costs were allocated to the exams.

During the period under analysis (2011) they were performed over 150,000 exams, further analysis showed that it was possible to create homogeneous groups of exams. These homogeneous groups are the following: Support to other services, Gastrointestinal imaging, Urologic imaging, Ultrasound, Mammography, X-ray, interventional/vascular radiology, Non contrast-enhanced magnetic resonance imaging (MRI), Contrast-enhanced MRI, Non-contrast-enhanced computed tomography, Contrast-enhanced computed tomography.

Activities, resources, cost drivers and cost objects are the elements necessary to implement an ABC model. Once identified the cost objects, the next step is to identify activities for each of the services provided. The process in the imaging service was divided into five parts namely, schedule the patient, receive the patient, make the exam, process the exam and prepare the report.

Resources were classified into: human resources, equipment, informatics resources, materials and others. With regard to human resources, they were considered physicians and all the staff at imaging area both administrative and production. In terms of informatics resources, they were considered the various information systems responsible to deliver or receive information. In this case, they were found a total of 8 different information systems, which in most cases have no communication with each other. In terms of equipment, they were considered all the machines and equipment (e.g. Scanner, Ultrasonography machine, Interventional radiology) used for the realization of the exams. Finally, among the materials found we must highlight maintenance equipment, administrative materials, hotel equipment, treatment equipment, electro-medical equipment, surgical items and medicines. Medicines are responsible for over 60 % of total material costs.

For the distribution of indirect costs, they were identified four cost drivers to relate the resources with the activities, and to relate the activities with the services. The selected cost drivers were the number of exams, labor hours, number of patients, machine hours, and square meters. There is a problem with identifying direct and indirect cost for each type of exam because there is no registration for each of them, so you had to use a driver for the respective assignment. In the cost model design process a problem with the assignment of direct materials was identified. In the traditional ABC model, direct materials are assigned directly to the products. For the estimation of costs the following model was used, taking into account the following parameters:

I *activities* $(i = 1, \ldots, I)$
J *Resources* $(j = 1, \ldots, j)$
K *Products* $(k = 1, \ldots, K)$
AR_{ij} *Total consupmtion resource for each activity*
TR_j *Predetermined Rate allocation for each resource*
P_{ki} *Normalized activity consumption for each product*
X_k *Cost of each product*
m *numbers of iterations* $(m = 1, \ldots, M)$
C_j *Total cost of resources j*
U_j *Cost of unused capacity for each resource*
e *Parameter estimated*

With this information, the traditional ABC model can be described as follows:

$$\begin{bmatrix} P_{11} & \cdots & P_{1I} \\ \vdots & \ddots & \vdots \\ P_{K1} & \cdots & P_{KI} \end{bmatrix} * \begin{bmatrix} AR_{11} & \cdots & AR_{1J} \\ \vdots & \ddots & \vdots \\ AR_{I1} & \cdots & AR_{IJ} \end{bmatrix} * \begin{bmatrix} TR_1 \\ TR_2 \\ TR_3 \\ \vdots \\ TR_J \end{bmatrix} = \begin{bmatrix} X_1 \\ X_2 \\ X_3 \\ \vdots \\ X_K \end{bmatrix} \qquad (1)$$

$$U_j = C_j - TR_j * \sum_{i=1}^{I} AR_{ij}, \quad \forall j \qquad (2)$$

$$\sum_{k=1}^{K} Pki = 1, \quad \forall i \tag{3}$$

$$TR_j = \frac{Estimated\ total\ cost\ of\ Resource\ j}{Estimated\ total\ activity\ base\ (labor,\ hours,\ kg,\ etc.)} \quad \forall j \tag{4}$$

With this model, we can calculate the cost of the products in a deterministic way. It is important to note that due to the large amount of information that we can have, once the calculation of the cost of products, a Pareto analysis must be performed to identify products, activities and resource worth continuing to study in depth and which can be used in the model with uncertainty. This allows us to reduce the information needed to support more efficiently the decision making. To build the ABC model with uncertainty, it is assumed that AR_{ij}, P_{ki} and TR_j are uncertain parameters. A sample AR_{ij}^e, P_{ki}^e and TR_j^e are generated for each input parameters AR_{ij}, P_{ki} and TR_j, using their probability density function (PDF). The value of X_k^e are the outcome variable, which is calculated considering:

$$\begin{aligned} X_k^e &= f(AR_{ij}, P_{ki}, TR_j), AR_{ij}^e = \left[AR_{ij,1}^e, AR_{ij,2}^e, \ldots, AR_{ij,M}^e \right], \\ P_{ki}^e &= \left[P_{ki,1}^e, P_{ki,2}^e, \ldots, P_{ki,M}^e \right], TR_j^e = \left[TR_{j,1}^e, TR_{j,2}^e, \ldots, TR_{j,M}^e \right] \end{aligned} \tag{5}$$

The procedure is repeated for m number of iterations. Finally, the outcomes are analyzed using statistic criteria, histograms, confidence intervals, etc.

4 Results and Discussion

One of the most complex problems to deal in terms of costing in the area of imaging is related to the consumption of materials, since it cannot be standardized to each patient. It basically depends on the patient's age and other personal characteristics. By applying the proposed model, they were found interesting results for the distribution of materials and consequently for the cost of exams by patient. Table 1 shows the unit cost for each exam in terms of materials and also the percentage that each type of material contributes to the unit cost. As can be seen, about 90 % of the costs are concentrated in materials processing, medicine and medical imaging, and these costs are basically concentrated in two types of examination, which are contrast-enhanced MRI and contrast-enhanced computed tomography. Doing a thorough analysis it was determined that the largest proportion of these costs are due to two types of materials (of about 300 that are in total) used in the imaging area which represent about 65 % of the total cost of materials throughout the service. The application of these materials depends on the age of the patient, the weight, and other parameters.

Table 1 Results of deterministic ABC

Percentage of the cost of materials for each type of exam with respect to the total cost					
Type of exam	Treatment material (%)	Medicines (%)	Electro-medicine material (%)	Other materials (%)	Total cost (%)
Support to other services	1.616	0.512	0.507	0.332	2.966
Gastrointestinal imaging	0.660	0.874	0.000	0.015	1.549
Ultrasound	0.014	0.427	0.102	0.966	1.509
Mammography	0.000	0.000	2.491	0.035	2.526
X-ray	0.059	0.396	2.581	4.389	7.425
Interventional/vascular radiology	3.824	0.046	0.000	0.224	4.095
Contrast-enhanced MRI	0.334	3.448	0.333	0.213	24.327
Non contrast-enhanced MRI	0.006	0.001	0.007	0.004	0.018
Contrast-enhanced computed tomography	0.460	48.173	0.459	0.297	49.390
Non contrast-enhanced computed tomography	1.323	0.066	1.538	0.995	3.922
Urologic imaging	0.374	1.313	0.000	0.586	2.273
Total	8.671	75.255	8.018	8.056	100.000

The consumption of materials may be approached with a triangular distributions of probability for each type of material, in order to measure the impact of such materials to the total cost and to the unit cost of such exams. This probabilistic model which was run in the Risk Simulator software version 9.0 is shown in Table 2. The expected variability of the total cost of materials is 15.40 %. Also given the current operating conditions, the maximum expected increase of the total cost is 41.07 % over the actual cost. Since the hospital is obligated to provide all health services demanded by patients' needs, this value indicates how important cost control and management is for the hospital and that measures should be taken in the logic of maintaining controlled activities that are critical and which can generate the increase in cost and can negatively influence the finances of the organization.

From the point of view of efficiency these results are very interesting because they suggest that it is possible that the efficiency of a hospital does not only be function of internal processes and how costs are managed and allocated, but that results are also significantly influenced by the type of patients served by the hospital. In this case, unit materials cost is not constant, because such variable cost may vary significantly depending on the patient's own characteristics.

Table 2 Results of probabilistic ABC

	Number of data points	Mean (%)	Median (%)	Standard deviation (%)	Variance (%)	Coefficient of variation (%)	Maximum (%)	Minimum (%)	Range (%)	Skewness (%)	Kurtosis (%)	25 % Percentile (%)	75 % Percentile (%)	Precision at 95 %
Variability of total cost	5,000	15.40	15.61	9.89	0.98	64.20	41.07	−15.39	56.46	−12.47	−56.15	8.14	22.96	0.0178
Variability of unit cost of contrast-enhanced MRI	5,000	23.37	23.91	15.67	2.46	67.05	53.64	−11.16	64.80	−8.39	−102.01	10.37	36.22	0.0186
Variability of unit cost of the contrast-enhanced computed tomography	5,000	14.85	15.45	16.99	2.89	114.43	47.75	−31.68	79.43	−25.75	−65.73	2.47	28.40	0.0317

5 Conclusions

Cost management in hospitals is very important as hospitals and generally public services have fixed operating budgets and the only variable that can be controlled to improve their financial performance is the production cost. It is interesting that in hospitals there is an implicit variable given by the type of patients that they attend. This variability can affect the efficiency of the hospital by an exogenous element thus not controllable by hospital managers and administrators. Furthermore, a better understanding of the process variability contributes for a better control of costs. The Monte Carlo simulation models proposed here may allow to some extent to predict the risk associated with the variability in costs and support the necessary steps which should be taken to better manage such risk, whether from the point of view of processes rationalization and of management.

References

Baker JJ, Boyd GF (1997) Activity-based costing in the operating room at valley view hospital. J Health Care Finance 24(1):1–9

Baxendale SJ, Dornbusch V (2000) Activity-based costing for a hospice. Strategic Finance-Montvale, 81:64–70

Chansaad A, Rattanamanee W (2012) A fuzzy time-driven activity-based costing model in an uncertain manufacturing environment. Apiems.net, 1949–1959

Davila A, Foster G, Oyon D (2009) Accounting and control, entrepreneurship and innovation: venturing into new research opportunities. Eur Account Rev 18(2):281–311. doi:10.1080/09638180902731455

Demeere N, Stouthuysen K, Roodhooft F (2009) Time-driven activity-based costing in an outpatient clinic environment: development, relevance and managerial impact. Health Policy (Amsterdam, Netherlands), 92(2–3):296–304. doi:10.1016/j.healthpol.2009.05.003

Hammersley JM, Handscomb DC (1964) Monte carlo methods, vol 1. Methuen & CO LTD, Springer, London

Kaplan R, Anderson S (2007) The innovation of time-driven activity-based costing. J Cost Manage 2(21):5–15

Laurila J, Suramo I, Brommels M, Tolppanen EM, Koivukangas P, Lanning P, Standertskjöld-Nordenstam G (2000) Activity-based costing in radiology. Application in a pediatric radiological unit. Acta Radiologica (Stockholm, Sweden: 1987), 41(2):189–195

Lawson R (2005) The use of activity based costing in the healthcare industry: 1994 vs 2004. Res Healthc Fin 10(1):77–94

Nachtmann H, Needy KL (2001) Fuzzy activity based costing: a methodology for handling uncertainty in activity based costing systems. Eng Econ 46(4):245–273. doi:10.1080/00137910108967577

Nachtmann H, Needy KL (2003) Methods for handling uncertainty in activity based costing systems. Eng Econ 48(3):259–282. doi:10.1080/00137910308965065

Rivero EJR, Emblemsvåg J (2007) Activity-based life-cycle costing in long-range planning. Rev Account Finance 6(4):370–390. doi:10.1108/14757700710835041

Ross TK (2004) Analyzing health care operations using ABC. J Health Care Finance 30(3):1–20

Innovation in Consulting Firms: An Area to Explore

Isaac Lemus-Aguilar and Antonio Hidalgo

Abstract Services are the most representative sector in developed economies due to their contribution to GDP and employment. Consulting firms are classified as part of the Knowledge Intensive Business Services (KIBS) and provide professional services to all types of organizations. Consulting firms usually innovate with their customers and suppliers in a nurturing environment for value co-creation. This environment is project-based, process-oriented and with intensive knowledge exchange among all stakeholders. Based on literature review, it has been found that despite the existence of frameworks for service innovation, none of them have specifically focused on consulting firms. Further implications on this issue are addressed for both academics and practitioners.

Keywords Innovation · Services · Consulting firms · Value co-creation

1 Introduction

There is a need for service innovation as an interdisciplinary science to help make innovation more systematic and sustainable; therefore, it has become a priority (Maglio and Spohrer 2013). Some decades ago, a country's economy relied on manufacturing, but nowadays services are driving productivity, economic growth and employment; forcing organizations to adopt a service-centric view in their existing business models (Salunke et al. 2011).

In order to survive, and organization has to evolve, in other words, it has to innovate. Different authors identify the benefits for an organization who innovates

---

I. Lemus-Aguilar (✉) · A. Hidalgo
Department of Industrial Management, Business Administration and Statistics,
Universidad Politécnica de Madrid, C/José Gutiérrez Abascal, 2, 28006 Madrid, Spain
e-mail: isaac.lemus@upm.es

A. Hidalgo
e-mail: antonio.hidalgo@upm.es

© Springer International Publishing Switzerland 2015
P. Cortés et al. (eds.), *Enhancing Synergies in a Collaborative Environment*,
Lecture Notes in Management and Industrial Engineering,
DOI 10.1007/978-3-319-14078-0_38

in services (Furseth and Cuthberson 2013; Ganz et al. 2011; Hidalgo and D'Alvano 2014; IfM & IBM 2007; Kidström et al. 2013; Maglio and Spohrer 2013; Mors 2010; Salunke et al. 2013). It does not matter in which part of the life cycle companies are (start-ups, developing and mature ones), nor their size (from small to medium and large enterprises) nor performance (leaders or followers in their market); they all need to become competitive in their own markets.

Current developed economies—as well as many developing ones—are based on services, representing in most case more that 70 % of the GDP and more that 80 % of total jobs in these economies (IfM & IBM 2007; OECD 2013). Taking into account both the importance of service in economy and the evolution that companies must pursue, Sanz and Jones (2013) argue that innovation is the mechanism that enables countries to transform challenges into opportunities. It is a source of integration, development, added value and new markets that generate income and wealth.

Consulting firms are organizations that belong to the services sector and are part of the Knowledge Intensive Business Services (KIBS) that have proven to be very innovative firms that boost economies. Consulting firms provide service innovation thought the co-production of the service that they deliver in interaction with its customers and suppliers. A preliminary literature review has been conducted about frameworks for service innovation focused in consulting firms and the findings are further developed and explained. Additional implications for research on this issue are addressed for both academics and practitioners as part of the conclusions of this paper.

2 Service Innovation and Consulting Firms

2.1 Service Innovation

Service innovation can be understood in different views. There are many key themes when reviewing service innovation since it incorporates knowledge and offerings, co-created by connected resources such as customers, suppliers and employees, which impact sales and costs enhancing competitive advantage. It also differs from manufacturing and product innovation. There are many theoretical perspectives but little attention to the mechanisms to combine resources in service firms leading to service innovation (Salunke et al. 2013).

Table 1 summarizes some recent perspectives about service innovation stated by diverse authors. Some researchers have a traditional novelty-oriented view (European Commission 2012b; Quintane et al. 2011); while others consider service innovation as a result of knowledge (Quintane et al. 2011; Salunke et al. 2011) or a service system (Ifm & IBM 2007; Maglio and Spohrer 2008; Vargo and Lusch 2004). There is also a relational approach emphasizing the role of customers and a network of actors as co-creators of service innovation (Chae 2014; Hidalgo and D'Alvano 2014). There was found no customized or exclusive service innovation standpoint in the context of consulting firms.

Table 1 Viewpoints on service innovation

Viewpoint	Definition	Authors
Novelty	Service innovation is a duplicable, new or significantly improved service concept and offerings.	European Commission (2012b), Quintane et al. (2011)
Knowledge	Service innovation is the extent to which new knowledge is integrated by the firm into service offerings, which directly or indirectly results in value for the firm and its customers/clients.	Quintane et al. (2011), Salunke et al. (2011)
Value network	Service value network is a set of activities where suppliers, service provider and customers integrate resources through service, and customer co-creates value in a specific cultural environment with service provider value proposal.	Chae (2014), Hidalgo and D'Alvano (2014)
Service system	Service system is a dynamic value co-creating configuration of resources, including people, technology, organization and shared information, connected internally and externally by value propositions, with the aim to meet customer's needs better than competitors.	IfM & IBM (2007), Maglio and Sphorer (2008), Vargo and Lusch (2004)

2.2 Consulting Firms

Wright et al. (2012) explain that consulting firms are organizations that act as change agents (both internal and external) and are often characterized as a key source of innovation, promoting "best practices" usually perceived as radical innovations. They also discuss that although consulting firms have been criticized because they standardize best practices, this knowledge standardization leads to innovation with their customers.

Anand et al. (2007) argue that the importance of consulting firms for knowledge-based innovation rely on: expertise and competence of their personnel as their main assets, widespread use of the partnership form of ownership, and inherent imperative for both organic growth and diversification on the context of innovation. In project-oriented environments such as consulting firms, there is a need to understand how managers are to create new resources combinations and strategically deploy outcomes that leads to service innovation (Salunke et al. 2013).

Consulting firms usually innovate with their customers and suppliers in a nurturing environment for value co-creation. This environment is project-based, process-oriented and with intensive knowledge exchange among all stakeholders. Consulting firms operate in different industry categories such as: management, financial services and accounting, human resources, information technology, design, sales and marketing, national security, litigation and economic, etc. (European Commission 2012a).

Consulting firms are classified as part of the Knowledge-Intensive Business Services (KIBS) and provide professional services to all types of organizations. Most studies on innovation in services have pointed to KIBS as the leading subsector regarding innovation and cooperation (Trigo and Vence 2012). KIBS are organizations that use and build knowledge as their primary raw material for their value-adding process (European Commission 2012a; Hidalgo and Albors 2008). Innovation, both incremental and radical, requires the conversion of knowledge into services as well as their successful diffusion into a society and/or market (Bouncken and Kraus 2013).

From an economic perspective, the Expert Panel in Service Innovation in the EU (Europe Innova 2011) states the benefits of KIBS and therefore service innovation as follows: regions with strong KIBS sectors exhibit the highest prosperity levels in Europe, and the presence of a strong KIBS sector positively affects regional innovation performance. In this context, the European Union has decided to develop an economy based on knowledge and innovation. One of the services sectors that have this transformative capacity are KIBS, because they collaborate closely with customers and across sectors. Spain, Italy and Germany account together for the 65 % top 23 largest KIBS regions and the 66 % top 9 employement KIBS regions (Europe Innova 2011).

3 Frameworks for Service Innovation

Preliminary literature review has identified different approaches on service innovation, both in the academic and practitioner area, but they all do not focus exclusively in the context of consulting firms. It has also not been tested if the proposed frameworks apply in consulting when it comes to co-elevate the innovativeness with their employers, customers and suppliers. Project-oriented firms are characterized by having long project life cycles where the provision of service often involves close collaboration with their clients, reflecting client input to innovation process (Salunke et al. 2011). Therefore, it is important that a model suitable for consulting firms considers an internal and external approach for service innovation.

Table 2 compiles some recent frameworks (denunciative but not limitative) on service innovation that could be used as a base for future research about consulting firms. Most of the authors propose a service innovation model based on their own perspective on service innovation. Some authors focus their model according to the source or reason of the innovation and the radicalism of the innovation (Chae 2012; Den Hertog et al. 2010). In contrast, other authors centre their model in the collaboration with networks with customers, suppliers and other institutions (Battisti 2012; D'Alvano and Hidalgo 2012), while others have a co-relational view (Furseth and Cuthberson 2013) encompassing elements that enable service innovation within and outside the firm.

Table 2 Recent frameworks for service innovation

Dimensions and capabilities (Den Hertog et al. 2010)	Orientations (Chae 2012)	Attributes (Battisti 2012)	Process (D'Alvano and Hidalgo 2012)	Elements (Furseth and Cuthberson 2013)
New service concept New customer interaction New business partner New revenue model New delivery system: personnel, organization, culture New delivery system: technological	Interactions (mutation, cross-over) Dimensions (customer-side, supply-side, geographical/institutional)	Network Diffusion Strategy	Scan Focus Resource Implement Learn Lead Co-operation	Value Business model Service system Customer experiences Technology Tangible assets Financial assets People Intangible assets

4 Some Examples from the Practitioner's Side

Consulting firms gain their competitive advantage from their ability to create and sustain knowledge resources and institutional capital based on legitimacy, reputation or relationships (Reihlen and Nikolova 2010). It is important that consulting firms realize this matter and incorporate innovation management into their strategy and operation, not only with their employees but also with their customers and suppliers.

Table 3 lists some real examples about innovation practices that are performed by consulting firms. Most of these firms are focusing on innovation co-creation through collaboration with their customers, integrating them into the project activities and having strong team interactions with continuous knowledge exchange. Some companies as is the case of KPMG, Fujitsu, Arup and Steria, have also recognized the importance to see innovation as a long term investment, so working with their clients in the long run will allow them to foster not only a good relationship but also the innovation outcome. It is also remarkable the case of companies that have created their own innovation framework for service delivery, making part of their value, culture and products the idea of thinking innovative and use this for the benefit of the organization, its suppliers and clients.

Table 3 Innovation practices examples in consulting firms (CBI 2008)

Company	Area	Innovation practices
Arup	Construction design and business consulting	Arup established a Design Technology Fund for internal research and a core innovation group that leads intra-collaboration and 'ideation'. It has external collaborative projects with other universities, institutions and governments. It works in the planning and design of the world's first eco-city Dongstan in China.
KPMG	Audit, tax and advisory services	KPMG has a Head of Innovation and an innovation process with a system based on 12 pillars: leadership, definition, investment, innovation communities, wider firm, ideas from anywhere, sharing culture, rewards, skills and behaviors, 'make it real', metrics and brand.
Fujitsu	IT management and outsourcing services	For Fujitsu, a successful innovation is only possible when it constantly collaborates with its customers through multiple point of contact. It has a culture based on the Fujitsu Services Innovation methodology. It takes a long term approach to innovation and measures several aspects of the services experience provided to customers to ensure this.
Steria	IT services and business consulting	Steria strives to close collaboration with customers. It acts as a hub in the innovation supply chain integrating customer needs with software developers to customize systems and software packages.

5 Conclusions

Consulting firms have a predominant role in service innovation research among other KIBS because this type of organizations have proven to be innovative and foster the economy. Besides, it is believed that consulting firms have a nurturing environment for service innovation to arise since the consulting service delivery is distinguished by being project-based, process-oriented, as well as intensive knowledge exchange and value co-creation among the consultants, customers and suppliers.

After preliminary literature review, there were identified some recent service innovation models but none of them focuses exclusively on service innovation in the context of consulting firms. Thus, there are still a field to study in order to have a clear comprehension on this matter.

From the academic perspective, it needs to be examined if consulting firms follow the same patterns as other service organizations, especially KIBS. Another interrogation is whether the frameworks developed for services innovation can be completely applied to consulting firms or if they need some customization in order to be more accurate. It is also important to understand the role of the consulting projects stakeholders in the innovation process (consultants, clients, suppliers, etc.). Finally, it would be relevant to understand how consulting firms manage innovation: strategy, structure, process, culture, learning and metrics.

From the practitioner side, consulting firms usually help their clients to innovation, but sometimes they do not apply the same approaches for service innovation within their own organization. From the real case examples, we can infer that some consulting firms have acknowledge the importance of innovation while working with their clients, but only few have immerse innovation into their strategy, structure, culture and process. It could be useful for consulting firms to take a look into current service innovation models and try to implement the best practices that they think that are suitable for their own company.

Acknowledgements This paper is produced as part of the EMJD Programme European Doctorate in Industrial Management (EDIM) funded by the European Commission, Erasmus Mundus Action 1.

References

Anand N, Gardner H, Morris T (2007) Knowledge-based innovation: emergence and embedding of new practice areas in management consulting firms. Acad Manag J 50:406–428
Battisti S (2012) Service innovation: the challenge of management in hypercompetitive markets. Int J Technol Market 7(2):99–118
Bouncken RB, Kraus S (2013) Innovation in knowledge-intensive industries: the double-edged sword of coopetition. J Bus Res 66:2060–2070
CBI (2008) CBI/QinetiQ report on innovation in UK service sector businesses. Excellence in Service Innovation, pp 38–39, 42–43, 64–65, July 2008
Chae BK (2012) An evolutionary framework for service innovation: insights of complexity theory for service science. Int J Prod Econ 135:813–822
Chae BK (2014) A complexity theory approach to IT-enabled services (IESs) and service innovation: Business analytics as an illustration of IESs. Decis Support Syst 57:1–10
D'Alvano L, Hidalgo A (2012) Innovation management techniques and development degree of innovation process in service organizations. R&D Manag 42(1):61–70
Den Hertog P, Van der Aa W, De Jong MW (2010) Capabilities for managing service innovation: towards a conceptual framework. J Serv Manag 21(4):490–514
Europe Innova (2011) Meeting the challenge of Europe 2020. The transformative power of service innovation: report by the Expert Panel of Service Innovation in the EU
European Commission (2012a) Knowledge-intensive (business) services in Europe. Directorate-General for Research and Innovation, Belgium
European Commission (2012b) The smart guide for service innovation. How to better capitalise on service innovation for regional structural change and industrial modernisation. Directorate-General for Enterprise and Industry, Belgium
Furseth PI, Cuthberson R (2013) The service innovation triangle: a tool for exploring value creation through service innovation. Int J Technol Market 8(2):159–176
Ganz W, Satzger G, Schultz C (eds) (2011) Methods in service innovation. Current trends and future perspectives. Fraunhofer Verlag, Germany
Hidalgo A, Albors J (2008) Innovation management techniques and tools: a review from theory and practice. R&D Manag 38(2):113–127
Hidalgo A, D'Alvano L (2014) Service innovation: inward and outward related activities and its mode of cooperation. J Bus Res 67(5):698–703

IfM & IBM (2007) Succeeding through service innovation: a discussion paper. University of Cambridge Institute for Manufacturing, Cambridge

Kidström D, Kowalkowski C, Sandberg E (2013) Enabling service innovation: a dynamic capabilities approach. J Bus Res 66:1063–1073

Maglio PP, Spohrer J (2008) Fundamentals of service science. J Acad Mark Sci 36:18–20

Maglio PP, Spohrer J (2013) A service perspective on business model innovation. Ind Mark Manage 42:665–670

Mors ML (2010) Innovation in a global consulting firm: when the problem is too much diversity. Strateg Manag J 31:841–872

OECD (2013) Science, technology and industry scoreboard 2013: innovation for growth. Publishing

Quintane E, Casselman RM, Reiche BS, Nylund P (2011) Innovation as a knowledge-based outcome. J Knowl Manag 15(6):928–947

Reihlen M, Nikolova N (2010) Knowledge production in consulting teams. Scand J Manag 6:279–289

Sanz L, Jones V (2013) Advances in business research in Latin American studies. J Bus Res 66:397–400

Salunke S, Weerawardena J, McColl-Kennedy J (2011) Towards a model of dynamic capabilities in innovation-based competitive strategy: insights from project-oriented service firms. Ind Market Manag J 40:1251–1263

Salunke S, Weerawardena J, McColl-Kennedy J (2013) Competing through service innovation: the role of bricolage and entrepreneurship in project-oriented firms. J Bus Res 66:1085–1097

Trigo A, Vence X (2012) Scope and patterns of innovation cooperation in Spanish service enterprises. Res Policy 41:602–613

Vargo SL, Lusch RF (2004) Evolving to a new dominant logic for marketing. J Market 68:1–17

Wright C, Sturdy A, Wylie N (2012) Management innovation through standardization: consultants as standardizers of organizational practice. Res Policy 41:652–662

Innovation Capability and the Feeling of Being an Innovative Organization

Marta Zárraga-Rodríguez and M. Jesús Álvarez

Abstract Companies today compete in a turbulent global environment characterized by considerable technological advance and a knowledge-based economy, and the basis for sustainable competitive advantage has shifted to innovation. Given that a capability is a source of competitive advantage under the Resource Based View (RBV) theory, the development of an innovation capability is a critical factor for success. The aim of this paper is to explore the degree to which certain companies have developed this innovation capability. We also explore the relationship between innovation capability and the feeling of being an innovative organization. Spain is no longer cost competitive, and this fact makes innovation a top priority for the tourism sector. In light of this state of affairs, the study focuses on Spanish hotels because it is expected that such companies have developed this innovation capability.

Keywords Innovation capability · Innovation practices · RBV · Hotels

1 Introduction

The tourism sector is very important to the Spanish economy and it has been very active in terms of competing with lower prices and more exotic and appealing destinations. Two decades ago the Spanish tourism sector decided to compete by improving the quality of service and management, i.e., by implementing strategies addressing quality. However, the tourism sector is aware of the fact that currently innovation should be considered a top priority. In order to be able to keep and

M. Zárraga-Rodríguez (✉)
ISSA-School of Management Assistants, Universidad de Navarra,
Edificio Amigos Campus Universitario, 31009 Pamplona, Spain
e-mail: mzarraga@unav.es

M.J. Álvarez
TECNUN, Universidad de Navarra,
Paseo Manuel Lardizabal, 13, 20018 San Sebastian, Spain

attract new customers it must develop its innovation capability. In this study, we present a set of innovation-related practices that have been culled from the literature and which serve as evidence of innovation capability development. Then we explore whether they are common practices among hotels and how they affect hotel managers' perception of being an innovative organization.

2 Framework and Objectives of the Study

Companies need to innovate in order to compete in today's turbulent environment. Innovation is particularly critical in service firms, which are marked-oriented. According to Agarwal et al. (2003), service firms need to continuously innovate in order to obtain a competitive advantage. Hotels are services firms within the tourism sector. Because tourists are looking for new experiences more than ever before, the success of a sector business like tourism relies on its ability to respond to these demands. The tourism sector is aware of the fact that it must follow the path of innovation in order to be able to keep and attract new customers. However there is a lack of innovation research in the hotel industry (Orfila-Sintes and Mattsson 2007).

There many studies dealing with the assessment of innovation performance (Ooi et al. 2012; Prajogo and Sohal 2002; Hoang et al. 2006; Abrunhosa and Sá 2007) but according to Börjesson and Elmquist (2011) there is a very little in-depth research on how organizational capabilities for innovation are developed in practice.

The RBV organizational theory states that the key to the strategic success of a company lies not only in the environment but also in the resources the organization has and in the strategic use the company makes of them; that is, in its organizational capabilities. A capability is a source of competitive advantage for the company that allows the generation of value and differentiation through the combined use of a series of resources (Peppard and Ward 2004; Ashurst et al. 2008). Innovation management can be viewed as a form of organizational capability (Lawson and Samson 2001) i.e., a company has innovation capability when innovation is a source of competitive advantage for the company.

It is difficult to observe capabilities, but we can detect practices that evidence their presence. According to Ashurst et al. (2008), practices are more concrete and observable than capabilities, and they are described as a set of socially defined ways of doing things in order to achieve an outcome. Hence, if we focus on practices that could be related to innovation capability we would gain deeper knowledge about how innovation capability is developed in practice.

When defining innovation practices, we take as a reference the work by Hu et al. (2008). They point out several innovation practices derived from a comprehensive review of the existing literature. Table 1 shows the set of innovation practices that serve as evidence of innovation capability development.

Although outcomes of the innovation process have been extensively studied in the literature, the singular pattern of tourism innovation compared with manufacturing and services in general has not been studied in depth, and there is an incomplete

Table 1 Innovation practices

Innovation capability related practices	Code
Innovation is not perceived as too risky and innovative proposals are readily accepted	P1
Resources (time, people, money, etc.) are devoted to the development of new services	P2
Managers actively promote and encourage the generation of innovative ideas	P3
Workers are encouraged to promote and develop ideas for new services (even if they don't succeed in the market)	P4
There is active collaboration between departments to develop new ideas	P5
Staff qualification and innovation management is intended	P6
The market is monitored to detect changes in demand and competitor innovations	P7
People in my organization are innovative	P8

understanding of how innovation processes take place in tourism organizations (Camison and Monfort-Mir 2011; Hjalager 2009). Practices related to innovation capability would be helpful to clarify the particular pattern of tourism innovation.

This study has two main objectives. First, we want to explore the innovation capability development in hotels in Spain. Second, we want to explore the relationship between innovation capability related practices and the feeling of being an innovative organization in this sector.

3 Research Methodology

A survey was conducted involving hotels located in the northern region of Spain. Two hundred hotels were selected from the official tourism webpage of three Spanish regions: the Basque Country, Navarra and Cantabria. Each hotel was phoned in advance, and we explained the aim of our research and requested their participation. The questionnaire was administered via a web page, which participants accessed with a link sent via e-mail. Over the course of the 2 weeks following the first e-mail, reminder e-mails that encouraged participation were sent. The response rate was 19 %.

The items on the questionnaire were presented as statements, and respondents, i.e. hotel managers, had to indicate their agreement on scale of 1 (strongly disagree) to 6 (strongly agree). The statements aimed to measure the presence of innovation practices. Statements were developed from a literature review and the final instrument contained 9 items, 8 of which addressed innovation practice and one that measured the feeling of being an innovative organization (FIO). We used multiple regression analysis to examine the relationships among the collected data.

4 Results and Discussion

Table 2 presents a summary of the scores given to each statement by the respondents. All the means are between 3.92 and 4.63, which implies that a high level of innovation practices is common.

In order to explore the relationship between the innovation practices (the set of predictor variables) and the feeling of being an innovative organization (the dependent variable), we used multiple regression analysis.

Through the multiple regression analysis, we were able to obtain a model that allowed us to explain how the independent or predictor variables influenced the dependent variable. At first we considered the eight innovation practices as predictors. However, as the correlation coefficients were quite high, it was not necessary to introduce all the predictors in the model. Instead we used the best subsets regression method to determine which predictor (independent) variables should be included in the multiple regression model. This method allowed us to identify the best-fitting regression models that could be constructed with the specified predictor variables. This method involved examining all possible subsets of predictors, that is, all of the models created from all possible combinations of predictor variables, using R^2 to check for the best model. First, all models that had only one predictor variable included were checked, and the two models with the highest R^2 were selected. Then all models that had only two predictor variables included were checked and the two models with the highest R^2 were again chosen. This process continued until all combinations of all predictor variables had been taken into account. We computed this method with a statistical software program (Minitab).

Table 3 shows the results. Each line of the output represents a different model and the two best models for each number of predictors appear.

R^2 shows the % of the dependent variable that can be explained by the predictor variables in the model. The adjusted R^2 is a useful tool for comparing the explanatory power of models with different numbers of predictors. The adjusted R^2 will increase only if the new term improves the model more than would be expected by chance. Mallows' Cp is a statistic commonly used as an aid in choosing between competing multiple regression models. A Mallows' Cp value that is close to the number of predictors plus the constant indicates that the model is relatively precise and unbiased in estimating the true regression coefficients and predicting future responses. However, the Mallows' Cp should be evaluated in conjunction with other statistics included in the best subsets output, such as R^2, Adjusted R^2 and S, for more accurate decision-making.

Table 3 shows that moving from the best model with three variables to models with four variables slightly improves the model fit (adj. R^2). However, note that

Table 2 Quantitative results: mean ($\bar{x}$) and deviation (σ) of the questionnaire scores

	P1	P2	P3	P4	P5	P6	P7	P8	FIO
$\bar{x}$	4.13	3.92	4.63	3.94	4.13	4.17	4.56	3.92	4.13
σ	1.07	1.06	1.09	1.32	1.12	1.17	1.24	1.01	1.07

Table 3 Best subsets analysis

Vars	R^2	Adj. R^2	Mallows' Cp	S	P1	P2	P3	P4	P5	P6	P7	P8
1	39.7	38.5	20.4	0.83643								X
1	38.3	37.1	22	0.84622			X					
2	57.8	56.1	1.9	0.70701			X					X
2	54.8	53	5.3	0.73174	X							X
3	60.6	58.2	0.7	0.68996	X	X						X
3	58.4	55.8	3.2	0.70919			X		X			X
4	61.6	58.3	1.6	0.68859	X	X			X			X
4	60.9	57.6	2.4	0.69484	X	X					X	X
5	61.8	57.6	3.4	0.69477	X	X			X	X		X
5	61.7	57.6	3.5	0.69509	X	X			X		X	X
6	62.0	57	5.1	0.69992	X	X			X	X	X	X
6	61.8	56.8	5.3	0.70157	X	X	X	X			X	X
7	62.1	56.1	7	0.70706	X	X	X	X	X	X	X	X
7	62.1	38.5	20.4	0.83643								X
8	62.1	37.1	22	0.84622			X					

Note: Vars no. of predictors in the model, R^2 model fit, S standard error of the regression, X predictors that are present in the model

models with five variables have an Adj.R^2 value that is lower than the best model with four variables, suggesting that adding the fifth variable does not add much value to the model. It should also be noted that the lowest value of S corresponds to the best model with four variables and that the lower the value of S is, the more accurate the predictions made with the regression line are. Therefore, the best subset of variables would be given by these variables: P2, P3, P5 and P8.

Although the value of Mallows' Cp in the selected four-variable model is not adequate, considering the values of R^2, Adj.R^2 and S, we consider this model to be the best one.

4.1 Multiple Regression Analysis

Once we have selected the best subset, by using a multiple regression analysis we can obtain a model that allows us to explain how the independent or predictor variables influence the dependent variable.

A four-predictor multiple linear regression model is proposed with four predictor variables: practice 2 (P2), practice 3 (P3), practice 5 (P5) and practice 8 (P8). The equation for the proposed multiple linear regression model is illustrated as follows:

$$\text{FIO} = b_0 + b_2(\text{P2}) + b_3(\text{P3}) + b_5(\text{P5}) + b_8(\text{P8}) + e$$

Table 4 Multiple regression analysis of innovation practices (4) on the perception of feeling part of an innovative organization

Regression analysis results					
R	0.7849				
R^2	0.6161				
F-Statistic	18.8611 $P < 0.001$				
Adj. R^2	0.5834				
S	0.6885				
N	52.000				
Predictor	*Coef.*	*SE Coef.*	*T*	*P*	*VIF*
(Constant)	0.0476	0.50928	0.0935	0.92594	
P2	0.23608	0.11932	1.9786	0.05374	1.732
P3	0.34446	0.121	2.8468	0.00653	1.854
P5	−0.11243	0.103	−1.0917	0.28052	1.433
P8	0.51727	0.10913	4.7400	2.0144E-05	1.298

where FIO = feeling of being an innovative organization (dependent variable), b_0 = constant, e = error.

We used Minitab statistical software for the regression analysis, and Table 4 shows the output provided by the tool.

To ensure the non-existence of multicollinearity, we have to pay attention to variation inflation factors (VIF). According to Hair et al. (1987), VIF should be lower than 10. According to the VIF values in Table 4, with these 4 independent variables we ensure the non-existence of multicollinearity.

According to the value of Adjusted R^2 (0.583), the model has a reasonable fit. The Adj. R^2 value of 0.583 implies that 58.3 % of the feeling of being an innovative organization can be explained by the independent variables in the model.

Three innovation practices were found to have a positive and significant relationship with the feeling of being an innovative organization: practice P2 (P2; b = 0.23608269; p = 0.0537), practice P3 (P3; b = 0. 34445478; $p < 0.01$) and practice P8 (P8; b = 0.51726959; $p < 0.01$). Although $p > 0.01$ for P2, we consider it to be a significant predictor variable because the model fit is better when we include it. Also, it makes sense to consider that the assignment of resources to the development of new services could influence the feeling of being an innovative organization; according to the literature review the assignment of resources is key when planning, and this reinforces our decision to include accuracy in this practice as a predictor variable in the model.

We used residual plots to examine the goodness of model fit and to determine if the regression assumptions are being met.

5 Conclusions

As a general conclusion, we see that the hotels analysed seem to have developed innovation capability to the extent that many innovation practices are perceived as common practices. The mean of the score given to each dimension by company managers is always above 3.9; this was to be expected since the hotel sector is moving towards innovation, where companies are being driven by competition from others and the need to make continuous improvement in offers within an increasingly competitive market.

The study has allowed areas of improvement to be identified by pointing out practices that, if they were commonly implemented, would improve the innovation capability, which would in turn be reflected in a company's results.

We have found a positive and significant relationship between some innovation practices and the feeling of being an innovative hotel. To the extent that the organization assigns resources to the development of new services, managers actively promote and encourage the generation of innovative ideas, and people inside the organization are viewed as innovative, the feeling of being innovative organization will be higher.

Research limitation: the small size of the sample makes it impossible to make inferences about the population as a whole. Nevertheless, the information gathered by the survey provides interesting information about innovation in this sector and it gives us a good qualitative approach to the subject. More data needs to be collected, however, in order to have a more complete picture.

References

Abrunhosa A, Sá PM (2007) Are TQM principles supporting innovation in the Portuguese footwear industry? Technovation 28(2008):208–211

Agarwal S, Erramilli MK, Dev CS (2003) Market orientation and performance in service firms: role of innovation. J Serv Mark 17(1):68–82

Ashurst C, Doherty NF, Peppard J (2008) Improving the impact of IT development projects: the benefits realization capability model. Eur J Inf Syst 17:352–370

Börjesson S, Elmquist M (2011) Developing innovation capabilities: a longitudinal study of a project at Volvo cars. Creativity Innov Manag 20(3):171–184

Camison C, Monfort-Mir VM (2011) Measuring innovation in tourism from the Schumpeterian and the dynamic-capabilities perspectives. Tour Manag 33(2012):776–789

Hair JF Jr, Anderson RE, Tatham RL (1987) Multivariate analysis, 2nd edn. Macmillan, New York

Hjalager AM (2009) A review of innovation research in tourism. Tour Manag 31(2010):1–12

Hoang DT, Igel B, Laosirihongthong T (2006) The impact of total quality management on innovation: findings from a developing country. Int J Qual Reliab Manag 23(9):1092–1117

Hu MLM, Horng JS, Sun YHC (2008) Hospitality teams: knowledge sharing and service innovation performance. Tour Manag 30(2009):41–50

Lawson B, Samson D (2001) Developing innovation capability in organizations: a dynamic capabilities approach. J Innov Manag 5(3):377–400

Ooi KB, Lin B, The PL, Chong AYL (2012) Does TQM support innovation performance in Malaysia's manufacturing industry? J Bus Econ Manag 13(2):366–393

Orfila-Sintes F, Mattsson J (2007) Innovation behavior in the hotel industry. Omega 37 (2009):380–394

Peppard J, Ward J (2004) Beyond strategic information systems: towards an IS capability. J Strat Inf Syst 13:167–194

Prajogo D, Sohal AS (2002) The relationship between TQM practices, quality performance, and innovation performance. Int J Qual Reliab Manag 20(8):901–918

Minimizing Carbon-Footprint of Municipal Waste Separate Collection Systems

Giancarlo Caponio, Giuseppe D'Alessandro, Salvatore Digiesi, Giorgio Mossa, Giovanni Mummolo and Rossella Verriello

Abstract Environmental performance of municipal waste management systems plays a key role in the so called smart cities performance. The authors propose a reference framework for public decision-making in optimizing municipal waste separate collection systems. A Mixed Integer Nonlinear Programming (MINLP) model is built up to set the collecting system (CS) and the optimal level of collection of each fraction. The goal is to minimize carbon footprint of the whole CS. The model is applied to a full-scale case study. Results obtained stress out the effectiveness of the model. Moreover the Environmental Efficiency Effort index is introduced to test the model solution evaluating the attitude of citizens in participating and improving waste separate collection.

Keywords Decision-making · Municipal waste collection system · MINLP

G. Caponio · S. Digiesi · G. Mossa · G. Mummolo · R. Verriello (✉)
Dipartimento di Meccanica, Matematica e Management, Politecnico di Bari,
Viale Japigia 182, 70126 Bari, Italy
e-mail: r.verriello@gmail.com

G. Caponio
e-mail: g.caponio@poliba.it

S. Digiesi
e-mail: s.digiesi@poliba.it

G. Mossa
e-mail: g.mossa@poliba.it

G. Mummolo
e-mail: giovanni.mummolo@poliba.it

G. D'Alessandro
ASM s.r.l, 70056 Molfetta, BA, Italy
e-mail: dalessandro@asmmolfetta.it

© Springer International Publishing Switzerland 2015
P. Cortés et al. (eds.), *Enhancing Synergies in a Collaborative Environment*,
Lecture Notes in Management and Industrial Engineering,
DOI 10.1007/978-3-319-14078-0_40

1 Introduction

Societal challenges finalized to the 'well-being' of EU Citizens represent at the same time an ethic commitment and opportunities of growth for industry. Industrial and societal challenges are mutual dependent and provide significant impacts on the grand challenges of the EU 2020 strategy (Mummolo 2014). Accordingly, a sustainable urban development of the cities is expected to play a crucial role. In the last decade the 'concept' of a Smart City took place in EU. 'Smart city' is defined as an urban area where smart solutions are implemented to support a sustainable urban development. An Urban Control Centre is adopted for a remote monitoring of energy grids (electric, thermal), buildings, and system of mobility, city utilities, and waste management system. Logistics of the waste management system plays a key role in sustainability of CSs. Sustainable logistics is a wide field of investigation where environmental issues are jointly considered with economic issues (Digiesi et al. 2012, 2013, 2014). Moreover, a wide literature is available on methods and experience to tackle urban greenhouse gas (GHG) emissions (Dhakal 2010; Dodman 2011). In the view of a urban sustainable development, a strategic role in the governance of the City is played by waste management and treatments systems. The phenomenon, known as 'Urban/Industrial Symbiosis', is being receiving wide attention. Recyclable waste originated both in urban and industrial districts can be recovered by a contiguous industrial district for recycling, re-use or waste-to-energy transformation. The eco-town program in Japan is an outstanding example of symbiosis (Van Berkel et al. 2009). The program identifies 26 eco-towns for a global investment of 1.65 billion USD in 61 innovative recycling projects and 107 new recycling facilities. The world potential market is of interest for the European industry. As an example, automatic waste separation is expected to grow by the factor five by 2020 (European Commission, October 2012). Technology of waste separation can widely affect municipal waste collection services of a smart city.

The paper proposes a decision-making framework outlining organizational and technological options for designing logistics of a municipal waste separate collection system (MWSCS) (Sect. 2). Furthermore, a MINLP model aiming at minimizing the carbon-footprint of the system while respecting technical and organizational constraints is proposed (Sect. 3). A full-scale case study, developed in Sect. 4, refers to Bari, a middle size city in southeast of Italy. Results obtained stress out the capability of the framework and of the optimization model in supporting public decision-making.

2 A Decision-Making Model for Designing a MWSCS

Designing a MWSCS requires both experiential and theoretical knowledge. Technical, economic, and environmental performance are obtained after basic organizational choices are taken and a simple input-output economic model is adopted. Public actors are not always aware nor of solutions adopted or the way they have been found. To improve awareness of decision-making, a reference three-level decision making model is proposed.

2.1 Decision Level 1: Municipal Waste Stream Grouping

A MWSCS is usually carried out by single waste streams (i.e. one for each recyclable material like paper, glass, organic waste, plastics and metal cans, etc.). Citizens are requested to sort and stock each household waste stream; waste quality is strongly affected by citizens environmental awareness which, in turn, affects efficiency of recycling processes. Innovative solutions consist of automatic or semi-automatic systems to sort materials from a multi-stream, i.e. a mix of mainly dry materials (e.g. paper plus plastics and metal cans). This would limit citizens efforts at the same time ensuring high and constant quality of waste streams to be recovered. The single-stream option requires smaller both bins and transport means, whereas the number of collection routes and related GHG emissions increase. On the other hand, the multi-stream option requires bigger bins and transport means plus energy consumption for separating the multi-stream in un-mixed streams. Technology improvements are expected to lead in the very next future to different grouping options involving also organic and glass fractions.

2.2 Decision Level 2: Municipal Waste Collecting Systems

At the second decision level, different CSs are considered. In (D'Alessandro et al. 2012) a review of municipal waste management and recovery systems is provided. The 'door-to-door' (DtD) option is the best choice for quality of collected materials and users monitoring. The DtD option fits well with urban features including a widespread presence of independent houses with private backyards where collection is typically carried out using small dedicated containers. In case of small urban aggregates of residential buildings with private common waste storage areas, an "aggregate" (A) curbside CS can be adopted. The A system usually requires higher capacity of both transport and temporary storage equipment than the DtD system. In this case, individual bins, located inside of blocks shared areas, serve more than one family and streets allow bigger trucks to transit. Higher capacity of logistic means allows less collection time and shorter routes with reduction in related GHG emissions. The same line of reasoning applies for the proximity CS (P), a technical option usually adopted for more intensive urban developed areas with a significant concentration of users living in high, but not strictly contiguous, buildings. This option requires higher capacity of storage and transport means than the DtD option and lower effort in waste collection. However, the expected quality of each material tends to reduce since it highly depends on the behaviour of a great number of citizens less subject to a tight monitoring. Finally, the street CS (S) is adopted for a large number of users living in city suburbs showing large streets and contiguous residential building (usually blocks without common open areas). This requires very large logistic equipment (transport means and bins) which also allow reduction in the number of routes.

2.3 Decision Level 3: Collection Frequency

Frequency of waste collection is a decision variable strictly dependent on both the waste fraction nature (e.g. the organic fraction usually requires a higher collecting frequency than a dry fraction) and on the CS adopted in the upper decision level. Regulations and guidelines set the minimum collection frequency for each waste fraction.

3 Minimizing the Carbon-Footprint of Municipal Waste Separate Collection Systems: A MINLP Model

The proposed model outlines the complexity of decision-making mainly for not-technical decision-makers (public administrators). Searching for a technological option that fits an assigned objective function requires the adoption of a systematic approach capable to quantify both technical (e.g. management and /or environmental) and economic variables. Mathematical programming is a widely adopted modelling technique in optimizing waste management systems (Mummolo et al. 2008). Multiple-criteria decision analysis is a valid support decision tool when the aim is finding the most appropriate separate CS and selection technologies among a finite number of alternatives according to a given set of criteria (Banar et al. 2009). The logical structure of the problem suggests the formulation of a Mixed Integer Nonlinear Programming (MINLP) problem to support decision makers in finding out the MWSCS with the lowest Carbon-footprint (MWCC). The problem is formulated as follows:

$$\text{O.F.:} \quad \underset{\{N_{ik}, N_{jik}, nw_{jik}\}}{\text{Min MWCC}} \sum_k \sum_i \sum_j \left(H \cdot v_j \cdot W \cdot fe_{ji} \cdot nw_{jik} \cdot N_{ik} \cdot N_{jik} / P_{ji} B_j \right) \quad (1)$$

where the decision variables are: $N_{j,\,i,\,k}$: number of users served by the j-th collecting system and the i-th waste stream grouping for the k-th waste fraction; $N_{i,\,k}$: is a boolean variable equal to 1 or 0 according to whether $N_{j,\,i,\,k} \neq 0$ or $N_{j,\,i,\,k} = 0$, respectively, $\forall$ i, j, k; $nw_{j,\,i,\,k}$: weekly collection frequency of the k-th fraction, by the j-th collecting system according to the i-th waste stream grouping; **j = 1**: 'door-to-door'; **j = 2**: 'aggregate curbside'; **j = 3**: 'proximity'; **j = 4**: 'street'; **i = 1**: 'single stream'; $i = 2$: 'multi stream'; $k = 1$: organic; $k = 2$: glass; $k = 3$: paper; $k = 4$: plastics and metal cans.

Model parameters are: $fe_{j,\,i}$: emission factor of the transport mean to collect any waste fraction according to the i-th waste stream grouping and the j-th CS; W: number of weeks in the observation period (W = 52 in case of one a year); $P_{j,\,i}$: number of collecting cycles per shift, i.e. number of picking operations of unit loads characterizing the i-th waste stream grouping and the j-th waste CS. It is considered 1 unit-load per user; Bj: bin capacity (users); H: number of work-shifts per day; v_j: average speed of the transport mean adopted by the j-th CS; P_c: production per capita of urban waste; c: average number of members in a family (user); $\eta_{k,\,j,\,i}$: collecting efficiency of the k-th material collected by the j-th waste CS according to the i-th waste stream grouping; p: percentage of the k-th type of waste in the total municipal solid waste produced.

The object function (1) is subject to the following constraints:

$$\sum_i N_{ik} = 1, \ldots \forall k \qquad (c.1)$$

$$N_{2,1} = 0 \wedge N_{2,2} = 0 \qquad (c.2)$$

$$N_{i,3} = N_{i,4} \qquad (c.3)$$

$$\sum_i \sum_j N_{i,k} \cdot N_{j,i,k} = N_{TOT}, \forall k \qquad (c.4)$$

$$N_{j,i,k} \leq N_{jMAX} \cdot N_{j,i,k} \geq N_{jMIN}, \forall i,j,k \qquad (c.5)$$

$$N_{1,1,2} = N_{2,1,2} = 0 \wedge N_{3,1,2} \geq 0 \wedge N_{4,1,2} \geq 0 \qquad (c.6)$$

$$N_{j,2,3} = N_{j,2,4}, \forall j \qquad (c.7)$$

$$nw_{j,i,k} \geq nw_{j,i,kMIN} \wedge nw_{j,i,k} \leq nw_{j,i,kMAX}, \forall i,j,k \qquad (c.8)$$

$$\sum_i \sum_j \sum_k (P_c \cdot c \cdot p_k \cdot N_{i,k} \cdot N_{j,i,k} \cdot \eta_{j,i,k}) \geq SC_{min} \qquad (c.9)$$

Constrain c.1 states that only one waste stream grouping is admitted for the k-th waste fraction. According to c.2 the organic and glass fractions are collected by a single stream waste grouping. Logistic opportunities suggest, as evident in c.3, adopting the same i-th waste grouping system for paper, plastics and metal cans. Constrain c.4 ensures that the sum of the number of users served by the j-th CS and the i-th waste stream grouping system is equal to the total number of the users (N_{TOT}), for the k-th waste fraction. The number of users served by the i-th waste grouping and the j-th CS for the k-th fraction, as evident in c.5, must comply with upper and lower limits depending on the urban configuration (e.g. historical vs. suburbs areas). Constraint c.6 states that the glass fraction can be collected only by 'proximity' or 'street' CS. Obviously, as shown in c.7, the number of users served by the j-th waste CS for the multi-stream grouping is the same for paper, plastics and metal cans streams. Constrain c.8 allows a variation of the weekly frequency of collection in a defined set of values according to practical and official waste management guidelines and local regulations. The last constrain allows the achievement of a minimum level of separate collection, SC_{min}.

4 A Full Scale Case Study

The proposed model has been applied to minimize the MWCC of Bari, a middle size city located in Southern Italy, accounting for 147,811 families (users from the point of view of the MWSCS) each of them having on the average 2.33 members per family and a waste production per capita of 513 kg/year (2013 data). The model also identifies the optimal waste fraction mix of recyclable materials (organic, glass, paper, plastics, and metal cans) produced by families. The MWSCS considers two classes of collecting transport means: a small one having a 7 m^3 transport capacity, average speed of 10 km/h, and an emission factor of 0.260 kg CO_{2eq}/km; a big one with 15 m^3 transport capacity, average speed of 15 km/h, and an emission factor of 0.697 kg CO_{2eq}/km (INEMAR 2010). The distribution of waste fractions is: organic 23.4 %, glass 5.6 %, paper 15.3 %, plastics 12.4 %, metal cans 3.03 %. By analysing four different urban areas of Bari (historical centre, areas with independent houses with private backyards and small clusters of residential buildings, more intensive urban developed areas, suburbs) the lower and the upper number of users that can be served and the capacity of the collecting bins for the different waste CSs have been evaluated (see Table 1). With a 6-hours work-shift, the productivity P_{ji} (workload capacity) of the workers team for each grouping and CS has been evaluated. A measure of the degree of users commitment in waste classification and collection has been estimated for each grouping and CS. Data are listed in Table 2.

SC_{min} has been set to 36.0 % (it is assumed that waste materials like wood, textile, electric and electronic equipment are managed by specific CSs/routes and allow to increase the overall separate waste collection up to 50.0 %). Results are in Table 3. According to the model proposed, organic and glass fractions have to be collected by a single stream grouping system according to constraints c.2: the rationale for this choice relies on the need to keep a high quality of fractions collected. Differently, the model suggests the adoption of a multi-stream grouping system for paper, plastics and metal cans fractions since such a grouping system

Table 1 Number of users and bin load capacity for each waste collection system

	Door to door (DtD)	Aggregate (A)	Proximity (P)	Street (S)
Users (upper limit)	14,337	53,158	78,443	133,473
Users (lower limit)	7,098	2,467	2,467	55,510
Bin capacity (user)	–	8	15	40

Table 2 Productivity (Pji) and efficiency (E) for each grouping and collection system

		Single stream				Multi material			
		DtD	A	P	S	DtD	A	P	S
Organic	E	0.7	0.7	0.6	0.5	–	–	–	–
	P (unit/shift)	800	200	200	150	–	–	–	–
Glass	E	–	–	0.6	0.5	–	–	–	–
	P (unit/shift)	–	–	200	150	–	–	–	–
Paper	E	0.65	0.65	0.6	0.5	0.8	0.8	0.7	0.6
	P (unit/shift)	800	200	200	150	700	200	200	150
Plastic & Al. Cans	E	0.65	0.65	0.6	0.5	0.8	0.8	0.7	0.6
	P (unit/shift)	800	200	200	150	700	200	200	150

Table 3 Solution of the optimization problem: number of users served by different waste collection systems and Level of waste separate collection

		Users served (#)				Level of separate collection (%)
		DTD	A	P	S	
Organic	Single stream	7,098	6,760	78,443	55,510	13.4
Glass	Single stream	–	–	78,443	69,368	3.1
Paper	Multi stream	7,098	6,760	78,443	55,510	10.3
Plastic & Al. Cans	Multi stream	7,098	6,760	78,443	55,510	10.4

allows for a reduction in the number of collection cycles (and GHG emissions) if it is compared with the number required for two separate cycles of collections (one for each fraction).

Solution obtained also privileges the P CS for organic, paper, plastic and metal cans as one can see by comparing results with the upper and lower limits (Table 1) of the number of users for each CS: the DtD system is limited at the minimum values of users served. The street CS is evaluated as not convenient from environmental point of view: the number of users served by this system is kept at the minimum value for each fraction. P and S systems are considered for the glass fraction (see constraint c.6): however, the model solution suggests the glass collecting to be carried out by the P system meanwhile keeping as much low as possible the S option. The S option reveals as not environmental compliant since the emission factors of high capacity transport means while the adoption of DtD CS is kept low since the number of workshifts and work cycles tend to increase because of high waste collection distance to be covered. The overall carbon footprint is estimated to around 348 t CO_{2eq}/year. A further analysis is carried out to investigate results sensitivity on the model parameters. A pregnant role is plaid by the efficiency of street collection, i.e. by the responsibility of citizens in leaving separate household waste to street bins. The street CS provides high quality and quantity of separate fractions if citizens have a collaborative behaviour without the need of a tight monitoring. To this purpose, the Environmental Efficiency Effort (Eη) index is defined as: $\Delta(CO_{2eq}/t_{sc})/\eta$ where the numerator represents the variation of GHG emissions [kg] per ton of separate waste collected and the denominator is the collecting efficiency variation, i.e. the variation in the involvement of the citizens due to increased environmental consciousness in improving the 'Street' collection efficiency: the higher involvement of citizens is, the higher is the 'Street' efficiency. Eη index is evaluated by considering, for each waste fraction, an increase in the 'Street Collection' efficiency of $\Delta\eta = 0.1$. Reduction in unitary GHG emissions is mainly obtained by adopting the multi-material CS for paper, plastics and metal cans (E$\eta = 1.6[kgCO_2/ton]$) as well as in the single stream organic collection (E$\eta = 1.3[kgCO_2/ton]$); a lower contribution is due to glass fraction (E$\eta = 0.4[kgCO_2/ton]$). The overall level of separate CS is increased of more than 4,000 t/year without increasing the overall amount of GHG emission (348 t CO_{2eq}/year) since the CS remains unchanged.

5 Conclusions

The proposed MINLP model reveals effective in minimizing the carbon footprint of MWSCSs. The model has been applied to a full-scale case study concerning Bari, a middle size city in the southeast of Italy. GHG emissions depend on both transport and routes required by the different CSs. Results obtained outline 'Door-to-Door' and 'Street' collections systems as the options responsible for higher GHG emissions (CO_{2eq}) than the 'Aggregate' and 'Proximity' ones. DtD system adopts greener transport means but requires longer collecting routes; the opposite situation holds for the 'Street' option. CSs have a different efficiency in capturing separate waste streams of adequate quality. The bad performance of the 'Street' option is mainly due to a low environmental consciousness of citizens coupled with the lack of monitoring actions. A sensitivity analysis revealed the high impact of the 'Street' collecting option efficiency on urban carbon-foot print mainly when the multi-stream grouping option of paper, plastics and metal cans is adopted. Investments to increase public awareness of citizens by communication and education campaigns are highly recommended since they would exert a beneficial impact on collection efficiency and, consequently, on the Carbon footprint of the city. Further research will be oriented toward investigating the effects of integrated collection and recovery facilities systems on the carbon-footprint of a smart city.

Acknowledgments The paper has been written within the framework of the project PON04a2_E "Smart Energy Master per il governo energetico del territorio—SINERGREEN—RES NOVAE". The project is supported by the Italian University and Research National Ministry to promote "Smart Cities Communities and Social Innovation" program.

References

Banar M, Ozkan A, Kulac A (2009) Choosing a recycling system using ANP and ELECTRE III technique. Turkish J Eng Environ Sci 34(3):145–154

D'Alessandro G, Gnoni G, Mummolo G (2012) Sustainable municipal solid waste management and recovery. In: CN Madu and C Kuei (eds) Handbook of sustainability management. Imperial College Press, London (ISBN 978-981-4354-81-3)

Dhakal S (2010) GHG emissions from urbanization and opportunities for urban carbon mitigation. Curr Opin Environ Sustain 2(4):277–283

Digiesi S, Mossa G, Mummolo G (2012) A loss factor based approach for sustainable logistics. Prod Plan Control 23(2–3):160–170

Digiesi S, Mossa G, Mummolo G (2013) Supply lead time uncertainty in a sustainable order quantity inventory model. Manage Prod Eng Rev 4:15–27. doi:10.2478/mper-2013-0034 (ISSN: 2082-1344)

Digiesi S, Mossa G, Rubino S (2014) A sustainable EOQ model for repairable spare parts under uncertain demand. IMA J Manage Math doi: 10.1093/imaman/dpu004

Dodman D (2011) Forces driving urban greenhouse gas emissions. Curr Opin Environ Sustain 3 (3):121–125

EU Commission (2012) Communication from the Commission to the European Parliament, The Council, The European Economic and Social Committee and the Committee of the Regions, 'A European strategy for key enabling technologies–A bridge to growth and jobs'

INEMAR (2010) Inventario EMissioni Aria-Regione Lombardia, Fattori Emissioni medi da traffico per tipo veicolo e inquinante. http://www.inemar.eu/

Mummolo G (2014) Looking at the future of industrial engineering in Europe. In: Hernández Iglesias C, López-Paredes A, Pérez Ríos JM (eds) Managing complexity: challenges for industrial engineering and operations management. Lecture notes in management and industrial engineering, vol 2. Springer, Berlin (ISBN 978-3-319-04704-1)

Mummolo G, Gnoni M, Ranieri L (2008) A mixed integer linear programming model for optimisation of organics management in an integrated solid waste system. J Environ Plan Manage 51:833–845 (ISSN: 0964-0568)

Van Berkel R, Fujita T, Hashimoto S, Geng Y (2009) Industrial and urban symbiosis in Japan: analysis of the eco-town program 1997–2006. J Environ Manage 1544–1556

A MILP Model for the Strategic Capacity Planning in Consultancy

Carme Martínez Costa, Manuel Mateo Doll and Amaia Lusa García

Abstract This paper deals with the Strategic Capacity Planning problem in knowledge intensive organizations (KIOs), particularly in consulting firms. Considering different characteristics such as the organization structure, the workforce characteristics, the capacity requirements, the capacity decisions or the evaluation criteria in a consulting company, a model based on MILP is developed to be applied to a particular case.

Keywords Strategic capacity planning · Staff planning · KIO/KIF · MILP · Consulting firm

1 Introduction

The problem of strategic capacity planning of an organization is to determine the appropriated quantity and typology of the resources in a long term horizon, considering one or several criteria.

Supported by the Spanish Ministry of Economy and Competitivity (project DPI2010-15614).

C. Martínez Costa
Management Department. IOC Research Institute. ETSEIB Engineering School,
Universitat Politècnica de Catalunya, Av. Diagonal 647, p7, 08028 Barcelona, Spain
e-mail: mcarme.martinez@upc.edu

M. Mateo Doll
Management Department. ETSEIB Engineering School, Universitat Politècnica de Catalunya,
Av. Diagonal 647, p7, 08028 Barcelona, Spain
e-mail: manel.mateo@upc.edu

A. Lusa García (✉)
Management Department. IOC Research Institute. ETSEIB Engineering School,
Universitat Politècnica de Catalunya, Av. Diagonal 647, p11, 08028 Barcelona, Spain
e-mail: amaia.lusa@upc.edu

© Springer International Publishing Switzerland 2015
P. Cortés et al. (eds.), *Enhancing Synergies in a Collaborative Environment*,
Lecture Notes in Management and Industrial Engineering,
DOI 10.1007/978-3-319-14078-0_41

In a manufacturing industry, these resources are e.g. the necessary equipment in the plant, while in other kind of organizations, such as the knowledge intensive organizations (KIOs), the main required resources are the professionals.

Consultancies are an example of KIO. The term KIO was introduced by Starbuck (1992). These organizations are characterized by the fact that workers are not easily replaced—some of them are highly qualified. Their capacity depends on the size and composition of the workforce. The workforce capacity is related to the human resource with necessary skills to meet the future demand. It involves identifying the current and future skill types and numbers of employees required for the business needs, transferring, hiring and firing (Song and Huang 2008).

Strategic capacity planning is extremely important for every company (Luss 1982; Olhager et al. 2001; Geng and Jiang 2009). For some manufacturing industries and services (telecommunications, transport, electricity distribution, etc.), companies have to invest huge amounts of money in tangible assets with long payout times (Paraskevopuolos et al. 1991), and these decisions are often irreversible.

For this reason, most of the capacity planning models have been developed for organizations that have significant expansion costs (see Martinez-Costa et al. 2014b for a literature review on this topic for manufacturing companies).

Professional service organizations have two main characteristics: (1) high levels of customer contact/service customization; (2) flexible processes with low capital and high labor intensity (Lewis and Brown 2012). Based on a review of the recent professional service firms, Von Nordenflycht's (2010) identify knowledge intensity and low capital intensity as the central characteristics of consulting firms.

This paper is focused in a consultancy for which the main strategic resource is human capital and the workforce planning constitutes a major problem. The size of the workforce, their expertise and the efficient allocation of human resources to projects play a significant role in firm's performance. Consulting, like any professional service, is a people business (Richter et al. 2008). Professional workers show different skill sets, interests and expectations, so companies face difficult decisions on how plan capacity in order to meet demand, while balancing the aspirations of people (Dixit et al. 2009).

The number of consultancy firms is very high around the world and covers a wide spectrum of fields (strategic consulting, technology consulting, human resources consulting, IT consulting, engineering consulting, etc.). Following Richter et al. (2008), two major types of consulting firms can be identified: those organised as professional partnerships and those following the managed professional business model. These two models have differences in the internal organisation and the governance structures along with different values, human resources policies and practices. These authors provide an overview of the main characteristics of the two archetypes found in the services by other studies (Cooper et al. 1996 and others). This study is focused on the case of a management consulting firm organized as a professional partnership.

Consultancy is an attractive field for researchers. Despite the relevance of consulting service, there is a lack of research in quantitative tools to support strategic capacity decisions in these companies. In order to make a step forward, this paper presents a mathematical model for solving the strategic capacity planning problem for a basic situation of a consulting firm. As far as we know, this model is the first to address capacity long-term planning for a professional service organization.

The organization of the rest of the paper is as follows: Sect. 2 describes briefly the problem. Section 3 presents the mathematical model. Finally, conclusions are detailed in Sects. 4 and ends with references respectively.

2 Problem Description

In a consulting company, the strategic capacity planning problem can be basically defined as a staff planning problem. The problem consists mainly on deciding the number and type of workers to hire and promote for each category, and to fire, for every period of the planning horizon, in a long-term (up to 5 years). Besides, decisions about training and transfers (from one office or business unit to another) must be also considered.

For the sake of simplicity, a consulting firm with one business unit (e.g. management consulting) and one geographic area (e.g., a country) is considered. Adapting the model for a company with other business units (such as audit, tax, financial advisory, outsourcing, etc.) or for a multinational company with distinct geographical units (offices) it is quite straightforward (in that case transfers of professionals between business units or between cities or countries should be considered).

Even though a general methodology can be developed to deal with this problem (Martinez-Costa et al. 2014a), each case needs to be modelled ad hoc. The main characteristics of the problem are introduced below:

- Organization Structure: We suppose a matrix-type structure for service lines (services offered to costumers) and industries (client's industry or sector). Concerning the capacity planning, workers are assigned to a service line and an industry. We consider that if the demand of projects in any service line and industry is stable, it is supposed that the consultant will continue in the same position; otherwise, if the demand on working hours in its position decreases, it will be necessary to change to another service line or industry. On the other hand, work is organized by project teams, and a worker can be assigned to more than one project simultaneously.

- Workforce characteristics: The workforce is supposed to be organized by categories (e.g., analyst, consultant, manager, senior manager, and senior executive or partner), and all members of a category are supposed to be able to perform the same set of tasks (or assume similar responsibilities). Each project team has a standard composition of members by category. Besides the category (which is directly related to the tasks to be performed and, somehow, with the speed in performing those tasks), other characteristics that can be considered are, among others: the chargeability (proportion of time to invoice), the age of people (to know the time to the retirement), the staff turnover (could be different by category) and the probability of achieving the objectives and then being promoted.
- Career model: Promotion is based on the achievement of objectives and any worker can promote to the maximum category. It is assumed that the workers that reach the objectives are automatically promoted and, also, that workers can remain in the same category for an unlimited time.
- Incomes: Based on forecasted demand and strategic growth decisions (target billing) in certain areas and strategic decisions to enter in new business, a lower and an upper bound for the incomes are supposed to be known.
- Capacity Requirements (Demand): The capacity requirements to meet the demand are supposed to be dependent on the forecasted incomes. It is assumed that a preferable composition of the capacity requirements (in terms of categories, service lines and industries) can be defined.
- Service Level: There would be many ways of quantifying the service level of a consultancy firm. For example, it can be forced that the available capacity meets, at least, a certain percentage of the required capacity. It is assumed that the demand cannot be delayed.
- Workforce Costs: The variable costs to be considered may include wages, hiring, firing, retirement and training. A linear relation between the number (and type) of workers and the costs is considered. It is also assumed that the cost is the same for all workers of a category (antiquity is not considered and an average cost is assumed).
- Objective function: The objective is to maximize a function that includes the profit and the penalty of a workforce composition on a required capacity composition out of the desired bounds.

3 Model Formulation

This section describes the mathematical model for the strategic capacity decisions in a standard consulting firm. The data, parameters, decision variables and the rest of variables are detailed in Tables 1, 2, 3 and 4. Later, the model is presented.

Table 1 Data

Data	Description
T	Planning horizon
K	Set of categories in the consulting firm
I	Set of industries in which the consulting firm works
L	Set of service lines considered in the consulting firm
Γ_k^+	Set of categories to which it is possible to access from the category k. $[\forall k]$
Γ_k^-	Set of categories from which it is possible to access to the category k. $[\forall k]$
cl_{kt}	Labor costs in [mu/consultant] for a consultant of category k in the period t. $[\forall k, t]$
cq_{kt}	Cost in [mu/consultant] due to the training of a consultant who changes to the category k in the period t. $[\forall k, t \mid \Gamma_k^- \neq \{\emptyset\}]$
ch_{kt}	Cost in [mu/consultant] due to the training of a consultant hired for the category k in the period t. $[\forall k, t]$
cr_{kt}	Cost in [mu/consultant] due to the retirement of a consultant of the category k in the period t. $[\forall k, t]$
cf_{kt}	Cost in [mu/consultant] due to the firing of a consultant of the category k in the period t. $[\forall k, t]$
p_{kt}	Price in [mu/h] that will be paid by the clients for a working hour of a consultant of the category k in the period t. $[\forall k, t]$
$rr_{kk't}$	Proportion of consultants in the category k who can be promoted to a category $k' \in \Gamma_k^+$ in the period $t+1$. $[\forall k, t \mid \Gamma_k^+ \neq \{\emptyset\}]$
rt_{kilt}	Proportion of consultants in the category k, expert on service line l in industry i in the period t who leave the company due to turnover at the end of the period t. $[\forall k, i, l, t]$
R_{kilt}	Expected number of consultants in the category k retired at the end of the period t, who are expert on service line l and industry i. $[\forall k, i, l, t]$
h_{kilt}	Mean capacity [h] in the period t for a consultant in the category k, who is expert on service line l in industry i. $[\forall k, i, l, t]$

Model

$$[MAX]z = \sum_{k,t}\left[\sum_{i,l} p_{kt} \cdot D_{kilt} - \sum_{i,l} cl_{kt} \cdot W_{kilt} - \sum_{k' \in \Gamma_k^+, i, l} cq_{kt} \cdot Q_{kk'ilt}\right.$$

$$\left. - \sum_{i,l}(ch_{kt} \cdot H_{kilt} + cr_{kt} \cdot R_{kilt} + cf_{kt} \cdot F_{kilt})\right] - \sum_{k,t}\left[\lambda_{kt} \cdot \left(\delta_{kt}^+ + \delta_{kt}^-\right)\right]$$

$$- \sum_{i,t}\left[\sigma_{it} \cdot \left(\varepsilon_{it}^+ + \varepsilon_{it}^-\right)\right] - \sum_{l,t}\left[\tau_{lt} \cdot \left(\beta_{lt}^+ + \beta_{lt}^-\right)\right]$$

$$- \sum_{t}[\mu_t \cdot \delta_t + \rho_t \cdot (\varepsilon_t + \beta_t)] - \omega \cdot \Delta - \pi \cdot E \quad (1)$$

$$LI_t \leq \sum_{k,i,l} p_{kt} \cdot D_{kilt} \leq UI_t \quad \forall t \quad (2)$$

Table 2 Parameters

Parameter	Description
LI_t, UI_t	Lower bound and upper bound for the incomes for the period t. [$\forall t$]
LK_{lt}, UK_{lt}	Preferable bounds for the proportion of consultants in the category k in the period t. This condition is not rigid, but non-compliance is penalized. [$\forall k, t$]
LS_{lt}, US_{lt}	Preferable bounds for the capacity requirements on the industry i in the period t. This condition is not rigid, but non-compliance is penalized. [$\forall i, t$]
LF_{lt}, UF_{lt}	Preferable bounds for the capacity requirements on the service line l in the period t. This condition is not rigid, but non-compliance is penalized. [$\forall l, t$]
$\lambda_{kt}, \mu_t, \omega$	Penalties associated, respectively, to the discrepancy between the preferable and obtained number of consultants of the category k, in the period t [$\forall k, t$], the maximum discrepancy for each period t [$\forall t$] and the maximum discrepancy
σ_{it}, τ_{lt}	Penalties associated, respectively, to the discrepancy between the preferable and obtained capacity requirements in period t on the industry i [$\forall i, t$] and in the service line l, in the period t. [$\forall l, t$]
ρ_t, π	Penalty associated, respectively, to the maximum discrepancy between the preferable and obtained capacity requirements on industries and service lines in the period t. [$\forall t$] and the maximum discrepancy between the preferable and obtained capacity requirements

Table 3 Decision variables

Variable	Description
$W_{kilt} \in \mathbb{Z}^+$	Number of consultants in the category k, expert on service line l in industry i in the period t. [$\forall k, i, l, t$]
$H_{kilt} \in \mathbb{Z}^+$	Number of hired consultants in time t to work in the category k, in industry i and on service line l, from the labor market. [$\forall k, i, l, t$]
$F_{kilt} \in \mathbb{Z}^+$	Number of consultants who are fired which work in the category k and are expert in industry i and on service line l, up to time t. [$\forall k, i, l, t$]

$$W_{kilt} \cdot h_{kilt} \geq D_{kilt} \quad \forall k, i, l, t \tag{3}$$

$$W_{kilt} = W_{kilt-1} + \sum_{j \in \Gamma_k^-} Q_{jkilt} + H_{kilt} - \left(\sum_{k' \in \Gamma_k^+} Q_{kk'ilt} + R_{kilt} + T_{kilt} + F_{kilt} \right) \quad \forall k, i, l, t \tag{4}$$

$$rr_{kk't} \cdot W_{kil,t-1} - 1 \leq Q_{kk'ilt} \leq rr_{kk't} \cdot W_{kil,t-1} \quad \forall i, l, t, \forall k \in K | \Gamma_k^+ \neq \{\emptyset\}, \forall k' \in \Gamma_k^+ \tag{5}$$

$$rt_{kilt} \cdot W_{kil,t-1} - 1 \leq T_{kilt} \leq rt_{kilt} \cdot W_{kil,t-1} \quad \forall k, i, l, t \tag{6}$$

$$LK_{kt} \cdot \sum_{j,i,l} W_{jilt} - \delta_{kt}^- \leq \sum_{i,l} W_{kilt} \leq UK_{kt} \cdot \sum_{j,i,l} W_{jilt} + \delta_{kt}^+ \quad \forall k, t \tag{7}$$

Table 4 Other variables

Variable	Description
$T_{kilt} \in \mathbb{Z}^+$	Forecasted number consultants in the category k who are expert on service line l and industry i and leave the firm due to turnover in the period t. [$\forall k, i, l, t$]
$Q_{kk'ilt} \in \mathbb{Z}^+$	Number of consultants who are promoted in time t from the category k to the category k', which are expert in industry i and service line l. [$\forall i, l, t, \forall k \mid \Gamma_k^+ \neq \{\emptyset\}, \forall k' \in \Gamma_k^+$]
$D_{kilt} \in \mathbb{R}^+$	Required capacity [h] of consultants in the category k who are expert on service line l in industry i within the period t. [$\forall k, i, l, t$]
$\delta_{kt}^+, \delta_{kt}^- \in \mathbb{R}^+$	Positive and negative discrepancy respectively, between the preferable and the obtained number of consultants in the category k for the period t. [$\forall k, t$]
$\delta_t \in \mathbb{R}^+$	Maximum discrepancy between the preferable and the obtained composition of consultants in all categories for the period t. [$\forall t$]
$\varepsilon_{it}^+, \varepsilon_{it}^- \in \mathbb{R}^+$	Positive and negative discrepancy respectively, between the preferable and the obtained requirements on the industry i for the period t. [$\forall i, t$]
$\varepsilon_t \in \mathbb{R}^+$	Maximum discrepancy between the preferable and the obtained requirements of the all industries for the period t. [$\forall t$]
$\beta_{lt}^+, \beta_{lt}^- \in \mathbb{R}^+$	Positive and negative discrepancy respectively, between the preferable and the obtained requirements on the service line l for the period t. [$\forall l, t$]
$\beta_t \in \mathbb{R}^+$	Maximum discrepancy between the preferable and the obtained requirements of the all service lines for the period t. [$\forall t$]
$\Delta \in \mathbb{R}^+$	Maximum discrepancy between the preferable and the obtained composition of consultants in all the periods and categories
$E \in \mathbb{R}^+$	Maximum discrepancy between the preferable and the obtained composition of required capacity in all the periods, industries and services lines

$$LS_{it} \cdot \sum_{k,j,l} D_{kjlt} - \varepsilon_{it}^- \leq \sum_{k,l} D_{kilt} \leq US_{it} \cdot \sum_{k,j,l} D_{kjlt} + \varepsilon_{it}^+ \quad \forall i, t \qquad (8)$$

$$LF_{lt} \cdot \sum_{k,i,j} D_{kijt} - \beta_{lt}^- \leq \sum_{k,i} D_{kilt} \leq UF_{lt} \cdot \sum_{k,i,j} D_{kijt} + \beta_{lt}^+ \quad \forall l, t \qquad (9)$$

$$\delta_t \geq \delta_{kt}^+ + \delta_{kt}^- \quad \forall k, t \qquad (10)$$

$$\varepsilon_t \geq \varepsilon_{it}^+ + \varepsilon_{it}^- \quad \forall i, t \qquad (11)$$

$$\beta_t \geq \beta_{lt}^+ + \beta_{lt}^- \quad \forall l, t \qquad (12)$$

$$\left.\begin{array}{l} \Delta \geq \delta_t \\ E \geq \varepsilon_t \\ E \geq \beta_t \end{array}\right\} \quad \forall t \qquad (13)$$

Equation (1) presents the objective function, which can be divided in two parts: the maximum profit (incomes minus costs) and the search of a composition as close as possible to the ideal one established by the firm. Constraint (2) links the required capacity and the incomes and imposes the lower and the upper bound for the incomes. Constraint (3) establishes the minimum available capacity for each category, industry, service line and period. The balance of the consultants in each category, industry, service line and period is shown in (4). Constraint (5) determines the ratio of consultants who accomplish the objectives in a period and, therefore, must be promoted. Constraint (6) imposes the turnover. Constraints linked to the ideal composition of the workforce are: (7) for each category, (8) for each industry and (9) for each service line. Finally, constraints (10–13) evaluate the discrepancies for all categories, industries or services and periods.

4 Conclusions

This paper proposes a MILP model for long-term workforce capacity planning in consulting firms. The problem has been described and the decision variables and factors to be considered in the problem formulation have been identified. Empirical studies can be done in the future to test the effectiveness of the model. As future lines of research, the model can be expanded to address the problem in consultancies with several offices and/or strategic business units, including in these cases the transfer of professionals between offices and/or business units. The model could be also adapted to an up-or-out career model, in which workers cannot remain in the same category for an unlimited time. Finally, uncertainty can also be considered.

References

Cooper DJ, Hinings CR, Greenwood R, Brown JL (1996) Sedimentation and transformation in organizational change: the case of Canadian law firms. Organ Stud 17(4):623–647

Dixit K, Goyal M, Gupta P, Kambhatla N, Lotlikar RM, Majumdar D, Parija GR, Roy S, Soni S (2009) Effective decision support for workforce deployment service systems. IEEE Int Conf Serv Comput. doi:10.1109/SCC.2009.24

Geng N, Jiang Z (2009) A review on strategic capacity planning for the semiconductor manufacturing industry. Int J Prod Res 47(13):3639–3655

Lewis MA, Brown AD (2012) How different is professional service operations management? J Oper Manage. doi:10.1016/j.jom.2011.04.002

Luss H (1982) Operation research and capacity expansion problems: a survey. Oper Res 30 (5):907–947

Martinez-Costa C, Lusa A, Mas-Machuca M, De la Torre R, Mateo M (2014a) Strategic capacity planning in kios: a classification scheme. Ann Indu Eng 2012:191–198

Martinez-Costa C, Mas-Machuca M, Benedito E, Corominas A (2014b) A review of mathematical programming models for strategic capacity planning in manufacturing. Int J Prod Econ. doi:10.1016/j.ijpe.2014.03.011

Olhager J, Rudberg M, Wikner J (2001) Long-term capacity management: linking the perspective from manufacturing strategy and sales and operations planning. Int J Prod Econ 69(2):215–225

Paraskevopoulos D, Karakitsos E, Rustem B (1991) Robust capacity planning under uncertainty. Manage Sci 37(7):787–800

Richter A, Dickmann M, Graubner M (2008) Patterns of human resource management in consulting firms. Pers Rev 37(2):184–202
Song H, Huang HC (2008) A succesive convex approximation method for multistage workforce capacity planning problem with turnover. Eur J Oper Res 188:29–48
Starbuck WH (1992) Learning by knowledge-intensive firms. J Manage Stud 29:713–740
Von Nordenflycht A (2010) What is a professional service firm? Toward a theory and taxonomy of knowledge-intensive firms. Acad Manag Rev 35(1):155–174

Forecasting the Big Services Era: Novel Approach Combining Statistical Methods, Expertise and Technology Roadmapping

Iñaki Bildosola, Rosa Rio-Bélver and Ernesto Cilleruelo

Abstract This paper aims at proposing a novel approach to gathering and structuring information concerning an emerging technology, generating a relevant profile, identifying its past evolution, forecasting the short and medium-term evolution and integrating all of the elements graphically into a hybrid roadmap. The approach combines four families of technological forecasting methods, namely: Statistical Methods in terms of Bibliometrics and Data Mining; Trend Analysis; Descriptive Methods in terms of Technological Roadmapping; and Expertise. Its future application to forecast the evolution of emerging IT trends in current entities, which are creating the Big Services era, is proposed as future work.

Keywords Technology forecasting · Technology roadmapping · Bibliometrics · Data mining · Big services era

1 Introduction

Academic literature is filled with attempts to answer questions about the future of any technology, either proposing methodologies for such purpose or by applying any or some of these methods to a specific technology. In this sense, foresight

---

I. Bildosola (✉) · E. Cilleruelo
Foresight, Technology and Management (FTM) Group. Department of Industrial Engineering, University of the Basque Country UPV/EHU, alameda Urquijo s/n, 48030 Bilbao, Spain
e-mail: inaki.bildosola@ehu.es

E. Cilleruelo
e-mail: ernesto.cilleruelo@ehu.es

R. Rio-Bélver
Foresight, Technology and Management (FTM) Group. Department of Industrial Engineering, University of the Basque Country UPV/EHU, calle Nieves Cano 12, 01006 Vitoria, Spain
e-mail: rosamara.rio@ehu.es

© Springer International Publishing Switzerland 2015
P. Cortés et al. (eds.), *Enhancing Synergies in a Collaborative Environment*,
Lecture Notes in Management and Industrial Engineering,
DOI 10.1007/978-3-319-14078-0_42

initiatives now have such a high presence in so many activities that, for instance, the Foresight Monitoring Network (EFMN) was created in 2004, evolving into the European Foresight Platform (EFP), as a Europe-wide network focused on monitoring and collecting data on foresight exercises. In its final report (2009), it was noted that the number of publications collected increased from 100 to more than 2,000. There have been attempts to summarize and structure the field, such as Popper's Foresight Diamond (2008a), which identifies 33 foresight methods. Thus, technology forecasting experts, since the initial studies, agree that models should be used in combination. This is reasoned by the fact that this is the best way to attempt to offset the weaknesses of one forecasting method with the strengths of others (Martin and Daim 2012).

On that basis, this paper aims at proposing a novel approach to gathering and structuring information concerning an emerging technology, generating a relevant profile, identifying its past evolution, forecasting the short and medium-term evolution and integrating all of the elements graphically into a hybrid roadmap. For this purpose, it combines four families of technological forecasting methods, namely: Statistical Methods in terms of Bibliometrics and Data Mining; Trend Analysis; Descriptive Methods in terms of Technological Roadmapping; and Expert Opinion. The paper is divided as follows: in Sect. 2 the key literature is analyzed, where the motivation and applicability of technology roadmapping, bibliometrics and data mining is identified. Section 3 introduces the investigative approach, where the process of actual profile generation and forecasting is described through eight steps. Finally, the future work is presented, where the application of the novel approach is proposed for current four emerging information technologies, also known as Big Services era enablers (Zhang 2012).

2 Background

2.1 Technology Roadmapping

Technology Roadmapping (TRM) was initially developed by Motorola, applying the first TRM in the 70's for car radios (Willyard et al. 1987), linking technology investment to product evolution, becoming widespread with other firms. Due to its variability, no 'official guide book' outlines how to develop technology roadmapping. Nonetheless, the IFM's T-Plan has reached the highest degree of standardization (Phaal et al. 2001). In fact, more than 2,000 public-domain roadmap documents, which applied the T-plan directly, could be located on the internet (Phaal 2011). Since the first paper relating to roadmapping (1997), the number of research papers relating to this area is continuously increasing, as shown in Table 1 (Carvalho et al. 2013). The same paper underlines the fact that the majority of the studies applied qualitative approaches, in the form of case studies, supporting the notion that issues relating to TRM are still being consolidated.

Table 1 TRM publications evolution (Carvalho et al. 2013)

	Period					
	1997–2001	2002–2006	Tendency	2007–2011	Tendency	Total
Level of analysis						
LA1-strategy and business level	2	17	↗	19	↗	38
LA2-innovation and NPD level	2	17	↗	22	↗	41
Total	4	34		41		79
Method						
CR1: literature review	2	8	↗	6	↘	16
CR2: simulation or theoretical modeling	0	0		2	↗	2
ER1: survey	0	3	↗	4	↗	7
ER2: case study	2	23	↗	28	↗	53
ER3: action research	0	0	–	1	↗	1
Total	4	34		41		79

The foresight capability of the TRM is exemplified, for instance, by the United States government case, which is based on the Department of Energy's Industrial Technology Program (ITP). Since the inception of these partnerships programs between industries and government the results are more than 220 technologies which entered commercial markets, 215 patents, 51 R&D awards and 9.3 QUADS of energy saved (EERE 2010). The program was performed through specific technology roadmaps for each industry which can be found in (Phaal 2011) under the US DOE signature. Focusing on big services era enablers, a cloud computing roadmap is proposed in (NIST 2011). Here the construction of the roadmap is described and it can be appreciated that quantitative and, above all, qualitative methods are being used in this process, based on stakeholders' expertise through public working groups. Another Big Services era technology is analyzed, in a more theoretical attempt, in Roman et al. (2000), where a roadmap is not generated as such, rather relevant trends and key research areas are identified and examined. Several more papers can be found related to emerging ITs, however, as far as we know no further specific attempts could be found in the current academic literature to profile and generate a roadmap of the aforementioned IT technologies.

2.2 Bibliometric Analysis

When it comes to statistical methods, bibliometrics analysis (BA) is based on three principles: activity measurement by counting publications, impact measurement by counting subsequent citations of a publication and linkage measurement involving co-citations and keywords used from paper to paper (Kongthon 2004). In terms of forecasting, bibliometrics can be used to understand the past and even potentially to forecast the future, but it should be used in combination with other analytical methods (Daim et al. 2006).

2.3 Data Mining

Nowadays, when trying to gather information with the aim of discovering the knowledge of any field, the process must inevitably encounter dealing with at least one database, this process has been historically known as knowledge discovery in databases (KDD). Since bibliometrics counts publication (or patent) activity to detect trends and changing patterns, data mining carries this a step further to process the content of those papers (Kostoff and Geisler 1999). It has to be stated that data mining has to be applied after some preprocessing in order to make it more effective, i.e., the raw information is refined and reduced (e.g. via bibliometrics). Since data mining refers to the process of obtaining information from multiple different data formats, when these data are text-based the process is known as text mining. Targeting the profile of emerging technologies through text mining has been also suggested and applied, in fact the importance of text mining and its applicability is supported by the appearance and growth of dedicated software. Thus, text mining tools can be used to examine research trends and patterns in the fields of technology management using software developed specifically for these types of knowledge mining applications (Porter et al. 2003).

3 Research Approach

This paper aims at proposing a novel approach to achieve the following objectives jointly for a given emerging technology: (1) Gather and structure the information concerning it, (2) Generate a comprehensive profile, (3) Discover its past evolution and forecast the short-term and medium-term future of that evolution, identifying important milestones, clusters, gaps and paths, (4) Integrate all of the elements graphically in a hybrid roadmap. The complete process is based on a combination of statistical methods, expertise and technology roadmapping. The process description is presented below step by step, divided in the profile generation and the TRM.

3.1 Generating the Profile

The first three steps consist in generating the profile of the technology field, by applying the bibliometrics method to analyze the published conference and journal articles about the field, from its emerging year to current publications. The development of a research profile can be used not only to assist academics with their literature review and research, but also to help practitioners to identify who the experts are and where the knowledge and expertise can be accessed (Gerdsri et al. 2013). The profile generating process is designed based on the nine-step tech mining process used by Porter and Cunningham (2004) and the refining performed by Gerdsri et al. (2013). Information retrieval will be based on SCI and SSCI databases. The work is performed using bibliometrics with the VantagePoint software.

- Step 1 *Identifying Data Sources*: Top journals and conferences in each field will be listed in a table; they will build the original database.
- Step 2 *Refining Search and Retrieving Data*: From the initial database, a refining process will be performed. The process of Boolean refining will be summarized. The outcome is a refined database to be mined.
- Step 3 *Generating the Profile*: The profile will be based on previous data and it will identify and graphically represent the technology's features, such as the number of researchers and practitioners involved in it; the number of the publications regarding its academic or corporation origin; the publication mode such as journals, proceedings or books; and the most characteristic details.

3.2 Constructing the Roadmap

When it comes to design the roadmap, a two-layer representation has been selected, as we give importance to the disruption of technologies in their embryonic state and their evolution toward a real usable product. When it comes to the time axis, we are interested in the past (recent past since the target is an emerging technology), present, short and medium-term. Long-term is not considered since it only would be realistic within an entity. Thus, our goal is to provide a hybrid representation, based on the visualization frame provided by Zhang et al. (2013), but adapted to our case. In each step different elements will be added to the roadmap (Fig. 1):

- Step 4 *Ontology Generation*: The process is based on iterative K-means clustering of the Keywords, where each step generates a deeper subdivision of the clusters, creating more sub-clusters and more layers of the ontology and a hierarchical structure. The list of sub-technologies of interest is a starting point with which to draw the sub-layers of the roadmap (Yoon and Phaal 2013). The partial outcome of this step is the hierarchical structure of the technology and its graphical representation
- Step 5 *Clustering*: This is the identification of clusters within the technology, giving them the right name accordingly. The partial outcome of this step is a knowledge map where all the clusters are identified. This clustering of keywords

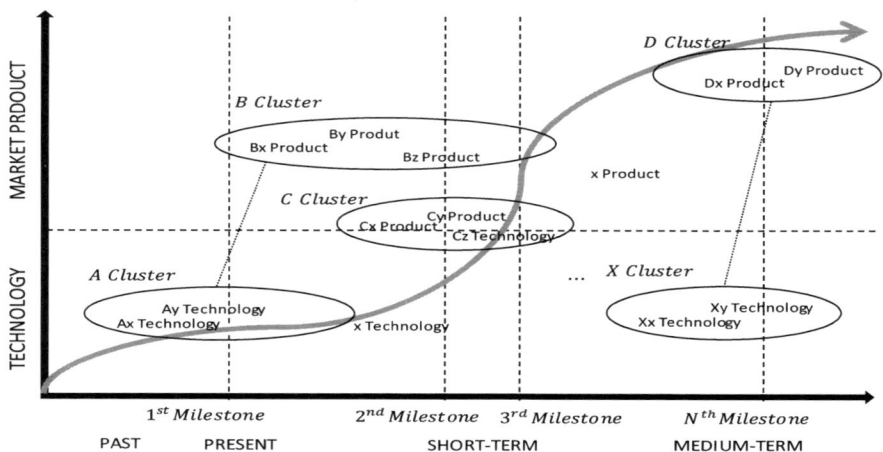

Fig. 1 Global technology roadmap, adapted from Zhang et al. (2013)

is based on a principal components analysis to group keywords that are frequently used together (Gerdsri et al. 2013), and differs from the previous step in its horizontal nature rather tan vertical. The clustering must be done time-based in order to position them in the correct place in the roadmap.

- Step 6 *Linking*: The links between the technology layer and the product layer are created by the Euclidean distance of the keyword-vectors between technology documents and product documents (Yoon and Phaal 2013). In this sense, before performing the step keyword vectors must be generated, consisting of the frequency of important words that frequently occur in documents. The usefulness of this element shall be the identification of the linkages of the technology with the products that have emerged from it.
- Step 7 *Trends identification*: The trend identification/extrapolation is considered as a forecasting quantitative method (Popper 2008b); here it is based on the outcome of analyzing previous steps. The outcome of this step is a year-based summary of the representative emerging keywords of the technology, presenting the growth rate of keywords. This trend can support the process of identifying potential technologies in the technology layer of roadmaps (Yoon and Phaal 2013). Thus, the results of this step are combined with the fifth and sixth, in order to adjust all the elements on the time axis, as its real placement will be when they are recognized as high-frequency elements.
- Step 8 *Reconstruction*: Several expertise-based methods are used in technology forecasting, such as expert panels, Delphi, key technologies and interviews. This is justified by several reasons: some of the key behavioral elements are included; the variety of inputs and thereby the quality of results will increase (in terms of richness of viewpoints, taking the expertise of stakeholders into account); it will

lead to broader support for the results and it may contribute to the democratic character of the process (TFAMWG 2004). In our case, once the profile is generated and the roadmap is constructed, the latter being especially subject to change, a survey-based assessment and statistical results analysis should be performed among experts and practitioners, because if high frequency modifications and appreciations occur, these should be considered, such as time-based displacements or clustering reconsiderations.

4 Future Work

Once the approach is presented, the next step shall be to test its applicability in real cases, namely certain specific technologies, as well as to infer whether another methodology also should be included to increase the process reliability. For this purpose, four emerging IT technologies have been selected, which are recognized as the new challenging IT trends for decision makers: cloud computing, big data, mobility and social business. Even if these technologies have their origin in the Computer Science and Networking Science fields among others, they are gradually reinventing IT systems and departments within enterprises, definitely gaining significant specific weight within the company. As enablers of the Big Services era, there is an indivisible relationship between these disrupting technologies, significantly 2012 was the first year when these technologies appeared in first place of the CIO's top three priorities (Gartner 2012).

In regard to the presented approach, it is worth to recognize that it is not free of modifications, which may arise in the process itself of implementation to the aforementioned technologies, as well as the outcome should be modified and finally endorsed based on expertise. In this sense, the inclusion of a Collective Intelligence method, such as Prediction Markets, is being considered, in a try to recover the experts' opinion in a valuable way for the *Reconstruction* step. These markets are used with the purpose of using the information content in the market values to make prediction about specific future events, its accuracy has already been discussed and proved (Berg et al. 2003). In this sense, the future work will take advantage from the synergy between the industrial fabric and the university at the local level in the Basque Country, in terms of emerging IT trends, which was generated by the work performed by Zabalza-Vivanco et al. (2013), which has been presented and illustrated in a forum as a Big Services era introduction to different decision makers.

Thereby, the expected results of the thesis work which follows this paper should be a complete profile for each of the aforementioned technologies, where the academic work will be presented in detail, regarding the nature, the sources, the number and the evolution of the publications. In addition to this, a complete roadmap will be constructed for each technology, with the purpose of identifying all the underlying sub-technologies and products, clustering all of them and naming the clusters logically, recognizing the possible linkages between the sub-technologies and products and identifying these elements as origin, present or future of the technology. In addition to this, all the estimated improvements shall be described comprehensively. In the end result, the aforementioned elements will be presented graphically as displayed in Sect. 3, aiming to provide a strategic vision of the evolution for each technology and a practical and solid tool, which can be used by practitioners, CIOs and decision makers to identify what the future holds for these technologies.

References

Berg J, Nelson F, Rietz T (2003) Accuracy and forecast standard error of prediction markets. Tippie College of Business Administration, University of Iowa

Carvalho MM, Fleury A, Lopes AP (2013) An overview of the literature on technology roadmapping (TRM): contributions and trends. Technol Forecast Soc Chang 80(7):1418–1437

Daim TU, Rueda G, Martin H, Gerdsri P (2006) Forecasting emerging technologies: use of bibliometrics and patent analysis. Technol Forecast Soc Chang 73(8):981–1012

Energy Efficiency and Renewable Energy (2010) Energy technology solutions: public-private partnerships transforming industry. U.S. Department of Energy

Gartner (2012) Gartner executive programs' worldwide survey of more than 2,300 CIOs shows flat it budgets in 2012, but it organizations must deliver on Multiple Priorities, 2012, from http://www.gartner.com/it/page.jsp?id=1897514

Gerdsri N, Kongthon A, Vatananan RS (2013) Mapping the knowledge evolution and professional network in the field of technology roadmapping: a bibliometric analysis. Technol Anal Strateg Manag 25(4):403–422

Kongthon A (2004) A text mining framework for discovering technological intelligence to support science and technology management. Doctoral dissertation, Georgia Institute of Technology

Kostoff RN, Geisler E (1999) Strategic management and implementation of textual data mining in government organizations. Technol Anal Strateg Manag 11(4):493–525

Martin H, Daim TU (2012) Technology roadmap development process (TRDP) for the service sector: a conceptual framework. Technol Soc 34(1):94–105

NIST (2011) NIST strategy to build a USG cloud computing technology roadmap. NIST working paper

Phaal R (2011) Public domain roadmaps. University of Cambridge

Phaal R, Farrukh CJ, Probert DR (2001) T-plan: fast start technology roadmapping: a practical guide for supporting technology and product planning. University of Cambridge

Popper R (2008a) In: Keenan M, Miles I, Popper R, Georghiou L, Cassingena Harper J (eds) Foresight methodology. Edward Elgar, Cheltenham

Popper R (2008b) How are foresight methods selected? Foresight 10(6):62–89

Porter AL, Cunningham SW (2004) Tech mining: exploiting new technologies for competitive advantage, vol 29. Wiley, Hoboken

Porter AL, Watts RJ, Anderson TR (2003) Mining PICMET: 1997–2003 papers help you track management of technology developments. In: Proceedings, Portland international conference on management of engineering and technology (PICMET)

Roman GC, Picco GP, Murphy AL (2000) Software engineering for mobility: a roadmap. In Proceedings of the conference on the future of software engineering, pp 241–258

TFAMWG (Technology Futures Analysis Methods Working Group) (2004) Technology futures analysis: Toward integration of the field and new methods. Technol Forecast Soc Chang 71:287–303

Willyard CH (1987) Motorola's technology roadmap process. Res Manag 30(5):13–19

Yoon B, Phaal R, Probert DR (2013) Structuring technological information for technology roadmapping: data mining approach. Technol Anal Strateg Manag 25(9):1119–1137

Zhang LJL (2012) Editorial: big services era: global trends of cloud computing and big data. IEEE Trans Serv Comput 4:467–468

Zabalza-Vivanco J, Rio-Belver R, Cilleruelo E, Acera-Osa FJ, Garechana G (2013) Decision tool based on cloud computing technology. In: 7th international conference on industrial engineering and industrial management, pp 137–145

Zhang Y, Guo Y, Wang X, Zhu D, Porter AL (2013) A hybrid visualisation model for technology roadmapping: bibliometrics, qualitative methodology and empirical study. Technol Anal Strateg Manag 25(6):707–724

Printed by Printforce, the Netherlands